Results and Problems in
Cell Differentiation

A Series of Topical Volumes in Developmental Biology

17

Editors

W. Hennig, L. Nover, and U. Scheer

Results and Problems in Cell Differentiation

Volume 1 · H. Ursprung (Ed)
The Stability of the Differentiated State

Volume 2 · J. Reinert, H. Ursprung (Eds)
Origin and Continuity of Cell Organelles

Volume 3 · H. Ursprung (Ed)
Nucleic Acid Hybridization in the Study of Cell Differentiation

Volume 4 · W. Beermann (Ed)
Developmental Studies on Giant Chromosomes

Volume 5 · H. Ursprung, R. Nöthiger (Eds)
The Biology of Imaginal Disks

Volume 6 · W. J. Dickinson, D. T. Sullivan
Gene-Enzyme Systems in Drosophila

Volume 7 · J. Reinert, H. Holtzer (Eds)
Cell Cycle and Cell Differentiation

Volume 8 · W. Beermann (Ed)
Biochemical Differentiation in Insect Glands

Volume 9 · W. J. Gehring (Ed)
Genetic Mosaics and Cell Differentiation

Volume 10 · J. Reinert (Ed)
Chloroplasts

Volume 11 · R. G. McKinnell, M. A. DiBerardino
M. Blumenfeld, R. D. Bergad (Eds)
Differentiation and Neoplasia

Volume 12 · J. Reinert, H. Binding (Eds)
Differentiation of Protoplasts and of Transformed Plant Cells

Volume 13 · W. Hennig (Ed)
Germ Line – Soma Differentiation

Volume 14 · W. Hennig (Ed)
Structure and Function of Eukaryotic Chromosomes

Volume 15 · W. Hennig (Ed)
Spermatogenesis: Genetic Aspects

Volume 16 · L. Nover, D. Neumann, K.-D. Scharf (Eds)
Heat Shock and Other Stress Response Systems of Plants

Volume 17 · L. Hightower, L. Nover (Eds)
Heat Shock and Development

L. Hightower L. Nover (Eds.)

Heat Shock
and Development

With 53 Figures and 13 Tables

Springer-Verlag

Berlin Heidelberg New York
London Paris Tokyo
Hong Kong Barcelona
Budapest

Professor Dr. LAWRENCE E. HIGHTOWER
Department of Molecular and Cell Biology
The University of Connecticut
75 North Eagleville Road, Storrs
CT 06269-3044, USA

Professor Dr. LUTZ NOVER
Institut für Biochemie der Pflanzen
Weinberg 3, D-4050 Halle (O)
Federal Republic of Germany

ISBN 3-540-53119-X Springer-Verlag Berlin Heidelberg New York
ISBN 0-387-53119-X Springer-Verlag New York Berlin Heidelberg

Library of Congress in Publication Data
Heat shock and development / L. Hightower, L. Nover, eds.   p   cm. –
(Results and problems in cell differentiation; 17) Includes bibliographical references and index.
ISBN 3-540-53119-X (Springer-Verlag Berlin Heidelberg New York: acid-free paper). –
ISBN 0-387-53119-X (Springer-Verlag New York Berlin Heidelberg: acid-free paper)
1. Heat shock proteins. 2. Chemical embryology. 3. Developmental genetics. I. Hightower, L. (Lawrence), 1946– .
II. Nover, Lutz. III. Series.
QH607.R4 vol. 17 [QP552.H43] 574.87'612 s–dc20 [574.2'2] 91-34150 CIP

© Springer-Verlag Berlin Heidelberg 1991
Printed in Germany

The use of general descriptive names, registered names, trademarks, etc. in this publication does not imply, even in the absence of a specific statement, that such names are exempt from the relevant protective laws and regulations and therefore free for general use.
Product liability: The publishers cannot guarantee the accuracy of any information about dosage and application contained in this book. In every individual case the user must check such information by consulting the relevant literature.

Production Editor: Martin Langner

Typesetting, International Typesetters, Inc. Makati, M. M., The Philippines
31/3145 - 5 4 3 2 1 0 – Printed on acid-free paper

Contents

Introduction. LUTZ NOVER and LAWRENCE HIGHTOWER 1

Part I
Heat Shock-Induced Developmental Abnormalties

1 Heat Shock Effects in Snail Development (With 6 Figures)

ELIDA K. BOON-NIERMEIJER

1	Introduction	7
2	Normal Development	8
3	Thermosensitivity During Development	10
3.1	Terms to Express Thermosensitivity	10
3.2	The Cleavage Period: Heat and the Cell Division Cycle	11
3.3	Death and Anomalies Induced by Heat During Development of *Lymnaea*	12
3.4	General Implications	14
3.4.1	Heat Shock as a Teratogen	14
3.4.2	Evidence for a Relationship Between Heat-Induced Anomalies and the Cell Division Cycle	16
3.4.3	Hypothesis Concerning Determinative Events	17
4	The Heat Shock Response During Development	18
4.1	Heat Shock Response and Thermotolerance	18
4.2	Definition of Thermotolerance	18
4.3	Thermotolerance in *Lymnaea:* Kinetics and HSP Synthesis	19
4.4	Heat-Induced Changes in Thermosensitivity and Gene Expression During Development	22
References	....	25

2 Environmentally Induced Development Defects in Drosophila (With 3 Figures)

NANCY S. PETERSEN and HERSCHEL K. MITCHELL

1	Historical Background	29
2	Phenocopy Induction	32

2.1   Conditions for Induction of Phenocopies 32
2.1.1   Sensitive Periods 32
2.1.2   Timing 32
2.1.3   Heating Conditions 33
2.2   Induction of Phenocopies in Recessive Mutant Heterozygotes 33
2.3   Effects of Heat Shock on Gene Expression 35
3   Phenocopy Prevention 36
3.1   Induction of Phenocopy Thermotolerance 36
3.2   Thermotolerance and Heat Shock Proteins 36
3.3   Molecular Models for Thermotolerance 39
4   Conclusions 40
References 41

3 The Use of Heat-Shock-Induced Ectopic Expression to Examine the Functions of Genes Regulating Development

GREG GIBSON

1   Scope 44
2   The Heat Shock Strategy 45
3   Applications 47
3.1   Minimal Induction: Sex Determination and Ageing 47
3.2   Early Development: Gradients and Metameric Stability 48
3.3   Establishing Identity: The Homeotic Genes 49
3.3.1   The Functional Structure of Homeotic Proteins 49
3.3.2   The Homeotic Regulatory Hierarchy 50
3.4   The Fates of Individual Cells: Photoreceptors 52
4   Conclusions 52
4.1   General Applicability to the Study of Development 52
4.2   Prospects 54
References 55

4 Thermotolerance and Heat Shock Response During Early Development of the Mammalian Embryo (With 7 Figures)

DAVID WALSH, KAREN LI, CAROL CROWTHER, DEBBIE MARSH,
and MARSHALL EDWARDS

1   Introduction 58
2   Developmental Defects Caused by Hyperthermia 58
3   The Heat Shock Response in Mammalian Embryos 59
3.1   Embryo Culture and Heat Shock Genes 60
3.2   Heat Shock and Neural Tube Closure 61
3.3   Cell Death of the Neuroectoderm 63
3.4   Induction of Heat Shock Proteins 63
3.4.1   Translation 63
3.4.2   Transcription 63

4   Thermotolerance and Heat Shock Protein Synthesis 66
5   Heat Shock and Cell Cycle Changes 66
6   Conclusion 68
References 69

5   Strain Differences in Expression
of the Murine Heat Shock Response:
Implications for Abnormal Neural Development (With 3 Figures)

MARK D. ENGLEN and RICHARD H. FINNELL

1   Introduction 71
2   The Heat Shock Proteins 71
3   Strain Differences in Heat-Induced Neural Tube Defects 72
4   The Murine Heat Shock Response 74
4.1   The Heat Shock Response in the Murine Embryo
        and Lymphocyte 74
4.2   A Genetic Basis for Strain Differences
        in the Murine Heat Shock Response 77
4.3   An In Vitro Model of the Murine Heat Shock Response 78
5   Conclusions 79
References 81

Part II
Cell-Specific and Developmental Control
of Hsp Synthesis

6   The Expression of Heat Shock Protein and Cognate Genes
During Plant Development

JILL WINTER and RALPH SINIBALDI

1   Introduction 85
2   Classes of Heat Stress Proteins and the Putative Functions
        of the Family Members 86
2.1   Hsp104 86
2.2   Hsp90 87
2.3   Hsp70 89
2.4   Hsp60 90
2.5   Low Molecular Weight Hsps (Hsp20 Family) 90
2.6   Other Heat Shock Proteins 93
3   Hsps and Hscs Expressed During Plant Development 93
3.1   Seeds and Seedlings 94
3.1.1   Hsps and hsp mRNAs During Seed Development 94
3.1.2   Heat Tolerance During Seed Germination and Endogenous Hsps 95
3.2   Roots and Leaves 96
3.3   Flowering 97

3.4   Pollen 97
3.4.1   Heat Stress During Pollen Development 97
3.4.2   Heat Stress During Pollen Germination 98
4   Conclusions 100
References 100

7 Expression of Heat Shock Proteins
During Development in *Drosophila* (With 3 Figures)

ANDRÉ P. ARRIGO and ROBERT M. TANGUAY

1   Introduction 106
2   Expression of Hsp83 106
3   Expression of Hsp70 and Its Cognates 107
4   Expression of the Small Hsps 108
4.1   Gene Structure and Control of Expression
        in the Absence of Stress 108
4.2   Tissue-Specific Expression of Hsp27 110
4.3   Tissue-Specific Expression of Hsp26 112
4.4   Tissue-Specific Expression of Hsp23 112
5   Cellular Localization and Function(s) of the Small Hsps
        During Development 113
6   Summary 116
References 116

8 Regulation of Heat Shock Gene Expression
During *Xenopus* Development (With 12 Figures)

JOHN J. HEIKKILA, PATRICK H. KRONE, and NICK OVSENEK

1   Introduction 120
2   Heat Shock-Induced Accumulation of Hsp
        and *Ubiquitin* mRNA in *Xenopus* Embryos is Developmentally
        Regulated 121
3   Pattern of Hsp and *Ubiquitin* mRNA Accumulation
        in Heat Shocked Embryos 123
4   Involvement of Cis- and Trans-Acting Factors
        in the Developmental Regulation of Hsp70 Gene Expression 124
5   Regulation of *Hsp30* Gene Expression During Development 131
6   Isolation and Sequence Analysis of *Hsp30* Genes
        from a *Xenopus laevis* Genomic Library 133
References 135

9 Heat Shock Gene Expression During Mammalian Gametogenesis and Early Embryogenesis

DEBRA J. WOLGEMUTH and CAROL M. GRUPPI

1	Introduction	138
1.1	Molecular Approaches to Studying Mammalian Germ-Cell Development	138
1.2	Brief Background and Classification of Hsp	140
1.3	Significance of Conservation of Coding and Regulatory Regions in *hsp* Genes	141
1.4	Rationale for Examining Hsp Expression and Function in Germ Cells	142
2	Heat-Shock Gene Expression and Function During Mammalian Spermatogenesis	142
2.1	Key Features of Mammalian Spermatogenesis	142
2.2	Expression of the *hsp70* Gene Family	143
2.3	Expression of *hsp90* Genes	144
3	Expression and Function of Hsp During Mammalian Oogenesis and Early Embryogenesis	144
3.1	Key Features of Mammalian Oogenesis and Very Early Embryonic Divisions	144
3.2	Expression of Hsp in Oocytes and Early Embryos	145
4	Summary and Speculation as to Function of Hsp in Mammalian Germ Cells and Embryos	146
4.1	Possible Functions of Hsp in General	146
4.2	Possible Functions of Heat Shock Genes in Male Germ Cell Differentiation	147
4.3	Molecular and Genetic Approaches for Identifying Function During Mammalian Gametogenesis	147
	References	149

10 Heat Shock Protein Synthesis in Preimplantation Mouse Embryo and Embryonal Carcinoma Cells (With 3 Figures)

VALÉRIE MEZGER, VINCENT LEGAGNEUX, CHARLES BABINET,
MICHEL MORANGE, and OLIVER BENSAUDE

1	Introduction	153
1.1	The Major Murine Heat-Shock Proteins	153
1.2	Heat-Shock Protein Expression During Gametogenesis	154
2	Heat-Shock Protein Synthesis in Unstressed Early Embryonic Cells	155
2.1	Heat-Shock Proteins, the First Major Products of Zygotic Transcription	155
2.2	High Spontaneous Expression of Hsps in the Preimplantation Mouse Embryo	156
2.3	Hsp Expression in Embryonal Carcinoma Cells	156
3	Transcription of Heat Shock Genes in Unstressed EC Cells	157
3.1	Transcriptional and Posttranscriptional Regulation of Spontaneous Hsp Synthesis in EC Cells	157

3.2   HSE-Binding Activity in EC Cells 157
3.3   An EIa-Like Activity in EC Cells 158
3.4   High Levels of B2 Transcripts
       in Undifferentiated Mouse Embryonic Cells 159
4   Defective Heat Shock Response in Early Embryonic Cells 160
4.1   Lack of Heat Shock Protein Inducibility
       in the Early Preimplantation Mouse Embryo 160
4.2   Inducible and Noninducible Embryonal Cell Lines 160
4.3   Noninducible EC Cells are Deficient in Transcriptional Transactivation
       of Heat Shock Genes by Stress 162
4.4   HSE-Binding Activity in Heat-Shocked EC Cells 163
5   Concluding Remarks 163
References 164

11   Transcriptional Regulation of Human *Hsp 70* Genes:
Relationship Between Cell Growth, Differentiation,
Virus Infection, and the Stress Response (With 5 Figures)

BENETTE PHILLIPS and RICHARD I. MORIMOTO

1   Introduction 167
2.   Factors Which Alter the Expression
       of *Hsp 70*, *Grp 78*, and *P72* 168
2.1   Factors Which Alter Expression of *Hsp 70* 168
2.1.1 Determinants of Basal Expression 168
2.1.2 Classical Stress-Response Inducers 170
2.1.3 DNA Viruses 170
2.1.4 Cell Cycle Regulation, Growth Factors 173
2.1.5 Agents Inducing Differentiation 173
2.1.6 Other Agents 175
2.2   Factors Which Alter the Expression of *Grp 78* 175
2.3   Factors Which Alter the Expression of *P72* 176
3   Mechanisms of Activation 176
3.1   Heat Shock Induction 177
3.2   Hemin Induction 180
3.3   Viral Induction 182
3.3.1 Adenovirus 182
3.3.2 Herpes Simplex Virus-1 (HSV-1) 183
3.3.3 Simian Virus 5 (SV5) 183
4   Concluding Remarks 184
References 184

12   Transforming Growth Factor-β
Regulates Basal Expression of the *hsp70* Gene Family
in Cultured Chicken Embryo Cells (With 5 Figures)

IVONE M. TAKENAKA, SETH SADIS, and LAWRENCE E. HIGHTOWER

1	Introduction	188
2	Biochemical and Biological Properties of TGF-β	189
3	TGF-β in Embryogenesis and Development	191
4	Heat Shock Proteins Are Induced During Embryogenesis and in Highly Mitogenic Cells	192
5	Regulators of Basal Expression of Heat Shock Gene Families in Unstressed Cells	194
6	TGF-β Rapidly Induces Hsc70 in Cultured Chicken Embryo Cells	195
7	The Hsc70 Molecular Chaperone Interacts with Diverse Polypeptide Sequences	200
8	Conclusion	204
References		205

13   Cell Growth, Cytoskeleton, and Heat Shock Proteins (With 2 Figures)

ICHIRO YAHARA, SHIGEO KOYASU, KAZUKO IIDA, HIDETOSHI IIDA,
FUMIO MATSUZAKI, SEIJI MATSUMOTO, and YOSHIHIKO MIYATA

1	Cyclic AMP and Expression of Heat Shock Proteins in the Budding Yeast	210
2	Heat Shock-Induced Reorganization of Cytoskeletal Structures	211
3	Hsp90 is an Actin-Binding Protein	213
References		215

14   Expression of Heat Shock Genes *(hsp70)*
in the Mammalian Nervous System (With 4 Figures)

IAN R. BROWN

1	Introduction	217
2	Early Studies on Brain Heat-Shock Proteins	218
3	Induction of Heat Shock Proteins in the Visual System	218
4	Analysis of *hsp70* mRNAs in the Mammalian Nervous System	219
5	Regional Differences in Expression of *hsp70* Genes in Brain Detected by In Situ Hybridization	219
6	Induction of an Hsp70 Gene at the Site of Tissue Injury in the Brain	222
7	Immunological Detection of Hsp70 in Brain Tissue	224
8	Tissue-Protective Effects of Heat Shock in the Nervous System	225
9	Conclusion	226
References		226

Introduction

L. Nover and L. Hightower

Though the roots of experimental stress biology at the cellular and organismic level can be traced back to the middle of the last century (Nover 1989), a decisive breakthrough came only in 1962 with the report on stress-induced changes of gene activity in *Drosophila* (Ritossa 1962) and the subsequent identification of the newly synthesized heat stress proteins (Tissieres et al. 1974) and mRNAs, respectively (McKenzie et al 1975; McKenzie and Meselson 1977). The selectivity of induction and the high rate of accumulation of Hsps facilitated the cloning and sequencing of the hs genes in *Drosophila* and the demonstration that all organisms react similarly when exposed to heat stress or chemical stressors (Ashburner and Bonner 1979; Schlesinger et al. 1982; Nover 1984). The explosive development of molecular stress research in the following 10 years illustrated that the stress response represents a characteristic network of dramatic but transient changes at many levels of cellular structure and function, including gene expression (Atkinson and Walden 1985; Tomasovic 1989; Georgopoulos et al. 1990; Nover et al. 1990; Nover 1991).

Besides the characterization of the hs genes and the mechanism of their induction, major interest concentrated on the heat stress proteins and their possible roles in induced stress tolerance. Rapidly, it became apparent that the major stress proteins are coded by five conserved multigene families (Lindquist and Craig 1988: Nover et al. 1990; Nover 1991), whose individual members are expressed independently in response to stress and/or developmental signals and/or to factors involved in cell cycle control. The multiplicity of heat stress and related proteins varies from cell type to cell type (Table 1). Though basically similar with respect to their function, distinct members of one family may be separated in their biological activity by confinement to different cell compartments. Stress proteins and their constitutive isoforms, respectively, are essential cell components under all conditions. Most of them or even all are evidently involved in the formation of transient protein complexes (Nover et al. 1990; Nover 1991). Members of the Hsp60 and Hsp70 families are ATP-binding proteins and may help in the folding of nascent or denatured proteins and in the assembly or disassembly of oligomeric protein complexes ("molecular chaperones", Deshaies et al. 1988; Ellis and Hemmingsen 1989; Goloubinoff et al. 1989; Rothman 1989). In the framework of this series of research monographs on results and problems in cell

Note: Abbreviations for stress proteins and genes. Heat stress proteins are abbreviated by *Hsp with numbers* indicating their Mr in kD, e.g., Hsp70 for a 70-kilodalton Hsp. The corresponding genes are given in italics, e.g. *hsp70*. Constitutively expressed members of the stress protein families are indicated by Hsc and *hsc* respectively, and similar rules apply to the glucose-regulated proteins of the ER (Grp78 for the protein and *grp78* for the gene). HSF, heat stress transcription factor; HSE, heat stress promoter element. Other, more specific, abbreviations are explained in the text.

Results and Problems in Cell Differentiation 17
Heat Shock and Development
Hightower and Nover (Eds.)
© Springer-Verlag Berlin Heidelberg 1991

Table 1. Survey of eukaryotic stress protein families (for details, see Nover et al. 1990; Nover 1991)

Hsp family	Genes	Cytoplasm	ER	Chloropl.[a]	Mitochondria[b]
Mammalian cells					
Hsp90	3	Hsp89α Hsp89ß	Grp94	-	-
Hsp70[a-]	7-8	Hsc72 Hsp70A Hsp70B	Grp78	-	Hsc70 (DnaK)
Hsp60	1	-	-	-	Hsp 60(GroEL)
Hsp 20	1	Hsp 25-27	-	-	-
Hsp10	1	-	-	-	Cpn10 (GroES)
Ubiquitin[c]		Ubiquitin	-	-	-
Plants cells					
Hsp90	1	Hsp80	?	-	-
Hsp70	5	Hsc70 Hsp70	Grp78	Hsc70$_c$	Hsc70$_m$
Hsp60	2	-	-	Hsc60$_c$	Hsc60$_m$
Hsp20	~20	15-20 Represent. of Hsps 15-19	?	Hsp21 Hsp22	?
Hsp10	1	-	-	Cpn10	?
Ubiquitin[c]		Ubiquitin			
***Saccharomyces cerevisiae* (yeast)**					
Hsp90	2	Hsp90 Hsc90	?	-	-
Hsp70	8	SSA1-4 SSB1,2	KAR2	-	SSC1
Hsp60	1	-	-	-	Hsp60
Hsp 20	1	Hsp26	-	-	-
Ubiquitin[c]		Ubiquitin	-	-	-

[a] Two to three additional sperm-specific proteins (see Chap 9).

[b] The organelles contain the prokaryotic type of stress proteins, i.e., DnaK-like (Hsp70) and GroEL-like (Hsp60). Different genes code for the representatives in plant chloroplasts and mitochondria, respectively.

[c] Besides the hs-inducible polyubiquitin gene, there are a number of constitutively expressed monoubiquitin and ubiquitin fusion genes.

differentiation, the editors wanted to emphasize the close relationship between stress proteins and developmental biology. On the one hand, heat and chemical stressors are well-known teratogens, because, due to the enormous amplification in numbers of cells, even minute changes in the expression, stability and/or intracellular localization of key proteins of development may exert dramatic effects on embryonic morphology. As is true for other parts of the stress response, a protection from teratogenic effects by mild pretreatments can be achieved and synthesis of stress proteins has been correlated frequently with this tolerance effect. These connections are discussed in detail for the snail *Lymnea stagnalis* (Boon-Niermeijer, Chap. 1), for *Drosophila melanogaster* (Petersen and Mitchell, Chap. 2), and for mammalian embryogenesis (Walsh et al. and Englen and Finnell, Chaps. 4 and 5).

Another remarkable part of research in developmental biology is outlined by G. Gibson (Chap. 3). The ready availability and regulatory properties of hs promoters made them prominent tools for gene technology, e.g., to study effects of ectopic and untimely expression of developmental genes. The number of such developmental gene constructs successfully transferred to the *Drosophila* germ line augments almost weekly and helps to elucidate the complex network of factors controlling development.

The second part of this book concerns the control of stress protein synthesis itself. This includes stress-independent expression under cell cycle control, or after virus infection, and following exposure to cytokines(Phillips and Morimoto, Takenaka et al., and Yahara et al., Chaps. 11–13). Members of the stress protein families, especially the Hsp90, Hsp70, and Hsp20 families, are detected as developmental markers during gametogenesis or embryogenesis in plants (Winter and Sinibaldi, Chap. 6,), *Drosophila* (Arrigo and Tanguay, Chap. 7), and mammals (Wolgemuth and Gruppi, and Mezger et al., Chaps. 9 and 10). Moreover, stress-inducibility of hs genes, seemingly common to all somatic cells, in fact shows cell-specific variation in highly specialized tissues, e.g., the mammalian brain (Brown, Chap. 14), or is even lacking in early embryonic stages or special developmental states (Winter and Sinibaldi, Heikkila et al., Mezger et al., Chaps. 6, 8, and 10). In some cases investigated in more detail, interesting peculiarities of the signal transduction mechanism involving the heat stress transcription factor (HSF) were found (Heikkila et al., Mezger et al., and Phillips and Morimoto, Chaps 8, 10 and 11).

We hope that by collecting information in a single monograph on the developmental systems that have formed the bases of major contributions in the heat shock field, the common themes will become more apparent. The literature covered here is typically scattered among developmental biology and teratology journals as well as molecular and cell biology journals. As the editors assembled the chapters to follow, we identified several common themes and we invite readers to extend the list: The capacities to induce heat shock proteins and thermotolerance are coordinately acquired during the blastula stage in organisms as diverse as frogs, snails, and flies; the high levels of heat shock proteins and/or cognates in pre-blastoderm embryos and embryonal carcinoma cells; the complex nature of acquired thermotolerance; and, the dependence of thermally induced abnormalities on both genetic and environmental factors.

And finally,we wish to point out the potential value of viewing the effects of heat shock proteins in the emerging functional context of these proteins and their cognates in cellular protein metabolism. It is becoming clear that these proteins normally play major roles in intracellular protein trafficking, polypeptide folding and assembly, membrane translocation of proteins, protein degradation, and antigen presentation. As part of stress responses, the induction of heat shock proteins appears keyed to cellular protein homeostasis and their functions may include removal of damaged proteins, possibly rescue and recycling of denatured proteins, and protection from lethal thermal damage. Indeed, a system of surveillance of the integrity of cellular proteins based on protein-protein interactions is coming into view that may have analogies to protein-based immune surveillance(Takenaka et al., Chap. 12).

At the very least, the emerging links to cellular protein metabolism can serve as a starting point for testable hypotheses for the roles of heat shock proteins in development. For example, studies of phenocopy prevention by acquired thermotolerance in *Drosophila* have implicated accelerated recovery of protein synthesis following heat stress as a primary protective effect (Petersen and Mitchell, Chap. 2). It has been suggested that the transforming growth factor-ß induces heat shock proteins that may act as molecular chaperones for other developmentally important proteins (Takenaka et al., Chap. 12), and that they may be involved in the assembly/disassembly of specialized structures during spermatogenesis and oogenesis (Wolgemuth and Gruppi, Chap. 9). The challenges for developmental biologists are to understand how certain members of heat shock protein families are noncoordinately induced during various stages of normal development and how heat shock proteins and their cognates participate in normal development and in the protection of embryos from teratogens.

This is the first monograph devoted entirely to an exploration of the roles of the heat shock proteins in development. Earlier review articles that synthesized specific aspects of the heat shock response among several experimental systems include chapters by Bienz and Pelham (1987), Bond and Schlesinger (1987), Lindquist and Craig (1988), and Nover (1991).

References

Ashburner M, Bonner JJ (1979) The induction of gene activity in *Drosophila* by heat shock. Cell 17:241–254

Atkinson BG, Walden DB (eds) (1985) Changes in eukaryotic gene expression in response to environmental stress. Academic Press, Orlando

Bienz M, Pelham HRB (1987) Mechanisms of heat shock gene activation in higher eukaryotes. Adv Genet 24:31–72

Bond U, Schlesinger MJ (1987) Heat shock proteins and development. Adv Genet 24:1–29

Deshaies RJ, Koch BD, Werner-Washburne M, Craig EA, Shekman R (1988) A subfamily of stress proteins facilitates translocation of secretory and mitochondrial precursor polypeptides. Nature 332:800–805

Ellis RJ, Hemmingsen SM (1989) Molecular chaperones: proteins essential for the biogenesis of some macromolecular structures. Trends Biochem Sci 14:339–342

Georgopoulos C, Tissieres A, Morimoto R (eds) (1990) Stress proteins in biology and medicine, Cold Spring Harbor Press, Cold Spring Harbor, New York

Goloubinoff P, Christeller JT, Gatenby AA, Lorimer GH (1989) Reconstitution of active dimeric ribulose bisphosphate carboxylase from an unfolded state depends on two chaperonin proteins and MgATP. Nature 342:884–888

Lindquist S, Craig EA (1988) The heat-shock proteins. Ann Rev Genet 22.631–677

McKenzie SL, Meselson M (1977) Translation in vitro of *Drosophila* heat-shock messages. J Mol Biol 117:279–283

McKenzie SL, Henikoff S, Meselson M (1975) Localization of RNA from heat-induced polysomes at puff sites in *Drosophila melanogaster*. Proc Natl Acad Sci USA 72:1117–1121

Nover L (ed) (1984) Heat shock response of eukaryotic cells. Springer, Berlin Heidelberg New York

Nover L (1989) 125 years of experimental heat shock research: historical roots of a discipline. Genome 31:668–670

Nover L (ed) (1991) Heat shock response. CRC Press, Boca Raton .

Nover L, Neumann D, Scharf KD (eds) (1990) Heat shock and other stress response systems of plants Springer Berlin Heidelberg New York

Ritossa F (1962) A new puffing pattern induced by heat shock and DNP in *Drosophila*. Experientia 18:571–573

Rothman JE (1989) Polypeptide chain binding proteins: catalysts of protein folding and related processes in cells. Cell 59:591–601

Schlesinger MJ, Ashburner M, Tissieres A (eds) (1982) Heat shock: from bacteria to man. Cold Spring Harbor Lab, Cold Spring Harbor, New York

Tissieres A, Mitchell HK, Tracy UM (1974) Protein synthesis in salivary glands of *D. melanogaster*. Relation to chromosome puffs. J Mol Biol 84:349–398

Tomasovic SP (1989) Functional aspects of the mammalian heat-stress protein response. Life Chem Rep 7:33–63

Part I

Heat Shock-Induced Developmental Abnormalities

1 Heat Shock Effects in Snail Development

Elida K. Boon-Niermeijer[1]

1 Introduction

In the study of the mechanisms underlying embryonic development, a variety of noxious treatments have been applied in the past. At the present, disturbing normal development by external influences remains a common method in experimental embryology. Heat shock has been used for a long time in this sense in a variety of organisms such as Amphibia, chicks, snails, *Drosophila*, and so on. These studies have in common that the defects are stage specific, depending on the period of development at which heat shock is applied. Which developmental biologist could foresee that heat shock would play such a pivotal role in molecular biology in the research on gene regulation? The discovery of the induction of a unique set of puffs by heat shock (Ritossa 1962) opened this field of investigation. Only about one decade later, it was demonstrated that a specific set of polypeptides was synthesized after exposure to heat, corresponding to a specific set of mRNAs. The response to heat is now known as a universal reaction of almost every living organism. It is even a response to stress in general. The heat shock genes and their products in different species show a high degree of homology. Their conservation during evolution suggests the early development of an essential function in all organisms (Schlesinger et al. 1982; Craig 1985; Lindquist 1986). This has been strengthened by the fact that their activity is not only influenced by external factors, but also fluctuates intrinsically, e.g., during the cell division cycle and embryonic development and differentiation. At the present, heat shock research is again concentrating on embryology. It is focused upon two main levels at which heat shock and development are cross-linked, i.e., at the levels of morphogenesis and gene regulation. Although these fields of investigation are still rather separate, they may converge in the future when our insights in both will have progressed.

Modern investigations into the effect of heat shock and other stresses on morphogenesis are those on the segmentation of somites in vertebrates and on the appearance of phenocopies in *Drosophila*. These studies give insight into the time, and perhaps the nature, of determinative events during normal development. This leads to the teratological studies concerning a better understanding of nongenetically determined congenital malformations during stress (drug)-sensitive phases of pregnancy.

Another level of interest are studies on the intrinsic change in thermosensitivity during development in relation to regulation of gene activity, especially that of heat

[1] Department of Molecular Cell Biology, University of Utrecht, Padualaan 8, P.O. Box 80.056, 3508 TB Utrecht, The Netherlands

Results and Problems in Cell Differentiation 17
Heat Shock and Development
Hightower and Nover (Eds.)
© Springer-Verlag Berlin Heidelberg 1991

shock genes or their cognates. The effect of external stress on thermosensitivity in relation to gene activity is the focus in this case. There is strong evidence that both the intrinsic thermosensitivity and the capacity for induced thermotolerance are developmentally regulated, i.e., stage dependent, as well as regionally specific.

The development of the fresh water snail *Lymnaea stagnalis* (Pulmonata) has been a subject of study for about 50 years (Raven 1942). The effect of heat shock on morphogenesis and on the early division cycles has been known for a long time (Visschedijk 1953; Geilenkirchen 1966; Verdonk and de Groot 1970; Boon-Niermeijer 1976). After the discovery of the heat shock response a more extensive study was carried out on the effect of heat and some other stresses on several aspects of the development of *Lymnaea* (Boon-Niermeijer 1987). In the present paper an overview is given of the data. Several properties of thermosensitivity are clearly distinguished. The problems in expressing thermosensitivity at different developmental stages are stated. The origin of morphogenetic disturbances is discussed in relation to the effect on the cell division cycle.

In a general comparison with data derived from other species, suggestions are made about the possible relationships among heat (stress, drugs), morphogenetic events and teratogenesis. It is postulated that morphogenetic events make use of the synchronization principle to determine groups of cells. Heat then interferes with mitosis and the cell cycle at those sites where the cells are in a particularly sensitive phase of the cycle. This leads to morphogenetic disturbances at those special sites.

Thermosensitivity, as well as thermotolerance, are developmentally regulated. During early cleavage, neither thermotolerance nor the heat shock response can be induced. The characteristics of thermotolerance during development from the late blastula onwards do not differ principally from those in nonembryological material. Two states of thermotolerance are distinguished, one independent of protein and Hsp synthesis, the other dependent on both.

2 Normal Development

Fertilized eggs of *Lymnaea* are deposited in egg masses, each containing about 80–150 egg capsules. As a rule, a capsule surrounds one egg. The egg capsules are freed from the egg mass, cleaned of adhering jelly, and kept in culture dishes in tap water. The embryos remain within their capsules throughout the experiments.

After deposition of the eggs, two polar bodies are extruded. The pronuclei fuse to form the zygote nucleus. Cell divisions take place about every 80 min at 25 °C, until after five cell division cycles the 24-cell stage is reached (Fig. 1). Cleavage is of the determinative type. The cell divisions take place according to a well-defined pattern, producing an identifiable, canonical three-dimensional array of blastomeres. The fates of the cells are predictable and traceable. The first two cleavages are meridional. At the successive divisions micromeres are formed at the animal side, alternating dexiotropically and laeotropically (Fig. 1). The cell cycles of the macro- and micromeres run in parallel, except that mitosis and cell division of the micromeres start 10–15 min later than in the macromeres. The 24-cell stage lasts about 3 h. During this period further differentiation will be determined. Cell lineage studies have revealed the fate of the cells up to larval and adult structures. At about 24 h after first cleavage,

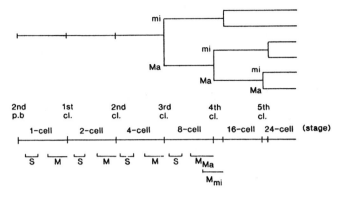

Fig. 1. Cell division schedule of *Lymnaea stagnalis*. Division rate of about 80 min at 25 °C up to the 24-cell stage. *cl* cleavage; *M* mitosis; *Ma* macromeres; *mi* micromeres; *p.b.* polar body; *S* S-phase (After Boon-Niermeijer 1975, 1976)

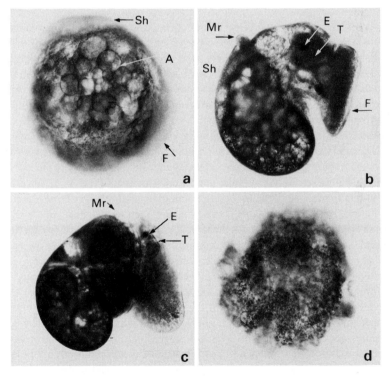

Fig. 2a-d. Larval stages of *Lymnaea*. *A* large endoderm cells; *E* eye; *F* foot; *Mr* mantle ridge; *Sh* shell; *T* tentacle; **a–c** Normal larvae 3, 5, and 7 days old; **d** 3-day-old trochophore shortly after heat shock (1 h, 40 °C), falling apart into loose, dying cells (Boon-Niermeijer and Van de Scheur 1984)

gastrulation starts, and at 30 h the embryo leaves the vitelline membrane and moves freely within the capsule by means of cilia. In the 3-day-old trochophore larva (Fig.2a) the shell gland has been developed. In the following days the shell grows at the edge by the activity of the mantle fold with which it is tightly connected (Timmermans 1969). Eventually the shell covers the visceral sac (Fig. 2b). The shell assumes an asymmetric cone. Typical large-celled larval structures form an extensive part of the body besides the adult small-celled structures, such as foot, cephalic plates with tentacles, eyes, etc. When the larva is five days old there is a regularly beating heart and a functional larval kidney. The larval structures degenerate gradually in the following two days. At the age of about eight days the young snail hatches (Fig. 2c). An extensive description of normal development has been published previously (Verdonk 1965; Van den Biggelaar 1971a, b; Raven 1975).

3 Thermosensitivity During Development

3.1 Terms to Express Thermosensitivity

In studies on thermosensitivity in cell lines, the most common parameter of expression is the percentage of survival assayed by the cloning capacity of the treated cells, i.e., the capacity to survive and to go on dividing. The advantage of this parameter is the all-or-none effect, but there are other possibilities, more subtle, but less unequivocal, e.g., change in shape of the cells due to the effect on the cytoskeleton, the effect on protein synthesis, etc. (Schamhart et al. 1984; Van Wijk et al. 1984; Wiegant et al. 1985).

Expression of thermosensitivity of multicellular organisms in terms of survival differs fundamentally from the cell line studies. In an organism as a whole, the all-or-none effect is much more complicated. Some cells may die, although the organism can survive, more or less hampered by the cellular loss. Some areas may be more sensitive than others, leading to localized effects. The function of special organs can be affected even without any loss of cells. These defects can all lead to the eventual death of the organism.

During embryonic development, these factors play an even greater role, since their proportional significance is not the same throughout development. For instance, the death of one cell has far-reaching consequences for the development of a four-cell embryo, but probably not for a gastrula. During the stages of organ formation and differentiation, the thermosensitivity of the different organs and the degree to which they are indispensable for the survival of the organism as a whole will be crucial. The consequence of this is that the time after heat shock at which survival is determined is essential. For developing embryos, it is not only a matter of surviving or not at the stage of heat exposure, but, after all, also a matter of developing into a viable organism. The latter needs much more than just surviving the moment of heat shock, since heat can also induce disturbances, which, in the long run, lead to death of the organism. In summary, the outcome of survival, if not scored at the stage of heat treatment, will depend completely on the developmental pathway remaining between the heated stage and stage at the moment that survival is determined.

While the survival studies are quantitative in principle, a qualitative aspect in the mechanism of thermosensitivity expression is introduced by the stage-specific sensitivity for the induction of discrete anomalies. This mechanism has been used in teratologic studies in vertebrates and phenocopy studies in *Drosophila*. In these cases, the interest is focused upon unravelling the time and nature of determinative events crucial for developmental processes.

3.2 The Cleavage Period: Heat and The Cell Division Cycle

During the early cleavage phase of *Lymnaea*, from the 1-cell to the 16-cell stage, thermosensitivity has been investigated using as parameters the duration of the cell cycle and the effect on morphogenesis (Geilenkirchen 1966; Boon-Niermeijer 1976). It has been determined as follows: groups of synchronous eggs are exposed to a heat shock of 37–38 °C for 10 min at successive intervals. The time of the next cell division and of the one after that is scored and the embryos are reared subsequently for another six days. Thermosensitivity was measured in two ways: (1) by means of the extension of the cell division cycle; and (2) by the effect on subsequent development. In both cases it changes periodically depending on the phases of the cell division cycles. Figure 3 shows the data for the period between the 8- and 16-cell stage. They are representative for the other cell cycles.

There are three peaks in the extension of the cell cycle: (1) at the initiation of a cell cycle; (2) during the G2-phase; and (3) at prometaphase. Early in metaphase the sensitivity of the cells decreases abruptly and cell division cannot be postponed any longer. The decay of sensitivity both at the onset of prophase and early in metaphase is so abrupt that it results in great time differences between eggs of one and the same, initially synchronous group, and even between equivalent blastomeres of the same egg. Apparently, defined transition points in the cell cycle are affected.

The effect of heat on later stages of development is also periodic and related to the cell division cycles. Developmental anomalies are maximally induced just before a cell division. This time coincides with the moment that the cell division cannot be postponed any longer (Fig. 3). The type of anomalies is stage dependent. Nondiscrete, severe anomalies arise after heat shock up to the eight-cell stage. Gastrulation is affected and at best hydropic bells arise without any resemblance to a snail. On the contrary, heat shock at the transition from the 8- to the 16-cell stage (i.e., fourth cleavage) can induce discrete anomalies, i.e., shell malformations at the morphogenetically sensitive period of the macromere cycle, and head malformations at the sensitive period of the micromere cycle, which occurs 10–15 min later (note that both larval and adult head structures originate from the descendants of the first micromeres). The occurrence of the discrete anomalies coincides well with the prospective significance of the cells heat-shocked just prior to their division.

Further support for this view comes from cytological observations on embryos heat-shocked during meta-anaphase of the successive cell division cycles (Boon-Niermeijer 1976). Feulgen-stained, whole mounts of eggs display a large number of cytologic anomalies resulting from an incomplete separation of chromatin. The well-defined cleavage pattern of *Lymnaea* allows the tracing of further events in the descendants of cells heat-shocked at meta-anaphase. Remarkable findings are tetrasters, chromosome bridges and loose chromosome fragments, fragmented nuclei, etc. Some cells die after one or two cycles following heat shock. They contain pycnotic, dotted chromosomes, are protruding from the surface of the embryo, and are eventually

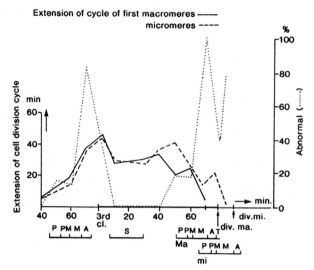

Fig. 3. Thermosensitivity from the 4- to the 16-cell stage, expressed in extension of the cell division cycles of macro- and micromeres (*left*) and in percentage of abnormal embryos (*right*). *P* prophase; *PM* prometaphase; *M* metaphase; *A* anaphase; *T* telophase; *div. Ma, div. mi* division of macro- and micromeres, respectively (After Boon-Niermeijer 1976)

extruded from the living embryo after about 1 day. Sometimes there is a deviation from the normal division chronology. Individual cells escape the cell division-inhibiting influence which is present during the normal 24-cell stage (see Fig. 1) and go on dividing at the division rate of 80 min which governs the preceding period.

Together, these data led to the view that developmental anomalies in *Lymnaea* derive from irreversible injuries of mitosis and/or division chronology, not from disturbance of other morphogenetic events. The discreteness of the anomalies is due to the different prospective significance of the affected cells, and not to the initial injuries which are universal for all cells which are in mitosis at the period of treatment.

3.3 Death and Anomalies Induced by Heat During Development of *Lymnaea*

The apparently contradictory results in studies on thermosensitivity throughout development in various species may be largely due to the use of different criteria by the various investigators in evaluating thermosensitivity. In *Lymnaea*, the simplest parameter of survival has been used: an early killing-assay, i.e., lethality either instantly or within 1 day without any sign of further development. Death can be easily determined, because dead embryos disintegrate as a whole (see Fig.2d). At the earliest stage investigated, the four-cell embryo, the heat exposure has to be confined to a definite period of the cell cycle (G2) to avoid the immediate disruption of mitosis and, as a consequence, cell death and/or abnormal development (Sect. 3.2). Survival at the four-cell stage is much more difficult to distinguish than at later stages, because usually the cells do not disintegrate and viability has to be assessed by trypan blue exclusion.

Exposures of 1 h to 40 °C reveal characteristic ups and downs in the percentage of killed specimens during development from the four-cell stage to the young snail (Fig. 4). The most sensitive phases are the trochophore/veliger larvae of 3–4 days and the young snail of 7 days. Survival assays of continous exposure to 43.6 °C confirm the unequivocal increase in thermosensitivity from the onset of development to the trochophore stage (Fig.6a). It is conspicuous that the four-cell stage is much less sensitive, in terms of early killing, than later stages.

Another way to evaluate thermosensitivity in *Lymnaea* is to monitor the effect on development shown by anomalies (or eventually death due to these anomalies) in those embryos which are not killed immediately (Fig. 4). Expressed in that way, thermosensitivity is highest at the earliest stage. No normal development occurs when embryos younger than 2 days are treated. Two, 3- and 4-day-old specimens surviving heat shock show unimpaired development except for a slight retardation. That means that they are rather insensitive. Sensitivity increases again when development continues: a high percentage of the 5- and 6-day-old larvae display a variety of anomalies. Heart beat and ciliary movement are interrupted and dead cells are extruded from the small-celled adult structures of foot, head, and mantle ridge. Larvae which survive have foot and shell anomalies, which are relics of the many dead cells lost. The larvae are often greatly swollen (hydropic).

A comparison between the two foregoing ways to express thermosensitivity makes clear that the conclusions depend on the method used, and that, using different methods, the results may seem contradictory. However, it should be realized that two different phenomena are studied. Therefore, when comparing results on thermosensitivity during development of different species, it is essential to consider the criteria used.

There is at least one study which confirms that early developmental stages are relatively insensitive to direct killing. *Drosophila* preblastoderm embryos survive a heat shock of 2 min at 43 °C better than blastoderm embryos (Eberlein and Mitchell 1987). Because of the very short exposure time, the chance to hit cells in mitosis is rather limited. In this case, thermosensitivity thus reflects the sensitivity of the material itself without the complications of mitotic disturbances. Confirmation of our results comes also from experiments with *Xenopus*. Oocytes appear to be extremely resistent to lysis by heat (King and Davis 1987). However, early cleavage stage embryos are highly sensitive, but the authors ascribe this to the high division rate

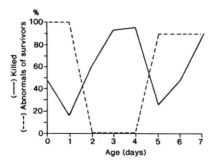

Fig. 4. Thermosensitivity during development of *Lymnaea* expressed in percentage of killed individuals (*full line*) and in percentage of abnormals of the surviving fraction (*broken line*) (After Boon-Niermeijer and Van de Scheur 1984)

(Davis and King 1989). Apparently, it was not excluded that the cells were in mitosis during heat exposure, and thus many cells may die due to mitotic disturbances. It should be worth the effort to repeat the experiments and see what happens if it is made sure, by precise timing and selection, that only interphase cells are treated.

Many more studies have been done on thermosensitivity during development using the criteria of normal vs. abnormal development, death before hatching, developmental arrest, etc. These criteria yield consistent results. A high sensitivity at the earlier stages has been shown for sea urchins (Roccheri et al. 1981), *Drosophila* (Graziosi et al. 1980; Bergh and Arking 1984); *Xenopus* (Bienz 1984; Heikkila et al. 1985; Nickells and Browder 1985), and mouse (Muller et al. 1985). However, in all these studies no care was taken to avoid mitosis during heat exposure, i.e., increased thermosensitivity due to mitotic abnormalities is included.

In summary, in investigations on thermosensitivity during development one has to realize what is the question, which are the proper criteria, and how results should be interpreted.

3.4 General Implications

3.4.1 Heat Shock as a Teratogen

Heat shock interferes with normal development in a great variety of species. A common feature is that the specificity of the induced anomaly depends on the developmental stage. This is the more striking, because many other adverse influences administered at the same stage evoke the same anomaly. To explain this phenomenon, the effect of heat shock and other insults has been related to an interruption of determinative events which take place at discrete stages during development. By definition (see Slack 1983), determination is a single event at which a cell becomes irreversibly committed to a particular fate. Heat shock experiments inducing anomalies are thus usually designed to answer the question: when does determination occur? Only those cells undergoing the determinative event at the time of the shock will be sensitive to disturbance. The position of the induced anomaly reflects the time at which critical developmental decisions are made. In this line of thought, the nature of the determinative event is not relevant.

The phenomenon of phenocopy induction in *Drosophila* represents this type of investigation. Phenocopies are malformations which arise after heat exposure at discrete developmental stages. They look like specific mutants, but are not inheritable. The phenotype of the mutant as well as of the phenocopy should arise from the absence of a specific gene expression in a specific time interval, either because the normal allele is missing or because heat interferes with its transcription. As we know, heat shock does interfere with transcription of genes active at the moment of heat shock, at the same time inducing the activity of the heat shock genes (Sect. 4). Phenocopies are produced only under conditions that repress virtually all transcription and translation. On the other hand, phenocopy production can be prevented by preexposure to a mild heat treatment, i.e., by induction of thermotolerance (Sect. 4). Conditions that prevent phenocopies are those which induce heat-shock protein synthesis without inhibiting ongoing cellular protein synthesis (Mitchell and Petersen 1982; see Chap. 2) The above data seem to prove that the suppressed gene expression is the cause of phenocopy production. The evidence needed, however, is the unequivocal demonstration of a correlation between lack of a specific transcript or protein and the occurrence of the abnormal phenotype and its reappearance as a result of thermotolerance respectively.

The embryonic stress hypothesis of teratogenesis, put forward by German (1984), is also based on the same principle. German used the example of heat as a human teratogen. The hypothesis proposes that many human developmental errors are due to the failure of essential loci to be transcribed into mRNA and translated into protein at some critical moment during embryonic life, because the gene expression machinery of the cell is occupied by heat shock protein synthesis. In this view the induction of the heat shock response, i.e., synthesis of Hsps and their messengers, should be rather threatening than protecting. The induction of Hsp synthesis is regarded to be the proof for teratogenic action (German 1984; German et al. 1986). However, instead of explaining the origin of anatomical malformations by the accumulation of heat shock mRNAs and proteins, the other aspect of the heat shock response should be considered, i.e., the suppression of normal gene expression. It is more plausible that abnormalities are caused by *the inhibition of the synthesis of particular messengers and proteins*, taking place at the moment of heat exposure and involved in the process of organogenesis. This is just the evidence which is lacking. It should be mentioned that Walsh and coworkers (1987) obtained evidence that the accumulation of Hsps per se was not teratogenic in the developing rat embryo, suggesting that it may be the absence of a required gene product and not the presence of Hsps that is teratogenic (see Chap. 4).

Another field of research on heat shock effects on the timing and nature of determinative events is the formation of segments, in vertebrates quite clearly expressed in the process of somitogenesis. In several amphibian species (Elsdale et al. 1976; Elsdale and Davidson 1986, 1987; Pearson and Elsdale 1979), in chick (Veini and Bellairs 1986; Primmett et al. 1988, 1989), and in zebrafish (Kimmel et al. 1988) heat interferes with the formation of somites. The most characteristic feature is that behind the somite last formed at the moment of heat exposure, a number of normal somites are still formed before abnormal ones appear. A similar phenomenon is observed in a totally different organism, the short germ embryo of the locust (Mee and French 1986a,b). In this species heat shock interferes with the process of segment formation, and again, after heat shock a constant number of normal segments develops before the first abnormal one appears.

It is important that the type of the segmental anomalies are all the same, whether they are induced in the xth or in the yth segment. Because of their different places in the body plan, however, they will lead to malformations in different structures in the later embryo. In other words, the initial effect of heat seems to be identical, but the morphogenetic outcome is different due to the different prospective significance of the damaged segment region.

The data have been discussed in relation to two models: (1) the clock and wave front model of Cooke and Zeeman [1976 (Pearson and Elsdale 1979; Elsdale and Davidson 1986)]; and (2) the progress zone model of Summerbell et al. (1973; Mee and French 1986a). A further discussion goes beyond the scope of this paper. Suffice it to say that the dependence of the specificity of the anomaly on the developmental stage can be explained by both models. They are not able, however, to explain the recent data of Primmett et al. (1989) on repetitive anomalies. In the next section it will be argued that these findings throw new light upon the possible nature of the regulation of determination which fits very well with the findings in *Lymnaea* (Boon-Niermeyer 1976).

3.4.2 Evidence for a Relationship Between Heat-Induced Anomalies
and the Cell Division Cycle

More and more evidence is accumulating to support the view that regulation of the cell division cycle plays an important role in determinative events. It comes mainly from two subjects of investigation, viz. (1) segmentation in vertebrates (amphibians and chick) and (2) neurogenesis in amniotes. The evidence is based largely on the outcome of heat shock experiments.

Although Veini and Bellairs (1986) already sugggested that "the same oscillating system which entrains the cells in somitomeres is involved in the response to heat shock", direct evidence to support this idea came some years later (Primmett et al. 1988, 1989.; Stern et al. 1988). In 2-day old chicken embryos, one single brief heat shock can generate repeated anomalies in the somites that develop after the shock. The first one arises six to seven somites behind the last somite formed at the moment of the shock. The next ones are again separated from each other by six to seven unaffected somites or a multiple of this. The six – to seven – somite interval corresponds to the value for the cell cycle duration of this tissue. A similar repetitive sequence of somite anomalies is induced by a variety of drugs that interfere with the cell cycle. On account of these data it is expected that a wave of cell divisions passes through the segmental plate. Indeed, the mitotic and ^{3}H-TdR-labeling indices display a pattern of high index separated from each other by regions, about six prospective somites long, of lower index (Stern and Bellairs 1984; Primmett et al. 1989). The cells which form the same somite are relatively synchronous over the range of at least two cell division cycles prior to segmentation. In the somite region synchrony is still maintained.

Evidence for the teratogenic effect of heat shock in homeothermic organisms, mammals and birds, has been reviewed by Edwards (1986). Hyperthermia is particularly damaging to the central nervous system (see Chaps. 4 and 5). The incidence and type of defect depends on the stage of development at the moment of heat exposure. Besides the effect on neurogenesis, there are a variety of other, less frequently studied defects, e.g., vertebral anomalies. I suppose they reflect an effect on the segmentation process of the mesoderm, as has been discussed above. Edwards mentions that proliferating cells are particularly sensitive to temperature elevations, resulting in arrest of mitotic activity and immediate death of cells in mitosis with threshold elevations of 1.5–2.5 °C and delayed death of cells at higher elevations (3.5 °C).

The particularly high sensitivity of neurogenesis may be understood in terms of the significance of cell death in the central nervous system. Firstly, the extent of cell division synchrony within a tissue will determine its vulnerability. In day-21 guinea pig embryos there are discrete zones along the ventricle with a high mitotic index. At the same positions and in the same frequency of mitotic cells in the controls, cells with clumped chromatin are present immediately after heat exposure. Cytoplasmic debris are leaking into the ventricle within 1–3 h and dead mitotic cells are rapidly removed. At the same time, there is a lag period before mitotic activity is resumed. The stage specificity of the effect comes from the fact that, at different stages, in different parts of the brain Anlage neurones are proliferating. A second reason for the high sensitivity of neural tissue certainly is the limited possibility of extra cell proliferation. After heat-induced death of neuroblast cells there is no compensatory growth. There are indications that the number of neurones present at the end of neurogenesis, determines the extent of the subsequent proliferation of glial cells. This may eventually lead to a proportional miniature and functionally deficient brain (Edwards 1986).

Recent investigations on heat shock also illustrate the thermosensitivity of the developing head and brain in 9.5-day-old rat embryos (presomitic, at neural tube closure). This stage represents a rapidly proliferating, migrating, and differentiating population of neuroectoderm cells, and heat shock induces specific neuroectoderm cell death. This results in major developmental defects of the eye and forebrain region (Walsh et al. 1987, 1989, see Chap. 4).

Shiota (1988) also observes pycnosis and reduced mitotic activity in mouse embryos after heat shock. He concludes that the induced neural tube defects may result from a temporary cessation of cell proliferation and partial necrosis of embryonic neuroepithelium.

In summary, in *Lymnaea* and many other organisms, heat shock effects on the cell cycle precede and cause developmental abnormalities. These general effects become unique because of the different prospective significance of the affected cells. This view has the advantage of specific results being explained by a general principle. It meets, furthermore, a number of observations on the effects of heat shock on morphogenesis:

1. The ubiquity of the anomalies induced by a variety of adverse influences at a given developmental stage is explained by their common effect of killing mitotic cells and deregulating the cell cycle.
2. During the blastulation period of many species, a high thermosensitivity has been shown; nonspecific developmental anomalies arise. These data fit well with the high extent of synchrony in cell divisions during this period and the relatively long duration of mitosis with respect to the rest of the cell cycle. There is, therefore, a high chance that many cells are hit, and, at the same time, that the final destination of these disturbed cells is in quite different parts of the later embryo, which leads to a multiplicity of malformations. After blastulation, the diversity between cells increases and only groups of cells, destined to form a discrete part of the body, will be synchronous, and therefore will be sensitive simultaneously to exposure to heat or other influences.

3.4.3 Hypothesis Concerning Determinative Events

The concept that heat induces discrete anomalies by mitotic interruption at specific stages postulates that synchrony within groups of cells destined to form a discrete Anlage is a general phenomenon. This led me to formulate the hypothesis that *determinative events make use of the synchronization of discrete populations of cells*. This hypothesis brings back the problem of regulation of morphogenesis to the far more general principle of the regulation of the cell cycle. Furthermore, cells passing through cell division synchronously loose their mutual affinity at the same time, and, more important, restore their mutual affinity simultaneously after division. This promotes the cells to organize into a group.

4 The Heat Shock Response During Development

4.1 Heat Shock Response and Thermotolerance

When living material is exposed to external stress situations, it reacts with a very characteristic and universal response, called the heat shock response. It is a very complex event with impact on the total cellular physiology. The structural and functional systems of the cell change drastically (cf. Nover 1984, 1991; Atkinson and Walden 1985; Burdon 1986; Craig 1985; Lindquist 1986). An evident feature is the transient (enhanced) expression of specific genes encoding the so-called heat shock proteins (Hsps). This phenomenon is universal and conservative, as appears from the following data: (1) it occurs in organisms as diverse as bacteria, plants and mammals (Schlesinger et al. 1982); (2) Hsps are synthesized in response to a variety of stress conditions, such as heat, hypoxia, exposure to ethanol, arsenite, and other chemicals (Li and Laszlo 1985); (3) there is a high degree of homology between Hsps of equivalent molecular weights from different origins (Craig 1985). These facts argue for a basic cellular function, representing a fundamental property of life.

Exposure to heat, or stress in general, affects not only the cellular physiology, but it evokes also a decrease in thermosensitivity, called thermotolerance. In one way or another the material seems to be prepared for a next deleterious challenge. Thermotolerance is transient and fades away under normal conditions. Many investigations show a coincidence between the enhanced synthesis of Hsps and the expression of thermotolerance (Li 1985; Li and Laszlo 1985; Subjeck and Shyy 1986). This suggests that Hsps have a homeostatic, protective or repair function. Recent reviews are dealing with the putative way they do so (Schlesinger 1986; Bienz and Pelham 1987). At least some Hsps play a role in the protein degradation system which may become overloaded by abnormal proteins caused by the shock. In addition, Nickells and Browder (1988) report evidence that Hsp 35 of *Xenopus* is equivalent to GAPDH, a glycolytic enzyme that meets the increasing demands for ATP after heat shock, as a result of the induction of a shift in energy metabolism to anaerobic glycolysis.

In the meantime, there is a growing number of observations indicating that the correlation between thermotolerance and Hsps is not as straight forward as it appeared to be initially. Thermotolerance can arise in the absence of protein synthesis in mammalian cells (Landry and Chrétien 1983; Tomasovic et al. 1983; Van Wijk et al. 1984; Widelitz et al. 1986), in yeast (Hall 1983; Watson et al. 1984), and in *Tetrahymena* (Hallberg et al. 1985). Thermotolerance has also been shown to occur without any concomitant enhanced synthesis of Hsps in growing pollen tubes of plants (Altschuler and Mascarenhas 1982) and in mammalian cells (Haveman et al. 1986). In Section 4.3 below, data on *Lymnaea* will be discussed, showing evidence that two states of thermotolerance have to be distinguished, one not correlated, the other one correlated with Hsp synthesis.

4.2 Definition of Thermotolerance

Before describing the experiments on *Lymnaea* and comparing the outcome with the literature, it is necessary to give a clear definition of thermotolerance, since the term is used in quite different senses. In cell line studies, it is used for any stress-induced increase in heat resistance, which is transient and not genetically determined (Hahn 1982). Also the term "acquired thermotolerance" is used to emphasize that it is a response on external influences. Corresponding with its definition, thermotolerance is expressed as an externally induced change in thermosensitivity, whatever the parameter used to express thermosensitivity itself (see Sect. 3.1).

A quite different phenomenon, which should be clearly distinguished from the above formulated thermotolerance, is the change in thermosensitivity which occurs intrinsically in the material, without any influence from outside. This is the case, for instance, during the development of an organism, during the cell cycle, and at distinct differentiation states of the cells (Bond and Schlesinger 1987; Nover 1991). In a variety of species, like *Drosophila, Xenopus*, mouse and sea urchin, an increase in thermoresistance from the early cleavage embryo to the blastulation phase has been reported. In their review, Browder et al. (1989) call it thermotolerance and they argue as follows:

"the term thermotolerance can also be applied to the developmentally acquired ability of an organism to survive heat shock at a temperature that would be lethal at earlier stages. Thermotolerance, in this instance, is due to developmental events at ambient temperature rather than previous exposure to a sublethal heat shock".

There are three serious objections to this argumentation: (1) it deviates from the common definition of thermotolerance used for other than developmental material; (2) it excludes the possibility to describe an increase instead of a decrease in thermosensitivity during development; (3) it excludes the possibility to use the term thermotolerance for the change in thermosensitivity induced by external stress during development. We prefer, therefore, to apply the description of a developmentally regulated, intrinsic change in thermosensitivity, and to keep the term thermotolerance for externally induced decrease in thermosensitivity.

4.3 Thermotolerance in *Lymnaea*: Kinetics and Hsp Synthesis

In *Lymnaea*, several qualities of thermotolerance have been investigated in three day-old trochophore larvae. Pilot experiments proved this stage to be able to become thermotolerant (Boon-Niermeijer and Van de Scheur 1984) and determination of thermosensitivity is quick and easy (Sect. 3.3). In principle, the experimental set up is the following. Larvae are exposed to a conditioning treatment (CT) inducing thermotolerance. The CT itself is harmless. After an interval (I) at normal conditions, thermosensitivity is measured. This can be done either by exposure to a test treatment (TT) of fixed duration and temperature (Fig. 5B, C), or by means of a survival curve for a fixed temperature (see Fig. 6A). The curve reveals a LD_{50} -value, i.e., the heat dose in minutes that kills 50% of the larvae. Conditioning treatments with heat (Boon-Niermeijer et al. 1986, 1988a), 2,4-dinitrophenol (DNP) (Boon-Niermeijer et al. 1987), and ethanol (Boon-Niermeijer et al. 1988a) evoke characteristic kinetic patterns of thermotolerance, related to quantitative and qualitative changes of protein synthesis. On account of the data, we proposed the existence of two states of thermotolerance, the a- and ß-state. The a-state is evoked by mild and severe CTs. It arises very quickly, independently of protein synthesis and is not correlated with Hsp synthesis. The ß-state arises only after rather severe CTs. It takes more time to develop and is associated with the synthesis of Hsps. Figure 5 shows some data to substantiate this model.

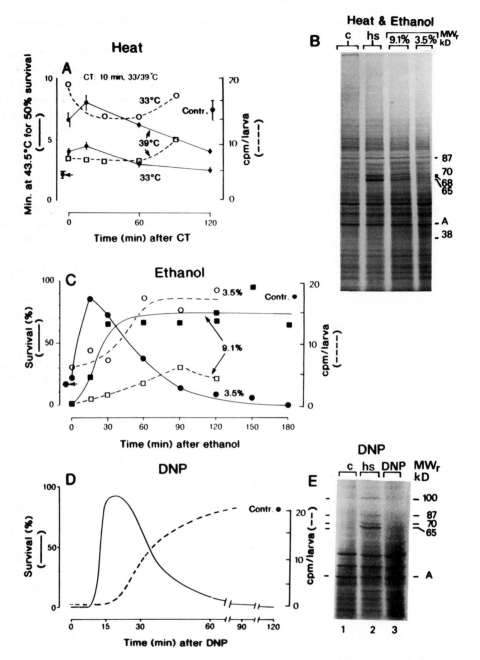

Fig. 5A-E. The effect of heat, ethanol, and 2, 4-dinitrophenol on thermosensitivity and quantitative and qualitative aspects of protein synthesis. Thermotolerance induced by preexposure to **A** *Heat* (10 min, 33 or 39 °C) , **C** *Ethanol* (45 min, 3.5 or 9.1%), or **D** *DNP* (10 min, 2.5 mM) was evaluated (*full lines*) as increase in LD_{50}-value [heat dose (min at 43.6 °C) killing 50% of the larvae] in **A**, or as percentage of survival after treatment at 43.6 °C during 3.5 min in **C** or 5 min in **D**. Value of control group indicated by *arrow* on the *left ordinate*. Effect of the same preexposures on ^{35}S-methionine incorporation (**A, C, D,** *broken lines*) and on the pattern of protein synthesis (**B, E,** 1 -h labeling with ^{35}S-methionine). M_r of some Hsps and actin A are indicated in kD

When the temperature rises from 25 to 33 or 39 °C during 10 min, the larvae become thermotolerant, as appears from the increase in LD_{50} -value above the control level (Fig. 5A). The quick increase is due to the α-state of thermotolerance, since the larvae are thermotolerant notwithstanding a considerable inhibition of the overall protein synthesis by treatment at 39 °C. The independence of the α-state on protein synthesis is confirmed by experiments with puromycin (Boon-Niermeijer et al. 1986). The longer maintenance of thermotolerance after 39 °C is ascribed to the ß-state. It coincides with the induction of synthesis of some Hsps (MWs 87, 70, 65 kD; Fig. 5B). This does not take place after a 33 °C-treatment (Boon-Niermeijer et al. 1986).

After preexposure to 3.5 or 9.1% ethanol during 45 min, larvae become thermotolerant within 15–30 min, as appears from the rise in survival after an exposure to 43.6 °C during 3.5 min (Fig. 5C). The quick rise is again due to the α-state, as it takes place without a significant level of protein synthesis and without the induction of Hsp synthesis (Fig. 5B). The more steady state of thermotolerance (ß-state) is induced by the high ethanol concentration, coinciding with the induction of synthesis of the Hsps 87, 70, and 65.

A 10 min exposure to 2.5 mM DNP evokes a very sharp increase in survival within an interval of 15 min (Fig. 5D). It appears to be a clear α-state of thermotolerance as it quickly fades away, and arises without the need of protein synthesis or enhanced Hsp synthesis (Fig. 5E). When the DNP conditions are more severe, no ß-state is induced, but treatment becomes noxious. In mammalian cells, DNP also induced thermotolerance but not the synthesis of Hsps (Haveman et al. 1986).

It is not clear to what extent the model of two states of thermotolerance has a general character, although several facts sustain this view. At the first place it seems to hold true also for a developmental stage of a quite unrelated species, the 9.5-day-old rat embryo (Walsh et al. 1987). The conditions are comparable with those for *Lymnaea*: a nonteratogenic heat shock (10 min at 42 °C) causes embryos to acquire thermotolerance against a severely teratogenic shock of 7.5 min at 43 °C during a 15 min recovery period at 38.5 °C. Although a heat shock response is evoked, as evidenced by the induction of a 71-kDa Hsp, the authors conclude:

" the rapid kinetics of acquisition of thermotolerance argues against a major role for newly synthesized Hsps in this response... The acquisition of thermotolerance in 9.5-gestational day rat embryos appears to be an example of acquired thermotolerance that does not require substantial Hsp synthesis."

There is a second reason sustaining the view that the existence of two states of thermotolerance is universal and not restricted to some defined material. The indications, that besides the correlated induction of thermotolerance and Hsp synthesis, thermotolerance can be induced without protein and Hsp synthesis, come from such diverse organisms as mammalian cells, yeast, *Tetrahymena*, and higher plants. Moreover, the same two states of thermotolerance as mentioned above were described recently for mammalian cells (Laszlo 1988).

It is unclear which processes are involved in the α–state of thermotolerance. One could imagine that functioning of the Hsps present already in the unstressed situation is sufficient. Their inactive state may change into a functional one by chemical modification, e.g., phosphorylation, methylation, ADP-ribosylation, by association, e.g., with the cytoskeleton, or by translocation to specific cell compartments, e.g., the nucleus (see Boon-Niermeijer et al. 1988a).

4.4 Heat–Induced Changes in Thermosensitivity
and Gene Expression During Development

The change in thermosensitivity during development is in itself an intriguing problem. What makes the material more sensitive, at one stage or another, to immediate killing by heat? The differences in competence to synthesize Hsps and to become thermotolerant are often pleaded as an explanation for stage-dependent thermo-sensitivity. But, as argued in Section 3.3 above, there is strong evidence that just at the early cleavage stage the material is more thermoresistant, not only in *Lymnaea*, but generally. At this stage, however, the embryos are not able to synthesize Hsps in response to heat exposure (cf. Bond and Schlesinger 1987; Browder et al. 1989). What does coincide, is the high constitutive level of synthesis of the Hsc70. It is the very first gene activity of the zygote (Bensaude et al. 1983; Wittig et al. 1983; Howlett 1986; see Chap. 10).

The change in thermosensitivity during development, as well as the change in competence to develop thermotolerance in relation to the quantitative and qualitative protein synthesis, has been investigated in *Lymnaea* (Boon-Niermeijer et al. 1988b). Thermosensitivity increases considerably between day 0 and day 1 as demonstrated by an impressive shift of the survival curve to shorter heat treatments (Fig. 6A). At day 0 the LD_{50} -value is 2–3.5 times as high as at the other stages (Table 1). The constitutive pattern of proteins synthesized at the subsequent stages shows some changes (Fig. 6C). At day 0 there is a relatively high synthesis of the proteins with apparent molecular weights of 70 and 103 kD, whereas the relative synthesis of 68 kD and 96 kD proteins increases after day 0. It is uncertain how the low thermosensitivity at day 0 should be explained on the basis of these data. The high constitutive expression of the 70 kD-protein may play a role. That this may be a general phenomenon is sustained by the fact that a similar high constitutive expression of a 70 kD-protein has been observed in cleaving mammalian embryos (cited above) and fertilized *Xenopus* eggs (Browder et al. 1987). The function of the contitutively synthesized 70 kD-protein should be similar in that case to that of Hsp70, viz. the repair of heat-induced cellular damage (Pelham 1984; Lewis and Pelham 1985; Bienz and Pelham 1987).

A high level of thermotolerance can be induced during development, except at the early cleavage stage. Table 1 shows the thermotolerance, induced by a conditioning treatment of 10 min at 39 °C, and measured 15 min later. Whatever the conditions used, no thermotolerance could be induced at day 0. With respect to the effect on

Table 1. Thermosensitivity during development in controls and after heat exposure

Age (days)	LD_{50} (min at 43.6 °C)		ΔLD_{50}[b] (min at 43.6 °C)
	Control		
	$-CT$[a]	$+CT$[a]	
0	7.5	7.8	0.3
1	3.7	9.5	5.8
2	3.0	10.1	7.1
3	2.1 ± 0.27	7.6	5.5

[a] CT conditioning treatment (10 min at 39 °C).
[b] Thermotolerance 15 min after CT is evidenced by the increase in LD_{50}-values of heat-pretreated material above the control level (ΔLD_{50}).

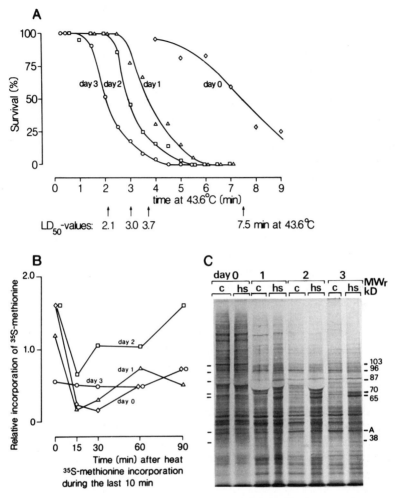

Fig. 6A-C. Thermosensitivity and the effect of heat on quantitative and qualitative aspects of protein synthesis during development of *Lymnaea*. **A** Thermosensitivity was evaluated in survival curves to 43.6 °C at the first four days of development. Each curve results in an LD_{50}-value, i. e., the heat dose killing 50% of the embryos (*arrows* along the abscissa). **B** Effect of 10 min at 39 °C on ^{35}S–methionine incorporation and **C** on the pattern of protein synthesis at the same stages as in **A**. **B** The control level at each stage was standardized to 1.0. The incorporation in the experimental groups was normalized to the control level of the corresponding stage. **C** Control (*c*) and heat shocked (*hs*) embryos were incubated in ^{35}S-methionine during 1 h and equal amounts of acid-insoluble radioactivity were loaded onto each lane for prominent bands changing their expression during development and/or by heat shock. M_r are indicated in kD; *A* actin (Boon-Niermeijer et al. 1988b)

overall protein synthesis, day 0 embryos respond in the same way as the later stages (Fig. 6B). An initial stimulation during the CT is followed by a sharp decrease during the first 15 min after CT and by a gradual repair. Only at day 3 is the initial stimulation absent. The protein synthesis patterns show that somewhere before day 1, but after the early cleavage stage, the embryos become able to execute the heat shock response, in the sense that the synthesis of several Hsps is induced (Hsps38, 65, 70, and 87 but not the 68 kD-protein, see Fig. 6C).

How do the data on *Lymnaea* fit with what is known on other species? How far have thermosensitivity and thermotolerance during development been studied in relation to constitutive and induced synthesis of Hsps? What makes a comparison rather difficult is, as pointed out above, the inconsequent use of the terms thermosensitivity and thermotolerance for different phenomena. This leads to a search for correlations between phenomena which in fact should not be compared at all. Therefore, in most of the reports on the early cleavage stage, thermosensitivity instead of thermotolerance has been related to the heat-induced synthesis of Hsps (reviewed by Browder et al. 1989). Only for the mouse has a study comparable to that on *Lymnaea* been carried out (Muller et al. 1985). The induction of thermotolerance and Hsp synthesis has been compared at the one-cell and the blastocyst stage. The one–cell embryo does not respond to preexposure to heat, whereas the blastocyst stage does, i.e., it becomes thermotolerant and synthesizes Hsps.

Concerning the acquisition of thermotolerance in later developmental stages, data are available on the frog (Pearson and Elsdale 1979; Elsdale and Davidson 1987). In later developmental stages of sea urchins (Sconzo et al. 1985), *Drosophila* (Mitchell and Petersen 1982; Petersen and Mitchell 1982), the brine shrimp *Artemia* (Miller and McLennan 1988a, b), and the rat (Mirkes 1987; Walsh et al. 1987, 1989), acquired thermotolerance arises together with the heat shock response. Walsh et al. (1989) confirm the data on *Lymnaea* concerning the quick rise of thermotolerance, and also the initial stimulation of overall protein synthesis (Fig. 6).

One would expect that the early stages, which are not able to respond to heat by synthesis of Hsps, might be able to develop the α-state of thermotolerance. This appears not to be the case. In fact, the conditions also at the later stages are such that the observed thermotolerance is due to the α-state, since it has been tested 15 min after the conditioning treatment. At that moment only the α-state has come to expression (Sect. 4.3 above). Also in rats the thermotolerance observed 15 min after pretreatment must be due to the α–state.

Recent observations throw new light on the pitfalls in the interpretation of qualitative patterns of protein synthesis induced by heat. Data may depend on the composition of the culture medium or the extraction method (Browder et al. 1987, 1989; Horrell et al. 1987; Davis and King 1989). Contradictory results on stage-dependent or tissue-specific expression of several Hsps may derive rather from different methods than from fundamentally different mechanisms.

In summary, it can be concluded that thermosensitivity as well as the capacity to develop thermotolerance are developmentally regulated. For both phenomena, the greatest transition takes place between the cleavage stage and the late blastula. In these transitions, a change in the regulation of gene expression might be involved, since in this developmental phase the activation of the embryonic genome starts. In this view, however, it is not clear why cleavage stage embryos are not able to develop the α–state of thermotolerance, which is expected not to depend on gene activity.

The data in the field of thermosensitivity and thermotolerance during development are rather fragmentary, chaotic, and seemingly contradictory. I plead, therefore, for a much stricter use of criteria, firstly in the evaluation of thermosensitivity, and secondly in the concept of thermotolerance. Only then can the comparison of phenomena be carried out properly and proper correlations can be made. Thermotolerance, not thermosensitivity, should be coupled to the heat-induced changes in the protein synthesis pattern. This is usually done in all studies except, surprisingly, in those on embryonic development.

It can be concluded that, during development from a transition point during blastulation onwards, the heat shock response and the induction of thermotolerance do not seem to differ fundamentally from those in non-embryological material. A rather unexplored field, however, which needs further investigation is the high intrinsic thermoresistance and the inability to develop additional thermotolerance at the early cleavage stage.

Acknowledgments. I thank Dr. R. Van Wijk and Dr. M.R. Dohmen for their valuable suggestions and criticism.

References

Altschuler M, Mascarenhas JP (1982) The synthesis of heat shock and normal proteins at high temperatures in plants and their possible roles in survival under heat stress. In: Schlesinger MJ, Ashburner M, Tissières A (eds) Heat shock from bacteria to man. Cold Spring Harbor Laboratory, New York, pp 321–327

Atkinson BG, Walden DB (eds) (1985) Changes in eukaryotic gene expression in response to environmental stress. Academic Press, Orlando

Bensaude O, Babinet C, Morange M, Jacob J (1983) Heat shock proteins, first major products of zygotic gene activity in mouse embryo. Nature 305: 331–333

Bergh S, Arking R (1984) Developmental profile of the heat shock response in early embryos of *Drosophila*. J Exp Zool 231:379–391

Bienz M (1984) Developmental control of the heat shock response in *Xenopus*. Proc Natl Acad Sci USA 81:3138–3142

Bienz M, Pelham HRB (1987) Mechanisms of heat shock gene activation in higher eukaryotes. Adv Genet 24:31–72

Bond U, Schlesinger MJ (1987) Heat shock proteins and development. Adv Genet 24:1–29

Boon-Niermeijer EK (1975) The effect of puromycin on the early cleavage cycles and morphogenesis of the pond snail *Lymnaea stagnalis*. Wilhelm Roux's Arch 177:29–40

Boon-Niermeijer EK (1976) Morphogenesis after heat shock during the cell cycle of *Lymnaea*: a new interpretation. Wilhelm Roux's Arch 180:241–252

Boon-Niermeijer EK (1987) Responses of a developing organism upon heat stress. A study on *Lymnaea stagnalis*. Thesis, State University of Utrecht, Utrecht

Boon-Niermeijer EK, Van de Scheur H (1984) Thermosensitivity during embryonic development of *Lymnaea stagnalis* (Mollusca). J Therm Biol 9:265–269

Boon-Niermeijer EK, Tuyl M, Van de Scheur H (1986) Evidence for two states of thermotolerance. Int J Hyperthermia 2:93–105

Boon-Niermeijer EK, Souren JEM, Van Wijk R (1987) Thermotolerance induced by 2,4-dinitrophenol. Int J Hyperthermia 3:133–141

Boon-Niermeijer EK, Souren JEM, De Waal AM, Van Wijk R (1988a) Thermotolerance induced by heat and ethanol. Int J Hyperthermia 4:211–222

Boon-Niermeijer EK, De Waal AM, Souren JEM, Van Wijk R (1988b) Heat-induced changes in thermosensitivity and gene expression during development. Dev Growth Differ 30:705–715

Browder LW, Pollock M, Heikkila JJ, Wilkes J, Wang T, Krone P, Ovsenek N, Kloc M (1987) Decay of the oocyte-type heat shock response of *Xenopus laevis*. Dev Biol 124:191–199

Browder LW, Pollock M, Nickells RW, Heikkila JJ, Winning RS (1989) Developmental regulation of the heat-shock response. In: DiBerardino M, Etkin LD (eds) Developmental biology: a comprehensive synthesis. Plenum, New York, pp 97–147 (Genomic adaptability in somatic cell specialization, vol 6)

Burdon RH (1986) Heat shock and the heat shock proteins. Biochem J 240:313–324

Cooke J, Zeeman EC (1976) A clock and wavefront model for control of the number of repeated structures during animal morphogenesis. J Theor Biol 58:455–476

Craig EA (1985) The heat shock response. CRC Crit Rev Biochem 18:239-280

Davis RE, King ML (1989) The developmental expression of the heat-shock response in *Xenopus laevis*. Development 105:213–222

Eberlein S, Mitchell HK (1987) Protein synthesis patterns following stage-specific heat shock in early *Drosophila* embryos. Mol Gen Genet 210:407–412

Edwards MJ (1986) Hyperthermia as a teratogen : a review of experimental studies and their clinical significance. Teratog Carcinog Mutagen 6:563–582

Elsdale T, Davidson D (1986) Somitogenesis in the frog. In: Bellairs R, Ede DA, Lash JW (eds) Somites in developing embryos. Plenum, New York, pp 119–134

Elsdale T, Davidson D (1987) Timekeeping by frog embryos, in normal development and after heat shock. Development 99:41-49

Elsdale T, Pearson M, Whitehead M (1976) Abnormalities in somite segmentation following heat shock to *Xenopus* embryos. J Embryol Exp Morphol 35:625–635

Geilenkirchen WLM (1966) Cell division and morphogenesis of *Limnaea* eggs after treatment with heat pulses at successive stages in early division cycles. J Embryol Exp Morphol 16:321–337

German J (1984) Embryonic stress hypothesis of teratogenesis. Am J Med 76:293-301

German J, Louie E, Banerjee D (1986) The heat shock response in vivo: experimental induction during mammalian organogenesis. Teratog Carcinog Mutagen 6:555–562

Graziosi G, Micali F, Marzari R, De Christini F, Savioni A (1980) Variability of response of early *Drosophila* embryos to heat shock. J Exp Zool 214:141–145

Hahn GM (1982) Hyperthermia and cancer. Plenum, New York

Hall BG (1983) Yeast thermotolerance does not require protein synthesis. J Bacteriol 157:1363–1365

Hallberg RL, Kraus KW, Hallberg EM (1985) Induction of acquired thermotolerance in *Tetrahymena thermophila*: effects of protein synthesis inhibitors. Mol Cell Biol 15:2061–2069

Haveman J, Li GC, Mak JY, Kipp JBA (1986) Chemically induced resistance to heat treatment and stress protein synthesis in cultured mammalian cells. Int J Radiat Biol 50:51–64

Heikkila JJ, Kloc M, Bury J, Schultz GA, Browder LW (1985) Acquisition of the heat-shock response and thermotolerance during early development of *Xenopus laevis*. Dev Biol 107:483–489

Horrell A, Shuttleworth J, Colman A, (1987) Transcript levels and translational control of hsp70 synthesis in *Xenopus* oocytes. Genes & Dev 1:433–444

Howlett SK (1986) The effect of inhibiting DNA replication in the one-cell mouse embryo. Wilhelm Roux's Arch Dev Biol 195:499–505

Kimmel CB, Sepich DS, Trevarrow B (1988) Development of segmentation in zebrafish. Development 104, suppl: 197-207

King ML, Davis R (1987) Do *Xenopus* oocytes have a heat shock response? Dev Biol 119:532–539

Landry J, Chrétien P (1983) Relationship between hyperthermia-induced heat-shock proteins and thermotolerance in Morris hepatoma cells. Can J Biochem Cell Biol 61:428–437

Laszlo A (1988) Evidence for two states of thermotolerance in mammalian cells. Int J Hyperthermia 4:513–526

Lewis MJ, Pelham HRB (1985) Involvement of ATP in the nuclear and nucleolar functions of the 70 kD heat shock protein. EMBO J 4:3137–3143

Li GC (1985) Elevated levels of 70,000 dalton heat shock protein in transiently thermotolerant Chinese hamster fibroblasts and in their stable heat resistant variants. Int J Radiat Oncol Biol Phys 11:165–177

Li GC, Laszlo A (1985) Thermotolerance in mammalian cells: a possible role for heat shock proteins. In: Atkinson BG, Walden DB (eds) Changes in eukaryotic gene expression in response to environmental stress. Academic Press, Orlando, pp 227–254

Lindquist S (1986) The heat shock response. Annu Rev Biochem 55:1151–1191

Mee JE, French V (1986a) Disruption of segmentation in a short germ insect embryo. I. The location of abnormalities induced by heat shock. J Embryol Exp Morphol 96:245–266

Mee JE, French V (1986b) Disruption of segmentation in a short germ insect embryo. II. The structure of segmental abnormalities induced by heat shock. J Embryol Exp Morphol 96:267–294

Miller D, McLennan AG (1988a) The heat shock response of the cryptobiotic brine shrimp *Artemia*. I. A comparison of the thermotolerance of cysts and larvae. J Therm Biol 13:119-123

Miller D, McLennan AG (1988b) The heat shock response of the cryptobiotic brine shrimp *Artemia*. II. Heat shock proteins. J Therm Biol 13:125–134

Mirkes PE, (1987) Hyperthermia-induced heat shock response and thermotolerance in postimplantation rat embryos. Dev Biol 119:115–122

Mitchell HK, Petersen NS (1982) Heat shock induction of abnormal morphogenesis in *Drosophila*. In: Schlesinger MJ, Ashburner M, Tissières A (eds) Heat shock from bacteria to man. Cold Spring Harbor Laboratory, New York, pp 337–344

Muller WU, Li GC, Goldstein LS (1985) Heat does not induce synthesis of heat shock proteins or thermotolerance in the earliest stage of mouse embryo development. Int J Hyperthermia 1:97–102

Nickells RW, Browder LW (1985) Region-specific heat-shock protein synthesis correlates with a biphasic acquisition of thermotolerance in *Xenopus laevis* embryos. Dev Biol 112:391–395

Nickells RW, Browder LW (1988) A role for glyceraldehyde-3-phosphate dehydrogenase in the development of thermotolerance in *Xenopus laevis* embryos. J Cell Biol 107:1901–1909

Nover L (ed) (1984) Heat shock response in eukaryotic cells. Springer, Berlin Heidelberg New York Tokyo

Nover L (ed) (1991) Heat shock response. CRC Press, Boca Raton, Fd

Pearson M, Elsdale T (1979) Somitogenesis in amphibian embryos. I. Experimental evidence for an interaction between two temporal factors in the specification of somite pattern. J Embryol Exp Morphol 51:27–50

Pelham HRB (1984) Hsp 70 accelerates the recovery of nucleolar morphology after heat shock. EMBO J 3:3095–3100

Petersen NS, Mitchell HK (1982) Effects of heat shock on gene expression during development: induction and prevention of the multihair phenocopy in *Drosophila*. In: Schlesinger MJ, Ashburner M, Tissières A (eds) Heat shock from bacteria to man. Cold Spring Harbor Laboratory, New York pp 345–352

Primmett DRN, Stern CD, Keynes RJ (1988) Heat shock causes repeated segmental anomalies in the chick embryo. Development 104:331–339

Primmett DRN, Norris WE, Carlson GJ, Keynes RJ, Stern CD (1989) Periodic segmental anomalies induced by heat shock in the chick embryo are associated with the cell cycle. Development 105:119–130

Raven C P (1942) The influence of lithium upon the development of the pond snail, *Limnaea stagnalis* L. Proc K Ned Akad Wet, Amsterdam 45:856–860

Raven C P (1975) Development. In: Fretter V, Peake J (eds) Functional anatomy and physiology (Pulmonates, vol 1). Academic Press, New York, pp 367–400

Ritossa FM (1962) A new puffing pattern induced by heat shock and DNP in *Drosophila*. Experientia 18:571–573

Roccheri MC, Di Bernardo MG, Giudice G (1981) Synthesis of heat-shock proteins in developing sea urchins. Dev Biol 83:173–177

Schamhart DHJ, Van Walraven HS, Wiegant FAC, Linnemans WAM, Van Rijn J, Van den Berg J, Van Wijk R (1984) Thermotolerance in cultured hepatoma cells: cell viability, cell morphology, protein synthesis, and heat shock proteins. Radiat Res 98:82–95

Schlesinger MJ (1986) Heat shock proteins: the search for functions. J Cell Biol 103:321–325

Schlesinger MJ, Ashburner M, Tissières A (eds) (1982) Heat shock from bacteria to man. Cold Spring Harbor Laboratory, New York

Sconzo G, Roccheri MC, Oliva D, La Rosa M, Giudice G (1985) Territorial localization of heat shock mRNA production in sea urchin gastrulae. Cell Biol Int Rep 9:877–881

Shiota K (1988) Induction of neural tube defects and skeletal malformations in mice following brief hyperthermia in utero. Biol Neonate 53:86–97

Slack JMW (1983) From egg to embryo: determinative events in early development. Cambridge University Press, Cambridge (Developmental and cell biology series, vol 13)

Stern CD, Bellairs R (1984) Mitotic activity during somite segmentation in the early chick embryo. Anat Embryol 169:97–102

Stern CD, Fraser SE, Keynes RJ, Primmett DRN (1988) A cell lineage analysis of segmentation in the chick embryo. Development 104, suppl:231–244

Subjeck JR, Shyy TT (1986) Stress protein systems of mammalian cells. Am J Physiol 250:C1–C17

Summerbell D, Lewis JH, Wolpert L (1973) Positional information in chick limb morphogenesis. Nature 224:492–496

Timmermans LPM (1969) Studies on shell formation in Molluscs. Neth J Zool 19:417–523

Tomasovic SP, Steck PA, Heitzman D (1983) Heat-stress proteins and thermal resistance in rat mammary tumor cells. Radiat Res 95:399–413

Van den Biggelaar JAM (1971a) Timing of the phases of the cell cycle with tritiated thymidine and Feulgen cytophotometry during the period of synchronous division in *Lymnaea*. J Embryol Exp Morphol 26:351–366

Van den Biggelaar JAM (1971b) Timing of the phases of the cell cycle during the period of asynchronous division up to the 49-cell stage in *Lymnaea*. J Embryol Exp Morphol 26:367–391

Van Wijk R, Otto AM, Jimenez de Asua L (1984) Effects of serum and growth factors on heat sensitivity in Swiss mouse 3T3 cells, J Cell Physiol 119:155–162

Veini M, Bellairs R (1986) Heat shock effects in chick embryos. In: Bellairs R, Ede DA, Lash JW (eds) Somites in developing embryos. Plenum, New York, pp 135–145

Verdonk NH (1965) Morphogenesis of the head region in *Lymnaea stagnalis* L. Thesis, University of Utrecht, Utrecht

Verdonk NH, De Groot SJ (1970) Periodic changes in sensitivity of *Lymnaea* eggs to a heat shock during early development. Proc K Ned Acad Wet C73:171–185

Visschedijk AHJ (1953) The effect of a heat shock on morphogenesis in *Lymnaea stagnalis*. Proc K Ned Akad Wet C56:590–596

Walsh DA, Klein NW, Hightower LE, Edwards MJ (1987) Heat shock and thermotolerance during early rat embryo development. Teratology 36:181–191

Walsh DA, Li K, Speirs J, Crowther CE, Edwards MJ (1989) Regulation of the inducible heat shock 71 genes in early neuronal development of cultured rat embryos. Teratology 40:321–334

Watson K, Dunlop G, Cavicchioli R (1984) Mitochondrial and cytoplasmic protein syntheses are not required for heat shock acquisition of ethanol and thermotolerance in yeast. FEBS Lett 172:299–302

Widelitz RB, Magun BE, Gerner EW (1986) Effects of cycloheximide on thermotolerance expression, heat shock protein synthesis, and heat shock protein mRNA accumulation in rat fibroblasts. Mol Cell Biol 6:1088–1094

Wiegant FAC, Tuyl M, Linnemans WAM (1985) Calmodulin inhibitors potentiate hyperthermic cell killing. Int J Hyperthermia 1:157–169

Wittig S, Hensse S, Keitel C, Elsner C, Wittig B (1983) Heat shock gene expression is regulated during teratocarcinoma cell differentiation and early embryonic development. Dev Biol 96:507–514

2 Environmentally Induced Development Defects in *Drosophila*

Nancy S. Petersen[1] and Herschel K. Mitchell[2]

1 Historical Background

The interactions of genes and environment during development have been fascinating, controversial subjects for more than a century. One of the earliest reports of environmental effects on insect development was made by George Dorfmeister in 1854. He demonstrated that exposure to extreme heat or cold could change the pattern on the butterfly's wings. More recent studies on environmental effects on insect development have primarily been done in fruit flies. *Drosophila* is used as a model system for studying developmental genetics because of its short life cycle and small genome size. These same attributes also make it a good model system for studying environmental effects on gene expression during development. Goldschmidt was the first to extensively study environmental influences on development in *Drosophila*. He showed that many different developmental defects which resemble mutant defects can be induced by heating during the pupal period. Unlike mutations, however, these defects are not inherited. Some examples of heat induced defects compared to mutant defects are shown in Fig. 1. Goldschmidt coined the name "phenocopy" to describe environmentally induced developmental defects. This name was chosen in order to emphasize the resemblance of the environmentally induced defects to mutant phenotypes (Goldschmidt 1935). Goldschmidt was impressed by the fact that chemicals as well as heat induced the same types of defects if they were given during the same sensitive period. He concluded that mutations, chemicals, and heat, might all affect a different initial chemical process, but the end result was the same because they diverted development from a normal path onto an abnormal path, and he felt that there are a limited number of abnormal developmental pathways which are not lethal.

"All our experiments indicate that a shock treatment within certain time limits during this sensitive period not only affects the immediate development of the organ but forces later developmental processes into such channels that the end product, the phenotype, is indistinguishable from that of a mutant. Normal environmental conditions and the usual or 'wild type' genetic constitution of organism will send the developing organ down one road: on the other hand, either an appropriate shock treatment or a different hereditary factor will send it down a new path which ends in a common result although the causes are different... . Mutant and phenocopy, then, look alike because changed genetic action as well as action by a phenotypic agent is limited to definite tracks. There can be no doubt that in the last analysis the primary change produced both by phenocopic agent and mutated hereditary material must be of chemical nature" (Goldschmidt 1949).

[1] Department of Molecular Biology, University of Wyoming, Laramie, Wyoming, USA
[2] Biology Division, California Institute of Technology, Pasadena, California, USA

Results and Problems in Cell Differentiation 17
Heat Shock and Development
Hightower and Nover (Eds.)
© Springer-Verlag Berlin Heidelberg 1991

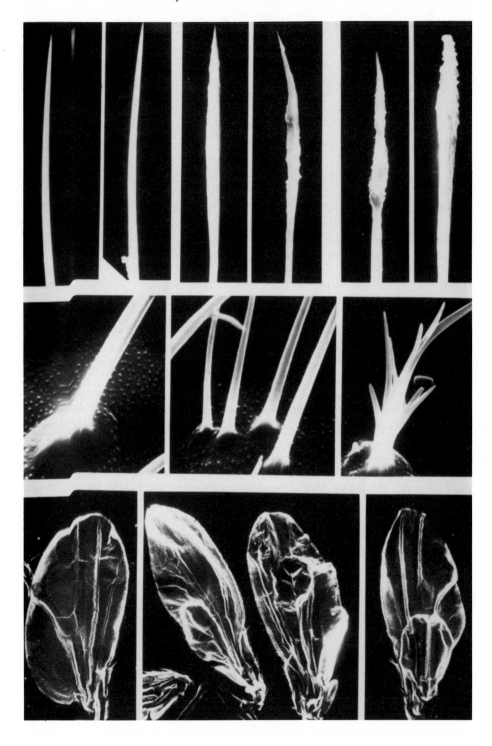

Goldschmidt's experiments and theories were not appreciated by most geneticists, and were largely ignored until the mid-sixtieth century, when Mitchell and Milkman started their studies on phenocopies. Milkman studied the crossveinless phenocopy. He showed that the severity of the crossveinless defect depended on the genetic background of the flies, providing evidence for interaction of genetics and environment in phenocopy induction (Milkman 1966). Milkman also proposed that the molecular mechanism behind phenocopy induction was protein denaturation. The phenomenon of phenocopies also fascinated Hadorn (Gloor 1947; Hadorn 1961) who interested Mitchell in the subject in the 1950s. Mitchell initially studied the blond phenocopy, which is due to the failure of phenol oxidase to be activated at the correct time in development and results in pale yellow rather than black bristles on the thorax of the fly (Mitchell 1966). The blond phenocopy closely resembles the phenotypes of the blond and straw mutations. The sensitive period for induction of the blond phenocopy by heat in normal pupae is 2–10 h earlier than the time of normal activation of phenol oxidase. In fact, phenol oxidase is present in extracts of both the mutant and the heat shocked pupae, indicating that processes more complicated than simple enzyme denaturation are at work. This observation led Mitchell to propose that heat affects the expression of a gene which codes for a protein involved in the phenol oxidase activation cascade (Mitchell 1966; Seybold et al. 1975).

Chemical treatments also can result in phenocopies. These are somewhat more difficult to study than heat induced phenocopies because of the problems in introducing chemicals through the pupal case, and because once introduced, a chemical cannot be removed. Rizki (1960) and Fristrom (1965) showed that a phenocopy of the mutant cryptocephal can be induced by feeding larvae N-acetyl glucosamine. Cryptocephal is a mutation in which the head appears to be missing because it fails to evert during pupation. The phenocopy is caused by N-acetyl glucosamine incorporation into the cuticle, resulting in a thicker cuticle layer which physically prevents the eversion of the head. This type of phenocopy appears to be an example of a very specific effect of a chemical on a specific process rather than a general mechanism. Ethanol also causes phenocopies in *Drosophila*. Larvae fed on 4–14% ethanol develop into abnormal adults with missing or malformed legs, wings, mouth parts or halteres (Ranganathan et al. 1987). Because more than one type of defect can be induced, heat and ethanol seem be examples of phenocopies which may be explained by a general mechanism. Now, due to progress in molecular biology, methods are available to determine how these phenocopies are induced and whether or not there is a basic general mechanism involved. This chapter will focus on how recent research has been able to answer some questions and make progress toward answering others.

Fig. 1. Mutants and phenocopies. Three examples illustrate similarities and differences between mutants and their phenocopies. In each example, the normal structure is on the *left*, the mutant structure is in the *middle*, and the phenocopy is on the *right*. *Top row*, the mutant *javalin* has arrowheadlike tips on the scutellar bristles rather than the normal points. In this case the spear tip phenocopy closely resembles the mutant. *Middle row*, the *multiple wing hair* mutant has multiple hairs per cell compared to the normal one hair per cell. The branched hair phenocopy is quite different having multiple branches on a single hair. *Lower row*, the phenocopy of the mutant *balloon* has a blister at the base of the wing. The mutant has very similar blisters but they are found in various positions

2 Phenocopy Induction

2.1 Conditions for Induction of Phenocopies

2.1.1 Sensitive Periods

Phenocopies in *Drosophila* can be induced in early embryos and in pupae. These are times when rapid changes in gene expression and cell shape are taking place. Phenocopies induced in embryos have very short sensitive periods, an example being the bithorax phenocopy (resembling the mutant for which it is named) which can be induced during a 10-min sensitive period at the cellular blastoderm stage by heat or ether (Gloor 1947; Santamaria 1979). Short high temperature shocks (2–3-min heat shocks at 43 °C) can also induce pair rule phenocopies in mid-blastoderm embryos, and germ band shortening in gastrulation stage embryos (Eberlein 1986; Eberlein and Mitchell 1987). However, conditions for getting 100% of each type of phenocopy have not been worked out due to difficulties in staging. In pupae, by contrast, sensitive periods for induction of phenocopies are usually 1–2 h long and it is relatively easy to find conditions where 100% of the pupae develop the same defect. Phenocopies in pupae are induced by short high temperature treatments of 30–40 min at 40–40.5 °C which are often lethal to some fraction of the pupae. Interestingly, these pupae do not die immediately; they continue to develop, but fail to eclose. If they are dissected from the pupal case, they have the same kind of defect as the survivors; often they are more severely affected.

2.1.2 Timing

The time when the pupae must be heated in order to produce a particular defect is called the sensitive period for the defect, and will be given in this paper in hours after white prepupa formation for development at 25 °C. Thirty-five sensitive periods for pupal defects have been described by Mitchell and Petersen (1982). The importance of timing of phenocopy sensitive periods can be illustrated using bristle defects. During mid-pupal period when thoracic bristles are developing, a heat shock within a 2 h period from 35–37 h, will produce hooked bristles; 2 h later from 39 to 42 h, smooth bristles (missing normal fluting) are induced. From 43–45 h, heat shocks result in bristles with a spear tip and a smooth base (Mitchell and Lipps 1978). Clearly, each of these sensitive periods reflects a different normal stage in bristle development which is sensitive to heat.

Similar structures on different parts of the body have different phenocopy sensitive periods. For example, many parts of the *Drosophila* body have bristles and/or tricombs (cell hairs). Phenocopies affecting these structures have different sensitive periods on the different parts of the body. The sensitive period for the hook phenocopy, which results in characteristic sharp bend in the bristles, is 35–37 h for the large bristles on the back of the thorax, 43–44 h for the bristles on the head, and 46 h for the bristles on the abdomen (Mitchell and Lipps 1978; Mitchell and Petersen 1982). The most easily induced phenocopy affecting hairs (tricombs) results in branched rather than straight hairs. This phenocopy also can be induced at different times on different parts of the body, on wings from 38–40 h, on the thorax from 43–46 h, on the legs from 41-45 h, and on the abdomen from 46–50 h. The differences in timing of these sensitive periods correspond to differences in the timing of the pattern of protein synthesis associated with epithelial cell differentiation and hair construction (Mitchell and Petersen 1983).

2.1.3 Heating Conditions

While the timing of the treatment determines the type of defect induced, the intensity and duration of the heat treatment affect the severity of the effects observed. Flies are normally raised at 18–25 °C. As mentioned above, heat treatments which induce phenocopies are usually on the verge of being lethal. The longer the exposure time and the higher the temperature, the more extreme the defect and the more animals affected. For the hook bristle phenocopy mentioned previously, this means that as the severity of the heat shock increases, the hook goes from being a mild bend in the bristle (40.2 °C) to a sharp split looking bend (40.6 °C), and from affecting one or two scutellar bristles on each fly under mild conditions to affecting all scutellar bristles under more extreme conditions. Even milder treatments, less than 40 °C, usually do not induce phenocopies. In fact, temperatures in the 33–37 °C range which are above the normal growing temperatures, but are not immediately lethal, induce the development of thermotolerance (discussed below) which will prevent the induction of phenocopies during a subsequent high temperature treatment.

2.2 Induction of Phenocopies in Recessive Mutant Heterozygotes

Besides phenocopies induced in wild-type flies, phenocopies of a recessive mutant can be induced in some recessive mutant heterozygotes. In mutant heterozygotes of *forked, singed*, and *multiple wing hair* (mwh), an additional phenocopy can be induced which resembles the mutant phenotype (Petersen and Mitchell 1987). Each mutant has a different sensitive period during which its phenocopy can be induced in the mutant heterozygote, and this phenotype is not induced in wild-type siblings by heating during this period. The multiple wing hair phenocopy is illustrated in Fig. 2. This phenocopy is particularly striking because it results in two to five hairs per cell. Normally there is one hair per cell. The multiple wing hair phenocopy can be induced by heating *mwh*/+ pupae between 32 and 33 h. Heating during the same period does not induce the phenocopy in wild-type flies. The same phenomenon is observed in *forked* or *singed* mutant heterozygote (Mitchell and Petersen 1985). Since both *forked* and *singed* are on the X chromosome, when wild-type females crossed to mutant males the resulting females are heterozygous for the mutation and sensitive to heat, while their male siblings are wild-type and resistant. An interpretation of this phenomenon is that heating affects the expression of the normal allele of the mutant gene which is already expressed at about half the normal level because there is only one copy of the gene. The further reduction in gene expression caused by the heat treatment is enough to lower the level of the gene product below the critical level required for normal construction of the affected structure. If a gene product is required at a specific time in development, then delaying the expression of the gene could also result in a lower level

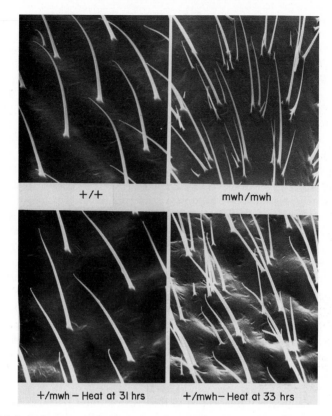

Fig. 2. Multiple hair phenocopy – the phenocopy of a recessive mutant. The multiple wing-hair mutation causes cells to produce several hairs per cell instead of a single hair. This mutation is recessive, so the mutant heterozygote has the normal (wild-type) phenotype. However, if the heterozygote is heated at 33 h of pupal development it will show the mutant phenotype. The illustration shows scanning-EM pictures of wing cells from normal adult flies (+/+), multiple wing-hair flies (*mwh/mwh*), and mutant heterozygotes which have been heated at two different times during pupal development. Only (+/*mwh*) flies heated during the sensitive period (32–33 h) show the mutant phenotype. Normal flies do not show the mutant phenotype even if they are heated during the sensitive period

of gene product at the critical time and would have the same effect as lowering the absolute amount of gene product made. This interpretation is in agreement with the earlier hypotheses of Mitchell that heat affects the level or timing of expression of a critical gene product. A possibly related phenomenon occurs in early embryos where heating mild alleles of the mutant *hairy* during blastoderm stage results in phenotypes characteristic of more extreme *hairy* alleles (Eberlein 1986). Further investigation of phenocopies induced in mutant heterozygotes may answer questions about how heat affects gene expression.

2.3 Effects of Heat Shock on Gene Expression

Phenocopy inducing heat shocks turn off all RNA and protein synthesis for a period of 15-20 h following the heat shock (Mitchell and Lipps 1978; Petersen and Mitchell 1982). This, along with the resemblance of many phenocopies to mutant defects, led to the suggestion that it is the synthesis of critical gene products required for bristle construction which is affected by heat. This hypothesis is further supported by the observation that heat can induce phenocopies in recessive mutant heterozygotes. All of these observations suggest that the molecular mechanisms involved in pheno- copy induction involve effects of heat on gene expression at some level. Furthermore, for both the blond phenocopy and the multiple wing-hair phenocopy, heat does not have its effect on the structure involved immediately. There is a delay between the heat shock and the appearance of the abnormality. For these reasons it is interesting to look at the effects of heat on mRNA and protein synthesis and processing during the phenocopy inducing heat treatment as well as during the recovery period at 25 °C.

The effects of heat on gene expression have been studied in detail during the sensitive period for induction of the branched hair phenocopy in wings. The sensitive period for this defect occurs at a time of rapid change in both mRNA levels and protein synthesis (Mitchell and Petersen 1981). A phenocopy-inducing heat shock (30–40 min at 40-40.1 °C) inhibits RNA and protein synthesis and causes a developmental delay of about 15 h (Mitchell and Lipps 1978; Petersen and Mitchell 1982). Protein synthesis and protein decay are completely inhibited for a period of several hours immediately following the heat shock (Petersen and Mitchell 1982; Petersen and Young 1989). Recovery starts gradually, first with synthesis of heat shock proteins, followed by a recovery of the normal developmental pattern of protein synthesis. During recovery from the heat shock, RNA synthesis is also inhibited, but concentrations of message stay relatively constant because mRNA decay is also inhibited (Petersen and Mitchell 1982; Petersen and Young 1991). Synthesis of heat shock mRNA resumes first, followed by the synthesis and decay of messages involved in the developmental program. The recovery of RNA and protein synthesis follows essentially the same pattern as at other times in development (Petersen and Mitchell 1981). When the heating is done during a sensitive period for induction of a phenocopy, a defect is induced. At other times the same heat treatment does not result in obvious defects. Apparently, whether or not a defect is induced depends on the particular developmen- tal process being interrupted. For the branched hair phenocopy, the delay in protein synthesis interrupts the order of the hair construction process by delaying the synthesis of cuticulin on the surface of the wings (Mitchell et al. 1983). The branches on wing hairs do not appear until about 15 h after the heat shock, when internal pressure involved in the next stage of development pushes membrane through the holes in the incomplete cuticulin. Thus the branched hair phenocopy is apparently due to failure to complete one process in development, cuticulin synthesis, before the next process, a shape change, begins (Petersen and Mitchell 1982). A more detailed understanding of how heat interferes with gene expression will come from the study of phenocopies induced in mutant heterozygotes where the gene whose expression is affected can be more positively identified. *Multiple wing hair* was a candidate for such a gene until it was shown that the phenocopy sensitive period for multiple wing hair in mutant het- erozygotes is clearly separate from that for branched hair (Petersen and Mitchell 1987).

The fact that a branched hair phenocopy can be induced in *multiple wing-hair* flies also indicates that these are independent events (Mitchell et al. 1990).

3 Phenocopy Prevention

3.1 Induction of Phenocopy Thermotolerance

The ability of flies at all development stages except early embryos (Walter et al. 1990) to survive at high temperatures can be dramatically improved by a short treatment of 30–60 min at 35 °C. This treatment induces the synthesis of heat shock mRNA and proteins which are thought to play a role in cell survival at otherwise lethal temperatures (Tissieres et al. 1974; Mitchell et al. 1979). The same treatment which enhances survival canal so prevent the induction of phenocopies if done immediately prior to a phenocopy inducing heat shock (Mitchell et al. 1979). Resistance to phenocopy induction is very short lived. A delay of 5 h between the thermotolerance inducing treatment and the heat shock results in a normal number of defective flies in the case of the hooked phenocopy. When done immediately prior to the phenocopy inducing treatment, however, the 35 °C-treatment is remarkably effective in preventing a wide variety of phenocopies (Petersen and Mitchell 1982; Petersen and Mitchell 1985). It is unclear whether the induced thermotolerance which prevents phenocopies is of the a or ß type described by Boon-Niermeijer in this Volume (Chap. 7) and earlier (Boon-Niermeijer et al. 1986). It is not possible to directly test for the necessity of protein synthesis in pupae and have animals continue to develop. The short time required for acquisition of tolerance suggests accumulation of newly synthesized heat shock protein may not be important. However, in larvae given the same thermotolerance inducing treatment to enhance survival, the decay of thermotolerance coincides with the decay of newly synthesized Hsp70 (Mitchell et al. 1985; N.S. Petersen unpubl.) Three examples of phenocopies which can be prevented by a 35 °C-thermotolerance inducing treatment are shown in Fig. 3. Phenocopies induced in heterozygote pupae are also prevented by the same thermotolerance-inducing treatment (30 min at 35 °C), suggesting that these phenocopies are basically similar to phenocopies induced in wild-type pupae (Mitchell and Petersen 1985; Petersen and Mitchell 1989). The effect of the preshock is not simply to alter the timing of the sensitive period, because the pretreatment is equally effective at the beginning and end of the sensitive period (Mitchell et al. 1979; Petersen and Mitchell 1989).

3.2 Thermotolerance and Heat Shock Proteins

The conditions for induction of phenocopy thermotolerance in *Drosophila* pupae are essentially the same as for survival thermotolerance. This suggests that phenocopy tolerance may be due to the same molecular mechanism(s) as survival thermotolerance, or to a subset of these. Conditions which induce both phenocopy thermotolerance and survival thermotolerance also induce the synthesis of heat shock proteins, suggesting a role for heat shock proteins in the development of thermotolerance

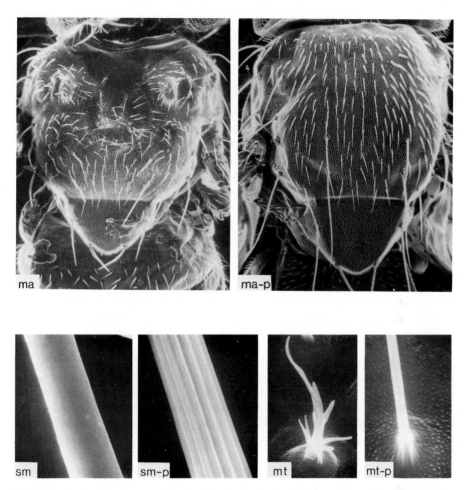

Fig. 3. Protection against phenocopy induction. If a heat shock of sufficient magnitude to induce a phenocopy is immediately preceded by a mild, thermotolerance-inducing, heat shock, the abnormality can be prevented. For each of the three pairs of structures shown, the defects caused by a single heat shock of 35 min at 40.5 °C are at the *left*. The same structures from animals which were pretreated (*-p*) first at 35 °C for 30 min and then at 40.5 °C for 35 min are shown at the *right*. The structures on the right in each pair are nearly normal. *Ma* and *ma-p* show the funny face phenocopy on the back of the thorax with a sensitive period at 30 h after white prepupa formation (at 25 °C). *Sm* and *sm-p* show smooth bristle phenocopy in scutellar bristles; sensitive period of 41–42 h. *Mt* and *mt-p* show the branched hair phenocopy on wings; sensitive period 38 h

(Mitchell et al. 1979). Further evidence for this is that chemicals which induce heat shock proteins can also confer thermotolerance in mammalian cell lines (Li and Lazlo 1985). More recently, direct evidence has implicated the synthesis of Hsp70 in the development of thermotolerance in mammalian cell lines (Riabowol et al. 1988; Johnston and Kucey 1988). In several cases, the synthesis of heat shock proteins other than and independent of Hsp70 have been shown to be required for the development of survival thermotolerance (Loomis and Wheeler 1982; Sanchez and Lindquist 1990). The development of thermotolerance in the absence of heat shock protein synthesis in cycloheximide treated cells has also been reported in a few instances (Hallberg et al.

1985; Widelitz et al. 1986). These differences in the requirements for heat shock protein synthesis under different conditions can be explained in several ways. Some have postulated two states of thermotolerance (Boon-Niermeijer Chap. 1; Lazlo 1988). Another way to look at the information available is to consider the many changes which are involved in the acquisition of thermotolerance, including changes in rRNA and mRNA synthesis and processing, changes in protein synthesis and processing, as well as changes in the cytoskeleton and in the cell membrane (Lindquist 1986). If thermotolerance involves several interacting proteins, then the apparent differences could be due to various components of the system being limiting under different conditions. Alternatively, if the most heat sensitive process (whose inactivation results in death) is different in individual cell types or cells under varying conditions, this could result in different proteins being required for the development of thermotolerance. In the most extreme form of this view, each heat shock protein would be involved in making a different cellular process thermotolerant and there could be more than two forms of thermotolerance. The reality will likely involve aspects of both models. Phenocopy prevention in thermotolerant pupae could also be due to different causes for different phenocopies. However, the fact that precisely the same thermotolerance-inducing conditions prevent a wide array of different phenocopies tends to argue that there is a common protection mechanism for phenocopies.

Survival thermotolerance in cell lines appears to be similar to phenocopy thermotolerance, and since survival thermotolerance involves the synthesis of heat shock proteins in at least some cases, it is useful to briefly discuss the possible roles of the heat shock proteins in this process. It is not possible to review this topic in this Chapter, so the reader is referred to reviews devoted to different aspects of this subject (Schlesinger et al. 1982; Neidhardt et al. 1984; Nover 1984; Craig 1985; Hightower et al. 1985; Petersen and Mitchell 1985; Lindquist 1986; Lindquist and Craig 1988). The most abundant and conserved of the eukaryotic heat shock proteins belong to two groups of proteins (1) the Hsp90 related proteins; and (2) the Hsp70 family. The fact that these proteins are synthesized in response to the same environmental stimuli in all procaryotes and eukaryotes, and that they are among the most conserved proteins known indicates the importance of their role in cell survival. The protein whose synthesis increases most with heat shock in eukaryotes is hsp70. Human hsp70 is 50% homologous to the *E. coli* protein, DnaK, which is also expressed following heat shock in *E. coli* (Bardwell and Craig 1984).

In many organisms including *Drosophila* there are non-heat inducible analogues of heat shock proteins called cognates or Hscs. In yeast, synthesis of some members of the Hsp70 family are essential for cell survival (Werner-Washburne et al. 1987). Progress has been made in understanding the roles of the non-heat induced members of the Hsp70 family (reviewed by Lindquist and Craig 1988). One of the more provocative recent developments is that a non-heat inducible homologue of Hsp70 (Hsc70) has been shown to enhance nascent protein transport across the endoplasmic reticulum (Chirico et al. 1988; Deshaies et al. 1988). Members of the Hsp70 family of proteins have also been reported to be involved in uncoating of clathrin vesicles and in the transport of proteins into lysosomes (Chappell et al. 1986, Chiang et al. 1989).

The major heat shock protein, Hsp70, or the closely related Hsc70 is normally present in large amounts very early in development in both mammalian embryos and in *Drosophila* (Chaps. 7–10). In the two-cell mouse embryo, Hsc70 is among the first zygotic genes to be expressed (Bensaude et al. 1983). In *Drosophila*, Hsc70 is the most abundant protein in oocytes and preblastoderm embryos (Palter et al. 1986). The fact that members of the Hsp70 family are normally expressed in large amounts at early stages of development suggests a role for the Hsp70 family in the process of activating gene expression as a cell goes from dormancy to a state of active growth.

The other major heat shock protein in eukaryotes, Hsp90, is synthesized normally in cells, but its synthesis is increased in heat shock cells under conditions where most cell protein synthesis is inhibited. A small fraction of cellular Hsp90 has been shown to be associated with steroid hormone receptors in the absence of hormone binding (Catelli et al. 1985).

The mechanism which ties together the diverse roles of members of the Hsp70 and Hsp90 families is the ATP-dependent unfolding or refolding of proteins. ATP-dependent unfolding of proteins is the proposed mechanism by which Hsc70 facilitates the transport across membranes. ATP-dependent unfolding of proteins has also been proposed to interfere with hormone receptor binding to DNA when Hsp90 is bound (Picard et al. 1988). A similar role has been proposed for these proteins following heat shock in the ATP-dependent refolding of proteins damaged by heat (Pelham 1986, 1988). Possible roles for heat shock proteins in prevention of phenocopies could be related to these functions; heat shock proteins could be involved in refolding nascent proteins or in transport and secretion of nascent proteins needed for critical developmental processes.

3.3 Molecular Models for Thermotolerance

Several molecular models have been suggested to explain heat-induced developmental defects and the mechanism by which they are prevented in thermotolerant cells. A long standing model is that heat shock destabilizes mRNA, and heat shock proteins prevent phenocopies by stabilizing critical mRNAs (Mitchell et al. 1979; Haass et al. 1989). This model is attractive because it would explain why heat induces different developmental defects at different stages in development. A major argument against this theory is that messenger RNA in general is stabilized following heat shock whether or not thermotolerance is induced (Farrell-Towt and Sanders 1984; Petersen and Mitchell 1981, 1982). In fact, specific pupal mRNAs decay more rapidly in thermotolerant animals than in animals receiving a phenocopy inducing heat shock (Petersen and Mitchell 1982; Petersen and Young 1991).

Another model for the induction of phenocopies holds that phenocopies are the result of a failure in mRNA processing following heat shock. In *Drosophila* cell lines, messages whose introns have not been removed can be transported into the cytoplasm and translated into abnormal proteins (Yost and Lindquist 1986, 1988). The abnormal proteins could cause phenocopies by interfering with normal development. Heat shock proteins might prevent phenocopy induction by enhancing the recovery of the splicing apparatus in the nucleus. This interesting model may explain some of the phenocopies of homeotic mutants which can be induced during *Drosophila* embryonic development. However, this type of model would not explain the induction of heterozygote phenocopies because the abnormal proteins are expected to produce a dominant effect. If the effect were dominant, heterozygote phenocopies should be induced in wild-type

Drosophila pupae. Furthermore, 2-D gel analysis of proteins made during recovery from heat shocks at 38 and 72 h has not shown any abnormal proteins made (Petersen and Mitchell 1987 and unpubl.).

Alternatively, phenocopies could be due to failure to synthesize enough of a critical protein, or failure in post-translational processing or transport (Petersen and Mitchell 1981; Petersen 1990). Many other models can be proposed since both thermotolerance inducing heat shocks and the more extreme phenocopy inducing heat shocks appear to affect many if not all of the different aspects of gene expression and even cell physiology in general. In order to determine what is the critical effect for phenocopy thermotolerance, it will be necessary to identify a gene whose expression is affected by heat, resulting in a phenocopy, and to compare the level of expression of that gene during normal development with the levels of expression following heat shocks which induce the phenocopy. The heterozygote phenocopies, where the gene whose expression is affected by heat shock can be identified, appear to be a good system in which to do this.

4 Conclusions

Phenocopies in *Drosophila* are developmental defects resembling mutant defects which are induced by environmental stress. Heat induced phenocopies have been studied most extensively. These can be induced by short high-temperature shocks which inhibit gene expression by inhibiting RNA and protein synthesis as well as RNA and protein processing and decay. The defects occur during recovery from heat shock following a developmental delay. A thermotolerance inducing treatment at a lower temperature immediately before the high temperature shock can completely prevent the occurrence of defects. The thermotolerance inducing treatment has many effects on cellular physiology including induction of heat-shock protein synthesis. The similarities between survival thermotolerance and phenocopy thermotolerance suggest that heat shock proteins may be involved in phenocopy thermotolerance.

The characteristics of *Drosophila* phenocopies lead to the hypothesis that phenocopies are due to direct effects of heat on the level and/or timing of gene expression. In thermotolerant animals, gene expression, including RNA synthesis and processing, as well as protein synthesis, recover much more rapidly than in nontolerant animals. The limiting process involved in the induction and prevention of phenocopies has yet to be determined. It is of interest to understand this process in detail because it appears to be very similar to conditions which induce birth defects in vertebrates. Similarities between *Drosophila* phenocopies and vertebrate birth defects include their critical dependence on the time of the insult and on the genetic background of the animal (Edwards 1969; Webster and Edwards 1984; Finnell et al. 1986,1988), the ability to prevent defects with a thermotolerance inducing treatment (Walsh et al. 1987), and the observation that many teratogenic chemicals as well as teratogenic heat shocks induce the synthesis of heat shocks proteins (Petersen 1990).

References

Bardwell JC, Craig EA (1984) Major heat shock gene of *Drosophila* and the heat inducible gene dnaK are homologous. Proc Natl Acad Sci USA 81:848–851

Bensaude O, Babinet C, Morange M, Jacob F (1983) Heat shock proteins, the first major products of zygotic gene activity in mouse embryos. Nature 305:331–332

Boon–Niermeijer EK, Tuyl M, Van De Scheur H (1986) Evidence for two states of thermotolerance. Int J Hyperthermia 2:93–105

Catelli MG, Binart N, Jung–Testas I, Renoir JM. Baulieu E–E, Feramisco JM, Welch WJ (1985) The common 90 kD protein component of non–transformed '8S' steroid receptors is a heat–shock protein. EMBO J 4:3131–3135

Chappell TG, Welch WJ, Schlossman DM, Palter KB, Schlesinger MJ, Rothman JE (1986) Uncoating ATPase is a member of the 70 kilodalton family of stress proteins. Cell 45:3–13

Chiang H–L, Terlecky SR, Plant CP, Dice JF (1989) A role for the 70–kilodalton heat shock protein in lysosomal degradation of intracellular proteins. Science 246:382–384

Chirico WJ, Waters MG, Blobel G (1988) 70K heat shock related proteins stimulate protein translocation into microsomes. Nature 332:805–810

Craig EA (1985) The heat shock response. CRC Crit Re Biochem 18:239–280

Deshaies RJ, Koch BD, Werner– Washburne M, Craig EA, Schekman RA (1988) A subfamily of stress proteins facilitates translocation of secretory and mitochondrial precursor peptides. Nature 332:800–805

Eberlein S (1986) Stage specific embryonic defects following heat shock in *Drosophila*. Dev Genet 5:179–197

Eberlein S, Mitchell HK (1987) Protein synthesis patterns following stage–specific heat shock in early *Drosophila* embryos. Mol Gen Genet 210:407–412

Edwards MJ (1969) Congenital defects in guinea pigs: fetal reabsorptions, abortions, and malformations following induced hyperthermia during early gestation. Teratology 2:313–328

Farrell–Towt J, Sanders M (1984) Noncoordinate histone synthesis in heat–shocked *Drosophila* cells is regulated at multiple levels. Mol Cell Biol 4:2676–2685

Finnell RH, Bennett GD, Karras SB, Mohl VK (1988) Common hierarchies of susceptibility to the induction of neural tube defects by valproic acid and its 4–propyl–4–pentenoic acid metabolite. Teratology 38:313–320

Finnell RH, Moon SP, Abbott LC, Golden JA, Chernoff GF (1986) Strain differences in heat–induced neural tube defects in mice. Teratology 33:247–252

Fristrom JW (1965) Development of the morphological mutant *cryptocephal* in *D. melanogaster*. Genetics 52:297–318

Gloor H (1947) Phenokopie–Versuche mit Aether an *Drosophila*. Rev Suisse Zool 54:637–712

Goldschmidt RB (1935) Gen und Ausseneigenschaft (Untersuchungen an *Drosophila)* 69:3869, 70–131

Goldsmidt RB (1949) Phenocopies. Sci Am Oct: 46–49

Haass C, Falkenburg PE, Kloetzel P–M (1989) In: ML Pardue, JR Feramisco, Lindquist S (eds.) Stress induced proteins Alan R Liss, New York, N.Y., pp 175–185

Hadorn E (1961) Developmental genetics and lethal factors Methuen, London UK, pp 236–256

Hallberg RL, Kraus KW, Hallberg EM (1985) Induction of acquired thermotolerance in *Tetrahymena thermophila* can be achieved without the prior synthesis of heat shock proteins. Mol Cell Biol 5:2061–2070

Hightower LE, Guidon PT, Whelan SA, White CN (1985) Stress responses in avian and mammalian cells. In: Atkinson BG, Walden DB, (eds.), Changes in eukaryotic gene expression in response to environmental stress. Academic Press, Orlando, pp. 197–210

Johnston RN, Kucey BL (1988) Competitive inhibition of *hsp70* gene expression causes thermosensitivity. Science 242:1551–1554

Lazlo A (1988) Evidence for two states of thermotolerance in mammalian cell. Int J hyperthermia 4:513–526

Li GC, Lazlo A (1985) Thermotolerance in mammalian cells. A possible role for heat shock proteins. In: Atkinson BG, Walden DB (eds) Changes in eukaryotic gene expresssion in response to stress. Academic Press, Orlando, pp 227–254

Lindquist S (1986) The heat–shock response. Ann RevBiochem 55:1151–1191

Lindquist S, Craig EA (1988) The heat–shock proteins. Ann Rev Genet 22:631–677

Loomis WF, Wheeler SA (1982) The physiological role of heat–shock proteins in *Dictyostelium*. In: Schlesinger M, Ashburner M, Tissières A (eds) Heat shock: from bacteria to man. Cold Spring Harbor Press, Cold Spring Harbor, p 353–359

Milkman R (1966) Analyses of some temperature effects on *Drosophila* pupae. Biol Bull 131:331–345

Mitchell HK (1966) Phenol oxidases and *Drosophila* development. Insect Physiol 12:755–765

Mitchell HK, Lipps L (1978) Heat shock and phenocopy induction in *Drosophila*. Cell 15:907–919

Mitchell HK, Petersen NS (1981) Rapid changes in gene expression in differentiating tissues of *Drosophila*. Dev Biol 85:233–242

Mitchell HK, Petersen NS (1982) Developmental abnormalities induced by heat shock. Dev Genet 3:91–102

Mitchell HK, Petersen NS (1983) Gradients of gene expression in wild–type and *bithorax* mutants of *Drosophila*. Dev Biol 95:459–467

Mitchell HK, Petersen NS (1985) The recessive phenotype of *forked* can be uncovered by heat shock in *Drosophila*. Dev Genet 6:93–100

Mitchell HK, Moller G, Petersen NS, Lipps–Sarmiento L (1979) Specific protection from phenocopy induction by heat shock. Dev Genet 1:181–192

Mitchell HK, Roach J, Petersen NS (1983) Morphogenesis of cell hairs in *Drosophila*. Dev Biol 95 :387–398

Mitchell HK, Petersen NS, Buzin C (1985) Self proteolysis of heat shock proteins. Proc Natl Acad Sci USA 82:4969–4973

Mitchell HK, Edens J, Petersen N (1990) Stages of cell hair construction in *Drosophila*. Dev Genet (in press)

Neidhardt FC, VanBogelen R, Vaughn V (1984)The genetics and regulation of heat–shock proteins. Ann Rev Genet 18:295–329

Nover L (1984) Heat shock response of eukaryotic cells. Springer, Berlin Heidelberg New York Tokyo

Palter KB, Wantanabe M, Stinson L, Mahowald AP, Craig EA (1986) Expression and localization of *Drosophila melanogaster* Hsp70 cognate proteins. Mol Cell Biol 6:1187–1203

Pelham H (1988) Coming in from the cold. Nature 332:776–777

Pelham HRB (1986) Speculations on the functions of the major heat shock and glucose–regulated proteins. Cell 46:959–961

Petersen NS (1990) Effects of heat and chemical stress on development. In: Scandalios G (ed) Advances in genetics:genomic responses to environmental stress. pp

Petersen NS, Bond B, Mitchell HK, Davidson N (1985) Stage specific regulation of actin genes in *Drosophila* wing cells. Dev Genet 5:219–225

Petersen NS, Mitchell HK (1981) Recovery of protein synthesis following heat shock: prior heat treatment affects the ability of cells to translate mRNA. Proc Natl Acad Sci USA 78:1708–1711

Petersen NS, Mitchell HK (1982) Effects of heat shock on gene expression during development: Induction and prevention of the multihair phenocopy in *Drosophila*. In: Schlesinger M, Ashburner M, Tissières A (eds) Heat shock from bacteria to man. Cold Spring Harbor Press, Cold Spring Harbor, pp 345–352

Petersen NS, Mitchell HK (1985) Heat shock proteins. In: Kerkut GA, Gilbert LI (eds) Comprehensive insect physiology, biochemistry, and pharmacology. Vol.X. Biochemistry. Pergamon Oxford, p 347–365

Petersen NS, Mitchell HK (1987) The induction of a multiple wing hair phenocopy by heat shock in mutant heterozygotes. Dev Biol 121:335–341

Petersen NS, Mitchell HK (1989) The forked phenocopy is prevented in thermotolerant pupae. In: Pardue ML, Fermisco J, Lindquist S. (eds) Stress induced proteins UCLA Symposia on Molecular and Cellular Biology, New series, Vol. 96, Alan R. Liss, New York, N.Y., pp 235–244

Petersen NS, Young P (1989) Effects of heat shock on protein processing and turnover in developing *Drosophila* wings. Dev Genet 10:11–15

Petersen NS, Young P (1991) Actin mRNA is stabilized by heat shock during *Drosophila* development. Cell Dev (submitted)

Picard D, Salser SJ, Yamamoto KR (1988) A movable and regulatable inactivation function within the steroid binding domain of the glucocorticoid receptor. Cell 54:1073–1080

Ranganathan S, Davis DG, Hood RD (1987) Developmental toxicity of ethanol in *Drosophila melanogaster*. Teratology 36:45–49

Riabowol KT, Mizzen LA, Welch WJ (1988) Heat shock is lethal to fibroblasts injected with antibodies to hsp70. Science 242:433–436

Rizki MTM (1960) Effects of glucosamine hydrochloride on the development of *Drosophila melanogaster*. Biol Bull 118:308–314

Sanchez Y, Lindquist SL(1990) HSP104 required for induced thermotolerance. Science 248:1112–1115

Santamaria P (1979) Heat shock induced phenocopies of dominant mutants of the *bithorax* complex in *Drosophila melanogaster*. Mol Gen Genet 172:161–163

Schlesinger MJ, Ashburner M, Tissières A (1982) Heat shock from bacteria to man. Cold Spring Harbor Press, Cold Spring Harbor, NY

Seybold WD, Meltzer PS, Mitchell HK (1975) Phenol oxidase activation in *Drosophila:* a cascade of reactions. Biochem Genet 13:85–108

Tissières A, Mitchell HK, Tracy UM (1974) Protein synthesis in salivary glands of *D. melanogaster* cells. J Mol Biol 84:389–398

Walsh DA, Klein NW, Hightower LE, Edwards MJ (1987) Heat shock and thermotolerance during early rat embryo development. Teratology 36:181–191

Walter MF, Petersen NS, Biessman H (1990) Heat shock causes the collapse of the intermediate filament cytoskeleton in *Drosophila* embryos. Dev Genet (in press)

Webster WS, Edwards MJ (1984) Hyperthermia and induction of neural tube defects in mice. Teratology 29:417–425

Werner–Washburne M, Stone DE, Craig EA (1987) Complex interactions among members of an essential subfamily of *hsp70* genes in *Saccharomyces cerevisiae*. Mol Cell Biol 7:2568–2577

Widelitz RB, Magun BE, Gerner EW (1986) Effects of cycloheximide on thermotolerance expression, heat shock protein synthesis, and heat shock protein mRNA accumulation in rat fibroblasts. Mol Cell Biol 6:1088–1094

Yost HJ, Lindquist S (1986) RNA splicing is interrupted by heat shock and is rescued by heat shock protein synthesis. Cell 45:185–193

Yost HJ, Lindquist S (1988) Translation of unspliced transcripts after heat shock. Science 242:1544–1548

3 The Use of Heat-Shock-Induced Ectopic Expression to Examine the Functions of Genes Regulating Development

Greg Gibson[1]

1 Scope

The beauty of the fruitfly, *Drosophila melanogaster*, as a system for the study of development derives principally from its amenability to genetic analysis (Morgan 1926). Yet there remain situations where classical genetics fails to address important questions: mutagenesis is by nature random and constrained. Once a gene has been identified, it is usually desirable to alter the sequence of its gene product, or the timing and location of its expression. To do so in a directed manner requires new technology. One successful approach has been the adoption of heat shock promoters to induce ectopic and high-level expression of developmentally active genes in transgenic flies.

With this technique, the molecular biologist can essentially make temperature-sensitive gain-of-function mutations to order, and use them to address a wide variety of developmental problems, often from a fresh angle. Its success builds largely upon the observations of numerous groups, which have revealed astounding degrees of functional specificity shown by genes controlling processes as diverse as sex determination, axis formation, segmentation, neurogenesis, and the establishment of segmental identity. Even though many of the genes examined encode different types of regulatory proteins which act in the nucleus, at the cell surface, or in signal transduction, it has been noted consistently that their ectopic expression has very limited (though usually rationalizable) effects on the determination of cell fates.

At the outset, an important theoretical limitation upon the heat shock technique should be considered: care must be taken in extrapolating from the phenotypic effects of ectopic and overexpression to wild-type gene function. Even though a gene may produce an effect, it cannot be assumed uncritically that this effect reflects the normal function of the gene. For example, ectopic expression of the segmentation and neurogenic *hairy* gene product very early in development has been found to affect sex determination, probably by interacting with proteins to which it is not normally exposed (D. Ish-Horowicz, pers. comm.). It is also conceivable that many developmental processes might depend upon tight spatial, temporal, and concentration control of gene expression. In such cases, "swamping the system" through overexpression may pre-

[1] Department of Developmental Biology, Beckman Center, Standford University School of Medicine, Palo Alto CA 94305, USA

Note: Abbreviations of genes.
Adh, alcohol dehydrogenase; *Antp*, Antennapedia; *Dfd*, Deformed; *EF1a*, elongation factor 1a; *ftz*, fushi tarazu; *hb*, hunchback; *hsp70*, 70 kD heat shock protein; *neo*, neomycin resistance; *rosy*, xanthine dehydrogenase; *Scr*, Sex combs reduçed; *sev*, sevenless; *tra*, transformer; *Ubx*, Ultrabithorax.

Results and Problems in Cell Differentiation 17
Heat Shock and Development
Hightower and Nover (Eds.)
© Springer-Verlag Berlin Heidelberg 1991

clude analysis of the problem. Clearly, though, in vitro and even classical genetic approaches are open to similar criticisms. With the availability of molecular probes with which to test hypotheses arising out of each experimental strategy, there is no reason to impose a blanket scepticism towards any one approach.

Indeed, the main argument to be presented here will be that the heat shock approach provides an important cell biological complement to more biochemical or molecular genetic strategies for studying gene function. Many results have conceived new interpretations of problems such as the organization of genetic hierarchies, gradient function, metameric stability, intercellular communication, and protein structure and function. Since there has been an explosion of data recently, this Chapter will concentrate on a few of the developmental problems which the technique has been used to address so far. Particular emphasis will be placed upon the homeotic genes to illustrate the range of questions which can be addressed. There will be no attempt to critically evaluate the precise experimental strategies employed, as these vary from project to project, reflecting the different natures of the genes involved. In the conclusion, a short discussion of a number of novel ways in which the basic technique might be expected to figure in future studies is presented.

2 The Heat Shock Strategy

The fundamental strategy, first employed by Lis and coworkers (1983), is conceptually simple. After cloning, structural analysis, and sequencing of the gene, all or a portion of it (usually a cDNA clone) is placed under the control of an inducible promoter. Most groups have employed approximately the first 50 base pairs of the *Drosophila hsp70* promoter, which contains all of the *cis*-acting elements required to induce transcription in all cells at elevated temperatures (Pelham 1982). This construct is then placed in a suitable transformation vector for P-element mediated transfer into the germline of recipient embryos (Rubin and Spradling 1982). Animals containing a single chromosomal copy of the transgene are selected, and through suitable genetic crosses it is usually possible to generate homozygous or balanced stocks for further analyses. The heat shock regime used to induce the transgene will depend upon the nature of the process being examined, but in general brief exposures of embryos (10–20 min) or larvae (30–60 min) to temperatures around 37 °C have been found to be sufficient if the animals are allowed to recover at 25 °C (see Lindquist 1986). Heat shocks are best administered by immersing embryos, or a humidified tube containing larvae, in a water bath at the required temperature, to ensure rapid heat transfer.

We and others have carried out careful control studies on flies transformed with an expression vector not carrying cloned DNA, and have found that under the relatively mild conditions typically employed, few developmental abnormalities are seen simply because of the increased temperature. However, it should be noted that it has been well documented that more severe heat shocks (delivered at 42 °C) can cause specific phenocopies of known mutations (Eberlein 1986), particularly those affecting bristle morphology (Mitchell and Lipps 1978). In addition, heat shocks very early in development (up to the syncitial blastoderm stage) do result in mild segmental fusions in up to 10% of embryos (Mlodzik et al 1990).

Although at least ten different types of heat shock transformation vectors have been used to produce transgenic flies, no systematic attempt has been made to analyze conditions for optimal expression. Various reports suggest that untranslated 3' mRNA signals can have a considerable impact on the levels and/or stability of the induced messenger RNA. Petersen and Lindquist (1989) have documented a potent RNA degradation signal in the 3' portion of the *hsp70* mRNA. This observation suggests that inclusion of these sequences will be desirable if rapid turnover of the gene product is sought, for example when the aim of the study is to examine the temporal requirements for the gene's activity. Care must also be taken in selecting 5' untranslated sequences, as these can cause strong repression of levels of protein production. Their effect is probably at the level of translational control, either as a result of altered accessibility of heat shock mRNAs to ribosomes (Klemenz et al. 1985), or of the presence of short upstream open reading frames (P. LeMotte, pers. comm.). It may also be possible to adjust levels of heat shock-induced transcription by varying the number of heat shock response elements in the promoter.

In most cases, the choice of a selectable marker for identification of transgenic flies is not a crucial factor, and a range of markers affecting eye color (*rosy* and *white*) and drug resistance (*Adh* and *neo*) are available. However, some genes, particularly those encoding homeodomain-containing proteins, do seem to cause a degree of discomfort to the host fly, and in this case a more sensitive marker gene, such as *white*, which allows identification of insertions in relatively inactive regions of the genome is recommended.

We have estimated, on the basis of Western analysis of protein levels after heat shock, that the variation in levels of expression of transgenes due to the site of insertion is usually within an order of magnitude (Gibson and Gehring 1988). Importantly, there is not necessarily a linear relationship between the amount of overproduced protein, and the expressivity of phenotypic effects. For this reason, it is normal to examine at least two independently obtained transgenic lines for each construct, particularly where quantitative comparisons are sought. In most cases reported so far, it seems that the heat shock promoter can be induced in all cells (including the pole cells) after the onset of zygotic transcription at the syncitial blastoderm stage of development (2 h after egg laying at 25 °C). Ish-Horowicz et al. (1989) have suggested that high levels of transcription cannot be induced until cellularization is complete, but it seems that development is sufficiently sensitive prior to this stage to allow low (and even immunochemically undetectable) levels of ectopically expressed protein to alter the course of development. In at least one case (A. Schier and W. Gehring, pers. comm.) the site of insertion has been found to produce an effect on the spatial distribution of an induced protein: a truncated Antenapedia protein was found to be expressed fortuitously in a "pair-rule"-type pattern of stripes after heat shock in the early embryo. This observation suggests that some of the effects on levels of expression may relate to the presence of nearby enhancer elements, and raises hopes that it may be possible one day to couple the heat shock promoter to tissue-specific enhancers and so induce spatially regulated gene expression.

For some purposes, such as the functional dissection of gene products, the rationale for using heat shock promoters stems in part from the difficulty of using the regulatory sequences of the gene itself. For example, in the case of the homeotic genes, these all cover tens of kilobases and are yet to be completely defined (Gehring and Hiromi 1986). Nevertheless, it is apparent that the heat shock approach introduces a number of other advantages, including the ability to address temporal aspects of gene function and cellular determination, to assay for residual in vivo functions even when rescue is not achieved, and to investigate epistatic relations between genes. In theory too, by using a normally quiescent promoter, it should be possible to obtain transgenic flies containing genes which would otherwise be dominant lethals if they disrupted wild-type gene function when expressed from their own promoter. Consequently, the heat shock system provides a convenient assay for measuring gene function prior to adoption of the normal promoter to ask more specific questions about rescue capabilities of the altered genes.

3 Applications

3.1 Minimal Induction: Sex Determination and Ageing

Ideally, it would be nice to be able to induce gene expression under conditions which fail to induce a stress response. In a couple of cases, it has been possible to achieve this goal by taking advantage of low levels of constitutive expression from the heat shock promoter.

Sex determination in *Drosophila* is under the control of a hierarchy of genes which act early in development to read the X:A chromosomal ratio and define cell states in a cell-autonomous manner (Baker and Belote 1983). The gene *transformer* (*tra*) occupies a critical step in the sex determination pathway, transmitting information from *Sex-lethal* to *doublesex* and eventually to genes involved in terminal differentiation. Two *tra* transcripts have been detected, one of which is female-specific, the other non-sex-specific. Alternative splicing is thought to lead to the production of a functional protein in females, while the non-sex-specific transcript fails to possess a long open reading frame (Boggs et al. 1987). By placing cDNAs derived from both of these transcripts under the control of the *hsp70* promoter, McKeown et al. (1988) were able to confirm this hypothesis. The female-specific mRNA is indeed sufficient to cause female determination, both in chromosomally male flies, and in females carrying a mutation at the *tra* locus. By contrast, over-expression of the non-sex-specific RNA under the same conditions (that is, growth at 25 °C) had no effect on the transgenic flies. Further, the experimental system has been used to demonstrate that the female *tra* messenger RNA does not derive from the non-sex-specific mRNA, and that the *tra* gene product is not necessary for splicing of its own primary transcript. These studies clearly illustrate the potential of the heat shock approach for addressing questions about molecular mechanisms in vivo.

Perhaps the most interesting results from McKeown et al.'s study concern its use to extend the classical genetic dissection of the sex determination pathway in *Drosophila*. By crossing the heat shock-*tra* construct into a variety of genetic backgrounds, the authors elegantly confirmed previous ideas about the order in which the sex determination genes act. They then invoked a simple argument to show that the sex determination pathway is essentially linear. Had the pathway been branched, it would have been predicted that ectopic expression of the female *tra* gene product would have been insufficient to convert chromosomal males into phenotypic females. However, since such a transformation was observed, they conclude that there are no genes downstream of *tra* which are independently regulated by another gene acting upstream of *tra* in the pathway. At least up until *tra*, the sex determination pathway must be linear, which is to say that the *Sex-lethal* gene product must act solely through *tra* to regulate genes further down in the genetic hierarchy.

A second example of the use of the *hsp70* promoter to produce low levels of constitutive transgene expression comes from a study of the ageing process. Biochemical studies in a number of species have shown a correlation between a decline in the levels of expression of the elongation factor EF1a, and the onset of senescene (Webster and Webster 1983). Transgenic laboratory flies carrying an extra copy of *EF1a* gene under the control of the *hsp70* promoter have been shown to live up to 20% longer than control flies carrying the *rosy* marker alone (and over twice as long as *rosy* flies). An interesting aspect of this study (Shepherd et al. 1989) was that the effect was enhanced by growing the flies at 29 °C rather than 25 °C. Growth at this temperature is thought not to interfere significantly with the normal metabolism of cells (as judged by the profile of protein production: Lindquist 1980), but it does lead to increased hsp70 transcription.

3.2 Early Development: Gradients and Metameric Stability

One of the most fascinating areas of current research into early development concerns the molecular mechanisms underlying the subdivision of the embryo into repeating units, or metameres. Extensive genetic screens have identified around 50 genes which are crucial for this process (Nusslein-Volhard and Wieschaus 1980). In brief, it is thought that a "cascade of gradients" of regulatory proteins are arranged in a complex hierarchy which divides the embryo along the anterior-posterior axis into successively smaller units (Ingham 1988; Struhl 1989b). There is strong evidence that a gradient of the maternally expressed gene *bicoid* organizes the anterior end of the embryo principally by activating the transcription of a class of target genes in a concentration threshold-dependent manner (Driever and Nusslein-Volhard 1988). In time, a series of "gap" genes, a number of which encode zinc-finger nuclear regulatory proteins, come to lie in adjacent domains along the length of the embryo; they are thought to activate, again in a threshold dependent manner, another set of nuclear regulatory genes, the "pair-rule" genes which define the boundaries of the first metameric units. The heat shock approach has recently made a number of contributions to the study of how "morphogens" act, and how the pair-rule genes help to define cellular identities with respect to segmental boundaries (Martinez-Arias 1989).

Ectopic expression of the gap gene *hunchback* (*hb*) generates embryos phenotypically equivalent to those lacking the genes which organize the posterior end of the embryo (Struhl 1989a). This observation suggested that the sole role of these posterior-group genes might be to repress maternal *hb* activity. Indeed, this hypothesis has been corroborated by a variety of genetic approaches, leading to a reappraisal of conceptions of how the two ends of the body are organized, and of how polarity is established in the early embryo. Given the presumed crucial role of local concentrations of regulatory proteins in defining boundaries of subordinate gene expression, it is worth considering the finding (Struhl 1989b) that ectopic over-expression of gap genes such as *Kruppel* and *hb* have remarkably localized effects on embryogenesis. While consistent with the idea that they operate as gradients, these results also emphasize the importance of cellular context for regulatory protein function. Little is yet known in detail about how the gap proteins regulate their target genes, but as the *cis*-acting control regions of the pair-rule genes start to be characterized (Goto et al. 1989; Harding et al. 1989), the availability of in vivo assays for gap protein function should prove to be an important complement to more biochemical studies.

That the heat shock approach, when combined with molecular analyses, has the potential to reveal much about the cell biology of segmentation, is illustrated by Ish-Horowicz et al.'s study (1989) of the effects of ectopic expression of the *fushi tarazu* *(ftz)* homeobox-containing pair-rule gene. Heat shock induction of *ftz* at the blasto-derm stage results in a near-reciprocal phenotype to loss of *ftz* activity: every odd parasegment appears to be deleted (Struhl 1985; Ish-Horowicz and Gyurkovics 1988). The most simple interpretation of this result, that cellular identity is determined by the combinatorial action of the various pair-rule genes and that disruption of the normal concentrations of their gene products gives rise to "nonsense" identities and cell death, is not supported by a rational study of alteration of gene expression in response to ectopic *ftz*. What is really affected is the stability of crucial boundaries which delineate the developing segments. Ectopic *ftz* was found to activate itself in one cell anterior to the normal *ftz* domain, thereby causing incorrect activation of *engrailed* and repression of *wingless* at the position where it is essential that a confrontation between these two gene products be established. Subsequently, this boundary degrades, and the even numbered parasegments fail to form. A shift in homeotic gene activity then occurs, which accounts for the deceptive "near-reciprocal" cuticular pattern. The heat shock approach should not be considered simply as a means for studying gene function, but rather as another system through which more detailed studies of cell biological prob-lems may be initiated.

3.3 Establishing Identity: The Homeotic Genes

3.3.1 The Functional Structure of Homeotic Proteins

The heat shock approach should also be of considerable use in addressing a variety of more molecular biological questions. For example, a number of groups have em-ployed it to study the functional structure of a family of homeodomain-containing proteins which control the identities of the individual body segments of *Drosophila*. The broad strategy is to construct deletion variants of, and hybrids between, different homeotic genes, and to compare their in vivo function (in terms of changes in larval and adult cuticular structures) with those of the wild-type cDNAs under the same experimental conditions. While it can be argued that the behavior of each gene when ectopically expressed is not identical to its normal behavior, it remains possible to compare gene products, since each of the homeotic genes examined so far, i.e. *De-formed* (*Dfd*: Kuziora and McGinnis 1988), *Sex combs reduced* (*Scr*: Gibson et al. 1990), *Antennapedia* (*Antp*: Schneuwly et al. 1987; Gibson and Gehring 1988), and

Ultrabithorax (*Ubx*: Mann and Hogness 1990; Gonzalez-Reyes and Morata 1990), behave quite differently under heat shock control. Furthermore, the transformations they produce are clearly related to those predicted from classical genetic analyses.

A consensus has emerged as a result of a large number of manipulations of these genes, which suggests that their functional specificity is almost entirely encoded within a discrete portion of each protein which includes the homeodomain. Deletion of the majority of the remaining sequences of each protein merely results in proteins with reduced potency: they produce qualitatively identical, but less severe, cuticular transformations (Gibson et al. 1990; Mann and Hogness 1990). In a sense, these "potentiation" sequences may be analogous to activation sequences identified in a number of transcription factors from mammalian cells and yeast (Ptashne 1988), but is important to note that their function almost certainly affects both the activation and repression of a series of target genes. Alternatively, they may act simply by influencing the stability of transcription complexes.

Homeodomain-swap experiments based on the *Dfd* gene have demonstrated that the homeodomain itself encodes a significant degree of in vivo functional specificity (Kuziora and McGinnis 1989). Much of this might be attributed to selection of specific target genes, presumably as a result of differential DNA-binding affinities encoded by the homeodomains. By the same token, studies on hybrids between *Scr, Antp,* and *Ubx* (Gibson et al. 1990; Mann and Hogness 1990) imply that sequences on both sides of each homeodomain do significantly influence the properties of the proteins. Moreover, different Ubx protein variants derived from alternatively spliced transcripts have been shown to induce transformations in different tissue types (Mann and Hogness 1990).

It is not yet clear how functional specificity relates mechanistically to protein structure. An attractive hypothesis is that residues within and adjacent to the homeodomain make specific contacts with ancillary transcription factors (Keleher et al. 1988; Stern et al. 1989). Such factors might modify the in vivo DNA-binding specificities of the homeoproteins, or determine whether they function as activators or repressors of transcription once bound to a promoter. The tissue-specific distribution of such factors would influence the range of cells in which the homeoproteins are capable of acting. Competition between homeoproteins would then reflect differences in DNA-binding affinity (Gibson and Gehring 1988), affinity for ancillary transcription factors (Gibson et al. 1990; Gonzalez-Reyes and Morata 1990), and combinatorial/additive interactions (Mann and Hogness 1990). These considerations imply that a proper understanding of homeotic protein function will require a variety of biochemical and molecular genetic approaches, both in vitro and in vivo.

3.3.2 The Homeotic Regulatory Hierarchy

The fact that the overexpression of any one of these genes has localized effects on development has led us to emphasize the stability of determined states, and the extent of buffering which apparently exists within cells (Gibson and Gehring 1988). It appears as if the transformations produced by the proteins are essentially all-or-nothing: cells switch fates rather than adopting intermediate phenotypes (Postlethwait and Schneiderman 1971). Weaker transformations caused by deletion proteins are interpreted as reflecting a decreased number of transformed clones, rather than less severe transformation of all cells in an organ. Much of the buffering probably reflects the involvement of an epistatic hierarchy amongst the genes of the *Antennapedia* and *bithorax* complexes, a finding which was not all together expected on the basis of genetic studies (Gonzalez-Reyes et al. 1990).

In light of the heat shock approach, it may be pertinent to reconsider some of the current notions of how segmental identity is translated into cellular differentiation. Garcia-Bellido's (1975) selector gene hypothesis postulates that an array of "realizator" genes lies downstream of the homeotic genes, and that it is their activity which directly establishes the size, shape, structure, and rate of division of individual cells. They must encode structural proteins such as components of the cytoskeleton. Evolutionary and ontogenetic perspectives both argue that many of the realizator genes should be targets for regulation by more than one homeotic gene. It is not known how direct such regulation may be.

One of the conceptual difficulties in thinking about homeotic protein function is understanding how they act globally to control the identities of whole segments, and yet precisely at the level of determination of fate of individual cells within a segment. Presumably they act in conjunction with other determinants of positional information to assign specific and heritable cellular identities. The buffering of cells against the effects of ectopically expressed homeotic proteins, as well as their capacity to undergo discrete switches in state of determination, both argue that the homeotic proteins either directly regulate a small number of target genes, or a larger number of target genes in a highly coordinated manner. Coordinated regulation, though, must be reconciled with the finding that altering just a few amino acids in one of the proteins can have a drastic influence on the types of transformations it induces (Kuziora and McGinnis 1989; Gibson et al. 1990; Mann and Hogness, 1990).

In addition, it should be considered that there appear to be narrow time windows within which the homeotic proteins act. For example, successively later induction of Antp protein in early third instar larvae leads to transformations of antennal structures in a distal-to-proximal direction over a 16-h period. Later heat shocks have no effect on structures transformed by earlier heat shocks (Gibson and Gehring 1988). This point raises the question of wether the requirement for homeotic gene activity is transient or maintained. Data from the analysis of a temperature-sensitive allele of *Ubx* (Kaufman et al. 1973), as well as mitotic clonal analyses of *Ubx* (Morata and Garcia-Bellido 1976), in the early 1970s both hinted that there may not be a requirement for *Ubx* gene activity after 8 h prior to pupation. Although the latter result has been interpreted in terms of the assumed "perdurance" of the homeotic protein until cellular differentiation (see Morata and Lawrence 1977), it is intriguing that detailed studies of *Scr* and *Antp* in heat shock assays has failed to reveal any transformations when they are induced after the mid-third instar larval period. In embryos too, the timing of heat shocks required to induce transformations precedes the first signs of thoraric and abdominal segmental differentiation by around 10 hrs, after which time induced Antp protein has long since been degraded (Gibson and Gehring 1988). Since these heat shock genes do not require autoactivation of an endogenous copy of the gene to produce their transformations (Gonzalez-Reyes et al. 1990; N.Urquia and G. Morata, pers. comm.) the question of whether or not homeotic protein function must be maintained throughout development should be considered open.

If the requirement is indeed transient, it might further be speculated that the coordinated regulation of target genes involves the regulation of a series of intermediate regulatory genes in a hierarchy leading eventually to the realizator genes.

3.4 The Fates of Individual Cells: Photoreceptors

The analysis of the genetic basis for the determination of fate at the level of individual cells has recently begun in a variety of invertebrate systems. The compound eye of *Drosophila* consists of an array of precisely ordered ommatidia, each of which contains a set of eight different photoreceptor cells, named R1 through R8 (Tomlinson 1988). A number of mutations which lead to transformations of one photoreceptor type into another have been isolated, and recently the heat shock approach has been used to investigate the nature of cell-autonomous gene function during eye morphogenesis.

The analysis of the *sevenless (sev)* gene is particularly interesting, as it demonstrates the tremendous functional specificity of a very different class of regulatory protein, a tyrosine kinase (Hafen et al. 1987). Ectopic expression of *sev* (Basler and Hafen 1989; Bowtell et al. 1989) has been shown not to cause any developmental abnormalities in a wild-type genetic background, a somewhat surprising finding given the fundamental role of this type of protein in signal transduction. However, it has been possible to achieve rescue of loss of *sev* function, by administering heat shocks in late third instar larvae as the morphogenetic furrow traverses the eye, organizing ommatidia assembly. Striking bands of rescued ommatidia in a mutant background can be produced, indicating that there is a very restricted temporal window in which *sev* is required. The spatial distribution of the gene product, by contrast, has been shown not to be critical. This result implies that *sev* activity is potentiated by receipt of a signal emanating from a neighboring cell, and that the timing of the arrival of the signal may be an important factor in distinguishing R7 from other photoreceptor cell types.

Ectopic expression of the homeobox gene *rough* has been achieved using two different promoters, derived from the *hsp70* and *sev* genes (Basler et al. 1990; Kimmel et al. 1990). In this case, the approach has led to a reconsideration of the apparent paradox that the deletion of a nuclear regulatory protein has non-cell autonomous phenotypic effects. The failure of R3 and R4 to develop correctly is now thought to result indirectly from improper specification of R2/5 identity in the absence of *rough,* which is thought to define a specific cell identity only after the decision to initiate neural development has been made.

4 Conclusions

4.1 General Applicability to the Study of Development

One of the major points emphasized in this review has been the empirical finding by a large number of groups that ectopic expression of genes generally causes only very limited and specific alteration of development. A priori, it might have been expected that the analysis of gene function using the heat shock approach would be of

Table 1. Summary of heat shock induced gene expression studies

Gene	Process	References
Caudal	Segmentation	Mlodzik et al. (1990)
Hunchback	Segmentation	Struhl (1989a)
Kruppel	Segmentation	Struhl (1989b)
Hairy	Segmentation	Ish-Horowicz and Pinchin (1987)
Fushi tarazu	Segmentation	Struhl (1985)
		Ish-Horowicz and Gyurkovics (1988)
		Ish-Horowicz et al. (1989)
Engrailed	Segmentation	Poole and Kornberg (1988)
Deformed	Homeotic identity	Kuziora and McGinnis (1988, 1989)
Sex combs reduced	Homeotic identity	Gibson et al. (1990)
Antennapedia	Homeotic identity	Schneuwly et al. (1987)
		Gibson and Gehring (1988)
		Gibson et al. (1990)
Ultrabithorax	Homeotic identity	Mann and Hogness (1990)
		Gonzalez-Reyes et al. (1990)
		Gonzalez-Reyes and Morata (1990)
Rough	Eye development	Basler et al. (1990)
		Kimmel et al. (1990)
Sevenless	Eye development	Basler and Hafen (1989)
		Bowtell et al. (1989)
D-ras	(Oncogenesis)	Bishop and Corces (1988)
Transformer	Sex determination	McKeown et al. (1988)
Elongation factor	Aging	Shepherd et al. (1989)
Shaker	Ion channels	Zagotta et al. (1989)
FLP recombinase	Recombination	Golic and Lindquist (1989)
P-transposase	Transposition	Stellar and Pirotta (1986)
	Splicing	Laski and Rubin (1989)
ß-galactosidase	Heat shock	Lis et al. (1983)
Adh	Heat shock	Klemenz et al. (1985)

limited application because of the expectation of nonspecific disruption of development. However, flies have shown a great deal of dignity in the face of repeated insult. Indeed, their generous display of developmental inertia is probably telling us much about both ontogenetic and evolutionary processes. For the experimentalist, it opens up the opportunity to study all aspects of development from novel angles, and as Table 1 illustrates, this opportunity is being seized.

Thus, the use of heat shock to study development of the peripheral nervous system has recently begun with yet another homeobox gene, *cut* (L. Jan, pers. comm.) and with the helix-loop-helix nuclear regulatory proteins of the *achaete-scute* complex (J. Modolell, pers. comm.). Behavioral studies have been initiated which involve ectopic expression of genes such as *period* (J. Hall, pers. comm.), a key regulator of circadian rhythm; and the *sex-peptide* (E.Kubli, pers. comm) which is transferred from the male to the female during copulation and modifies courting behavior. Similarly, dissection of the functions of various ion channels involved in transmission of nervous impulses has been initiated with studies of the neurophysiological properties of flies carrying heat shock-*Shaker* constructs (Zagotta et al. 1989). Bishop and Corces (1988) have studied the effects of ectopic expression of an activated form of the *Drosophila* homolog of a mammalian oncogene, the ras G-protein. The analysis of segmentation will undoubtedly benefit from studies of the effects of ectopic expression of members of the segment-polarity class of genes, such as *engrailed* (Poole and Kornberg 1988) and *wingless* (J. Noordermeer, P. Lawrence, and R. Nusse, pers. comm.) which include nuclear and cell-surface proteins. Furthermore, the heat shock approach has already proved its worth in study of the function of genes with non developmental roles, such as the *P-element transposase* gene (Stellar and Pirotta 1986; Laski and Rubin 1989).

4.2 Prospects

It remains to emphasize that the heat shock approach is by no means limited to the analysis of the structures and functions of previously identified genes. For example, attempts are under way to randomly generate conditional dominant (gain-of function) mutations, by P-element mediated integration of a heat shock promoter throughout the genome, in the hope that administration of heat shock will give rise to ectopic expression of regulatory genes. It may also soon be possible to couple the heat shock promoter to tissue-specific enhancer elements in such a way as to generate spatially restricted ectopic expression (see Parker-Thornberg and Bonner 1987, for an alternative strategy). Alternatively, it should be possible to use heated needles to deliver localized heat shocks (Monsma et al. 1988).

A very promising use of heat shock induced expression is in the detection of downstream (target) genes for a variety of regulatory proteins. Here, the strategy is to compare patterns of gene expression at certain time points after heat shock in strains which do and do not induce the gene of interest. Molecular strategies concentrate on plus/minus or subtractive screens for differentially expressed cDNA clones, and can be conducted in embryos, imaginal discs, or even cell culture lines. A more cell biological approach is to search for differences in gene expression in vivo as reported by so-called enhancer detector lines (O'Kane and Gehring 1987). A variation on this theme might be to search for genes whose expression is altered by loss of function of a gene. In this case, heat-shock induced expression of the *FLP recombinase* (Golic and Lindquist 1989) would be used to controllably delete a trans-gene which had previously been introduced into a fly which is genetically null for the gene of interest.

It is not yet clear whether heat shock strategies might prove to be useful in other organisms. Preliminary studies in the nematode *Caenorhabditis elegans* are encouraging, although it has not been established that the heat shock promoter is active in all cells of the developing worm (Stringham et al. 1990). In rodents, more problems might be encountered with the logistics of administration of heat shocks given the susceptibility of early embryos to stress (Walsh 1990, see Chap. 4). That ectopic expression can be used to produce developmental abnormalities has however been clearly shown both in *Xenopus* and the mouse, using microinjection and constitutive expression strategies (Ruiz i Altaba and Melton 1989; Kessel et al. 1990). Here, the possibility may soon exist to use heterologous hormone-inducible promoters to control ectopic expression of genes. It is to be hoped that, this time in terms of strategy, *Drosophila* again proves to be a worthy model for vertebrate development.

Acknowledgments. It is a pleasure to thank Walter Gehring for his encouragement to write this review, and for his support during the course of the work described here. I'd also like thank Marcel van den Heuvel, Deborah Andrew, Jasprien Noordermeer, Elizabeth Gavis, Richard Mann, Matthew Scott and Juan Botas for their comments on the manuscript, and all those mentioned in the text who supplied unpublished information. This work was supported by a grant (to Professor Walter J. Gehring) from the Swiss National Science Foundation, and by a post-doctoral fellowship from the Helen Hay Whitney Foundation.

References

Baker BS, Belote JM (1983) Sex determination and dosage compensation in *Drosophila melanogaster.* Annu Rev Genet 17:345–393

Basler K, Hafen E (1989) Ubiquitous expression of sevenless: position–dependent specification of cell fate. Science 243:931–934

Basler K, Yen D, Tomlinson A, Hafen E (1990) Reprogramming cell fate in the developing *Drosophila* retina: transformation of R7 cells by ectopic expression of rough. Genes Dev 4:728–739

Bishop JG III, Corces VG (1988) Expression of an activated ras gene causes developmental abnormalities in transgenic *Drosophila melanogaster.* Genes Dev 2:567–577

Boggs RT, Gregor P, Idriss S, Belote J, McKeown M (1987) Regulation of sexual differentiation in *D. melanogaster* via alternative splicing of RNA from the transformer gene. Cell 50:739–747

Bowtell DL, Simon MA, Rubin G (1989) Ommatidia in the developing *Drosophila* eye require and can respond to sevenless for only a restricted period. Cell 58:931–936

Driever W, Nusslein–Volhard C (1988) The bicoid protein determines position in the *Drosophila* embryo in a concentration dependent manner. Cell 54:95–104

Eberlein S (1986) Stage specific embryonic defects following heat shock in *Drosophila.* Dev Genet 6:179–197

Garcia–Bellido A (1975) Genetic control of wing disc development in *Drosophila.* CIBA Found Symp 29:161–182

Gehring WJ, Hiromi Y (1986) Homeotic genes and the homeobox. Annu Rev Genet 20:147–173

Gibson G, Gehring WJ (1988) Head and thoracic transformations caused by ectopic expression of Antennapedia during *Drosophila* development. Development 102:657–675

Gibson G, Schier A, LeMotte P, Gehring WJ (1990) The specificities of *Sex combs reduced* and *Antennapedia* are defined by a distinct portion of each protein which includes the homeodomain. Cell 62:1087–1103

Golic KG, Lindquist S (1989) The FLP recombinase of yeast catalyses site-specific recombination in the *Drosophila* genome. Cell 59:499–509

Gonzalez-Reyes A, Morata G (1990) The developmental effect of overexpressing a Ubx product in *Drosophila* embryos is dependent on its interactions with other homeotic products. Cell 61:515–522

Gonzalez-Reyes A, Urquia N, Gehring WJ, Struhl G, Morata G (1990) Are cross-regulatory interactions between homeotic genes functionally significant? Nature 344:78–80

Goto T, Macdonald P, Maniatis T (1989) Early and late patterns of even skipped expression are controlled by distinct regulatory elements that respond to different spatial cues. Cell 57:413–422

Hafen E, Basler K, Edstroem J, Rubin GM (1987) Sevenless, a cell–specific homeotic gene of Drosophila, encodes a putative transmembrane receptor with a tyrosine kinase domain. Science 236:55–63

Harding K, Hoey T, Warrior R, Levine M (1989) Autoregulatory and gap gene response elements of the even skipped promoter of *Drosophila.* EMBO J 8:1205–1212

Ingham PW (1988) The molecular genetics of embryonic pattern formation in *Drosophila.* Nature 335:25–33

Ish–Horowicz D, Gyurkovics H (1988) Ectopic segmentation gene expression and metameric regulation in *Drosophila.* Development (Suppl) 104:67–73

Ish-Horowicz D, Pinchin SM (1987) Pattern abnormalities induced by ectopic expression of the *Drosophila* gene hairy associated with repression of *ftz* transcription. Cell 51:405–415

Ish–Horowicz D, Pinchin SM, Ingham PW, Gyurkovics HG (1989) Autocatalytic *ftz* activation and metameric instability induced by ectopic *ftz* expression. Cell 57:223–232

Kaufman TC, Tasaka S, Suzuki DT (1973) The interaction of two complex loci, *zeste* and *bithorax* in *Drosophila melanogaster*. Genetics 75:299–321

Keleher CA, Goutte C, Johnson AD (1988) The yeast cell-type specific repressor a2 acts cooperatively with a non-cell-type-specific protein. Cell 53:927–936

Kessel M. Balling R, Gruss P (1990) Variations of cervical vertebrae after expression of a Hox-1.1 transgene in mice. Cell 61:301–308

Kimmel BE, Heberlein U, Rubin GM (1990) The homeodomain protein rough is expressed in a subset of cells in the developing *Drosophila* eye where it can specify photoreceptor cell subtype. Gene Dev 4:712–727

Klemenz R, Hultmark D, Gehring WJ (1985) Selective translation of heat shock mRNA in *Drosophila melanogaster* depends on sequence information in the leader. EMBO J 4:2053–2060

Kuziora MA, McGinnis WJ (1988) Autoregulation of a *Drosophila* homeotic selector gene. Cell 55:477–485

Kuziora MA, McGinnis WJ (1989) A homeodomain substitution changes the regulatory specificity of the Deformed protein in *Drosophila* embryos. Cell 59:563–571

Laski FA, Rubin GM (1989) Analysis of the cis–acting requirements for germ-line-specific splicing of the P-element ORF2–ORF3 intron. Genes Dev 3:720–728

Lindquist S (1980) Varying patterns of protein synthesis in *Drosophila* during heat shock: implications for regulation. Dev Biol 77:463–479

Lindquist S (1986) The heat shock response. Annu Rev Biochem 55:1151–1191

Lis JT, Simon JA, Sutton CA (1983) New heat shock puffs and ß–galactosidase activity resulting from transformation of *Drosophila* with an hsp70-lacZ hybrid gene. Cell 35:403–410

Mann R, Hogness DS (1990) Functional dissection of Ultrabithorax proteins in *D. melanogaster*. Cell 60:579–610

Martinez-Arias A (1989) A cellular basis for pattern formation in the insect epidermis. Trends Genetics 5:262–267

McKeown M, Belote JM, Boggs RT (1988) Ectopic expression of the female transformer gene product leads to female differentiation of chromosomally male *Drosophila*. Cell 53:887–895

Mitchell HK, Lipps L (1978) Heat shock and phenocopy induction in *Drosophila*. Cell 15:907–918

Mlodzik M, Gibson G, Gehring WJ (1990) Effects of ectopic expression of caudal during *Drosophila* development, Development 109:271–277

Monsma SA, Ard R, Lis J, Wolfner M (1988) Localized heat shock induction in *Drosophila melanogaster*. J Exp Zool 247:279–284

Morata G, Garcia-Bellido A (1976) Developmental analysis of some mutants of the bithorax system of *Drosophila*. Wilhelm Roux Arch Dev Biol 179:125–143

Morata G, Lawrence P (1977) Homeotic genes, compartments, and cell determination in Drosophila. Nature 265:211–216

Morgan TH (1926) The theory of the gene. Reprinted by Garland New York, 1989

Nusslein-Volhard C, Wieschaus E (1980) Mutations affecting segment number and polarity in the *Drosophila* embryo. Nature 287:795–801

O'Kane CJ, Gehring WJ (1987) Detection in situ of genomic regulatory elements in *Drosophila*. Proc Natl Acad Sci USA 84:9123–9127

Parker-Thornberg J, Bonner JJ (1987) Mutations that induce the heat shock response of *Drosophila*. Cell 51:763–772

Pelham HRB (1982) A regulatory upstream promoter element in the *Drosophila hsp70* heat shock gene. Cell 30:517–528

Petersen RB, Linquist S (1989) Regulation of hsp70 synthesis by messenger RNA degradation. Cell Regul. 1:135–149

Poole SJ, Kornberg T (1988) Modifying expression of the engrailed gene of *Drosophila melanogaster*. Development (Suppl) 104:85–94

Postlethwait JH, Schneiderman HA (1971) Pattern formation and determination in the antenna of the homeotic mutant Antennapedia of *Drosophila melanogaster*. Dev Biol 25:606–640

Ptashne M (1988) How eukaryotic transcriptional activators work. Nature 335:683–689

Rubin GM, Spradling AC (1982) Genetic'transformation of *Drosophila* with transposable element vectors. Science 218:348–353

Ruiz i Altaba A, Melton DA (1989) Involvement of the *Xenopus* homeobox gene *Xhox3* in pattern formation along the anterior–posterior axis. Cell 57:317–326

Schneuwly S, Klemenz R, Gehring WJ (1987) Redesigning the body plan of *Drosophila* by ectopic expression of the homeotic gene Antennapedia. Nature 325:816–818

Shepherd JCW, Walldorf U, Hug P, Gehring WJ (1989) Fruit flies with additional expression of the elongation factor EF-1a live longer. Proc Natl Acad Sci USA 86:7520–7521

Stellar H, Pirotta V (1986) P transposons controlled by the heat shock promoter. Mol Cell Biol 6:1640–1649

Stern S, Tanaka M, Herr W (1989) The Oct–1 homeodomain directs formation of a multiprotein–DNA complex with the HSV transactivator VP16. Nature 314:624–630

Stringham E, Jones D, Candido P (1990) Studies of hsp-16 and ubq-1 expression in the nematode *Caenorhabditis elegans* (Abstr). Worm Breeder's Gaz 11:(2):25

Struhl G (1985) Near-reciprocal phenotypes caused by inactivation or indiscriminate expression of the *Drosophila* segmentation gene ftz. Nature 318:677–680

Struhl G (1989a) Differing strategies for organizing anterior and posterior body pattern in *Drosophila* embryos. Nature 338:741–744

Struhl G (1989b) Morphogen gradients and the control of body pattern in insect embryos. CIBA Found Symp 144:65–91

Tomlinson A (1988) Cellular interactions in the developing *Drosophila* eye. Development 104:183–193

Webster GC, Webster SL (1983) Decline in synthesis of elongation factor one (EF–1) precedes the decreased synthesis of total protein in ageing *Drosophila melanogaster*. Mech Ageing Dev 22:121–128

Zagotta WN, Germeraad S, Garber S, Hoshi T, Aldrich RW (1989) Properties of ShB A–type potassium channels expressed in Shaker mutant *Drosophila* by germline transformation. Neuron 3:773–782

4 Thermotolerance and Heat Shock Response During Early Development of the Mammalian Embryo

David Walsh, Karen Li, Carol Crowther, Debbie Marsh, and Marshall Edwards[1]

1 Introduction

Through evolution, mammals have acquired the ability to maintain a relatively constant body temperature over a wide range of ambient temperatures. Temperature stability is important especially during embryogenesis when the developing organs of the embryo are undergoing rapid proliferation and differentiation. Development at this stage is extremely sensitive to temperature elevation.

Hyperthermia during pregnancy can result in a spectrum of adverse reactions, including embryonic or fetal death, prenatal growth retardation and developmental abnormalities. Hyperthermia is a recognized teratogen in many animal species including chickens, mice, rats, hamsters, rabbits, guinea pigs, sheep, pigs and monkeys (reviewed by Edwards 1986).

There are many known causes of hyperthermia in pregnant women. These include environmental exposure to heat, heavy exercise (particularly in a hot environment), febrile infections, drugs and some metabolic conditions. The incidence of febrile infections in women during the first 12 weeks of pregnancy (organogenesis) has been estimated to be between 2.5 and 8% (Layde et al. 1980; Lipson 1988). Considerable evidence has been obtained from prospective and retrospective surveys that indicates significant associations between birth defects and maternal hyperthermia (Edwards 1989).

2 Developmental Defects Caused By Hyperthermia

Heat shock during development induces morphological defects in both vertebrates and invertebrates (Petersen 1990). Many different developmental defects have been produced experimentally by moderate physiological temperature elevations. In all species, the central nervous system (CNS) appears to be very susceptible to damage by heat, particularly during neural tube closure (the initiation of the CNS). In human development, this event occurs between 3 and 4 weeks of pregnancy. The types and severity of defects depend on the species, the stage of development at which heat shock occurs and the dose of heat (dose being a product of the elevation above normal temperature and the duration of the elevation).

---

[1] University of Sydney, Department of Veterinary Clinical Sciences, Sydney, NSW, 2006, Australia

Results and Problems in Cell Differentiation 17
Heat Shock and Development
Hightower and Nover (Eds.)
©Springer-Verlag Berlin Heidelberg 1991

Embryonic tissues undergoing rapid cellular proliferation appear to be most at risk to hyperthermia. This probably accounts to a large extent for the high susceptibility of the developing eye and brain. However, actively proliferating adult tissues, such as testis and bone marrow, are also at risk (Edwards and Penny 1985). By comparison, non-proliferative, metabolically active tissues in either embryos or adult animals are relatively resistant.

Common abnormalities of the CNS caused by hyperthermia include neural tube defects (anencephaly, exencephaly, encephalocele), microphthalmia, microencephaly, neurogenic talipes and arthrogryposis. Major developmental deformities usually result from cell death. The actual mechanisms causing the cells to die are unknown. Recent studies have reported that interruption of the normal developmental program may trigger a set of "suicide' genes responsible for programmed and rapid cell death in developing nematodes (Ellis and Horvitz 1986; Yuan and Horvitz 1990). Vascular disturbances may also be involved in mammalian cell death (Nilsen 1985).

Using guinea pigs at 21 days of pregnancy (equivalent to weeks 7-8 in human pregnancy), the acute effects of transiently raising body temperature 2 to 3.5 °C on the embryonic brain have been studied (Edwards et al. 1974). Cells in S-phase and mitosis were most susceptible to heat, with lethal damage occurring after temperature elevations of 2 °C or more (Wanner et al. 1976; Upfold et al. 1989). Within minutes of achieving an elevation of 2 °C, cells showed clumping of chromatin, karyorhexis and subsequent death (by apoptosis) within 4–6 hr. Further mitotic activity resumed after about 8 h in a partially synchronized burst. The degree of synchronization appeared to increase the sensitivity of the embryo to a subsequent heat shock. If the second heat shock occurred after a lengthy delay (8–10 h), severe defects were seen (Edwards et al. 1974, 1976). Recent studies (Upfold et al. 1989) have shown that in the 21-day embryonic guinea-pig cell death is concentrated in the alar (sensory) regions of the brain. This subsequently affects the layer IV pyramidal neurons in mature adults. The whole brain weight of offspring heated at this stage is permanently reduced. Affected brains appear to be proportional miniatures of normal brains. They contain fewer cells and less myelin, although the amounts are appropriate for the size of the brain (Edwards et al. 1976). The ratio of glial to neuronal cells (usually 5:1) appeared unchanged.

3 The Heat Shock Response in Mammalian Embryos

Heat shock of mammalian cells, as with all other organisms studied, elicits an immediate cellular and molecular response. For example, in the neural plate of 9.5-day rat embryos, changes occur in the structure and shape of the neuroectoderm cells that result in alteration of closure of the neural tube. The nuclear and microtubule organizations are dramatically affected by heat shock. Rapid changes are also seen in the cell cycle immediately after heat shock, affecting the progression of the cells through the cell cycle (Walsh and Morris 1989). At the molecular level, the heat shock response is characterized by; (1) a rapid transcriptional activation of the heat shock gene families; (2) an immediate, selectively enchanced translation of the corresponding mRNAs; (3) simultaneous inhibition of all other protein syntheses; and (4) following the stress

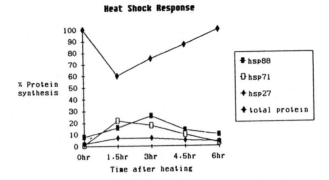

Fig. 1. Protein synthesis after a heat shock of 42 °C for 10 min, expressed as % of synthesis in control embryos. Embryos at 9.5 days were labelled with [^{35}S] methionine for 60 min at 38 °C before collection at specific time intervals after heating. The proteins were analyzed on SDS gels by fluorography and optical densitometry. A general inhibition of total protein synthesis was observed 1.5 h after heat shock together with an increase of Hsp synthesis

response, decay of heat shock protein (Hsp) mRNA and the degradation of the Hsp and associated recovery of normal cellular protein synthesis (Fig.1 Walsh et al. 1985).

An interesting aspect of the heat shock response is the ability of cells to acquire thermotolerance and protection from further heat shock exposure. This same mechanism may also be responsible for adaptation of the cells to higher temperatures. Thermotolerance occurs when cells are exposed to a sublethal heat shock that induces the heat shock response. These cells, when further challenged to a known lethal heat shock, appear to be thermally protected. Although the actual mechanism that allows cells to survive extreme heat shock is not clear, these cells appear to have the ability to recover rapidly. It has generally been acknowledged that Hsps play a role in providing acquired thermotolerance.

3.1 Embryo Culture and Heat Shock Genes

We have developed a whole embryo culture system that allows normal development of rat embryos to proceed over a 48-h period. Rat embryos are removed from the uterus of 9.5 days of gestation and cultured in heat-inactivated rat serum (New et al., 1973; Klein et al., 1980). At 9.5 days (early gastrulation), the embryo is at one of the most sensitive stages in development i.e., neural tube closure. The neural plate induces the neural groove to form the neural tube. After 24 h of culture, the neural fold in the midbrain region has fused and closure of the neuroectodermal ridge proceeds posteriorly and anteriorly. After 48 h of culture a well-developed forebrain, midbrain and hindbrain is evident as are the branchial arches, otic and optic placodes, a beating heart, forelimb buds and yolk sac circulation. (Fig. 2).

The embryo culture system offers several advantages for studying the effects of heat shock and thermotolerance on embryo development. Accurate staging of embryogenesis can be achieved. This enables precise developmental windows to be studied at

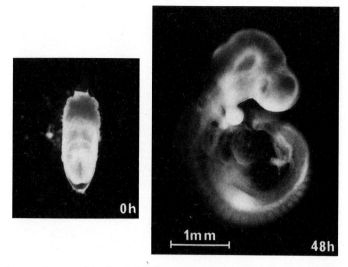

Fig. 2. Developmental stages of in vitro 9.5-11.5-day whole rat embryo in culture. This period of culture embraces most of the major events of organogenesis. The embryo develops from the relatively undifferentiated head-fold stage (0 h) to a stage with a beating heart, prominent optic vesicles and a closed neural tube at the end of culture (48 h)

both the cellular and molecular level. In addition, exact levels of heat can be applied to the embryos at specific developmental stages without interference from maternal factors.

3.2 Heat Shock and Neural Tube Closure

A series of heat shock regimes have been developed that induce the heat shock response and thermotolerance in the rat embryo (Walsh et al. 1987). When 9.5-day embryos were heat shocked at 43 °C for 7.5 min, normal development of the anterior neural plate was interrupted. This resulted in characteristic ocular, brain and facial malformations (Fig. 3). The 43 °C heat shock delayed neural tube closure and subsequent induction of the developing forebrain. By comparison, a mild heat shock of 42 °C for 10 min induced the heat shock response but produced no morphological deformities. There was, however, a slight delay (4 h or 2 somites) in growth of the embryo after 48 h of culture. The significance of this delay in terms of neurological development is unclear. In terms of overall growth, the stressed embryo in vivo when compared to controls appears to fully recover and to have overcome this delay by the time of birth. Preheating embryos at 42 °C for 10 min resulted in an acquired thermotolerance which protected against subsequent exposure to the teratogenic 43 °C heat. Thermotolerance was acquired within 15-min of the initial 42 °C heat shock and persisted for 6 h. The observation that a 15-min "rest" period (at 38.5 °C) is required to allow for development of thermotolerance suggests that cellular processes, such as protein synthesis, may be required for thermoprotection.

David Walsh et al.

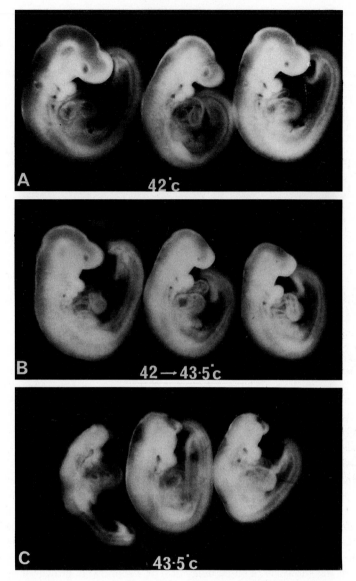

Fig. 3A-C. Rat embryos at 11.5 days cultured for 48 h following various heat treatments at the neural plate stage. **A** Embryos exposed to a mild heat shock of 42 °C for 10 min. Embryos had a well-developed forebrain, midbrain and hindbrain region. **B** Embryos subjected to 42 °C for 10 min to induce the heat shock response, followed by a recovery period of 60 min at 38.5 °C, then a further heat shock of 43.5 °C for 7.5 min. Although the embryos were smaller, caused by a delay in growth, no morphological defects were seen. **C** Embryos exposed at 43.5 °C for 7.5 min, displaying developmental defects of both the eye and forebrain

3.3 Cell Death of the Neuroectoderm

The neural plate of 9.5-day embryos consists of a layer of neuroectodermal cells that proliferate and fold to form the neural tube. These cells undergo interkinetic nuclear migration during the normal cell cycle of 6–7 h. The nuclei pass through Go and early S-phase at the basement membrane with mitosis occuring at the ventricular surface (see Fig. 6A). Embryos heat shocked at 42 °C showed no interruption to this cellular migration, no morphological defects and no cell death in the neuroectoderm. Heat shock 43 °C, however, caused developmental defects. These defects occurred as a result of massive cell death (apoptosis), particularly in the neuroectodermal ridges of the neural fold in the anterior neural plate and the heart primordia. This region of the neural plate consists of rapidly proliferating cells that, at this stage, are initiating and inducing the optic placodes (eyes), forebrain and heart. Cell death within the neuroectoderm occurred 2-3 h after the 43 °C heat shock but did not take place in thermoprotected embryos (42/43 °C). Cell death may be the result of damage to and inhibition of synthesis of essential proteins required for normal cell proliferation and function.

3.4 Induction of Heat Shock Proteins

3.4.1 Translation

[^{35}S]-methionine incorporation studies of Hsp synthesis after heat shock have shown an association between Hsp induction and a general reduction of normal protein synthesis after heat shock (see Fig. 1). The length and level of inhibition depend on the severity of the heat shock. In all cases, inhibition occurs for up to 1.5 h. Simultaneously, the Hsp71 was selectively expressed and translated (Fig. 4). The Hsps27, 71, 88 and the heat shock cognate Hsc73 (Walsh et al. 1987, 1989), which is constitutively expressed at this stage of development, are also synthesized and respond to heat shock. This selective pattern of protein synthesis gradually returned to the control pattern 7–8 h. Control embryos did not appear to synthesize the inducible Hsp71 (Fig. 4).

Cells with acquired thermotolerance appeared to recover and synthesize normal proteins within a shorter time frame after heat shock. Thermotolerant cells and cells exposed to a nonteratogenic heat shock show a regulated protein recovery over 8 h. The recovery of protein synthesis was also associated with the rapid decay of Hsp mRNA and degradation of Hsps. A teratogenic heat shock at 43 °C for 7.5 min appears to cause irreversible inhibition of protein synthesis. A teratogenic heat shock caused a 20 fold increase in expression of the *hsp71* gene (Fig. 5), but little of this mRNA appears to be translated.

3.4.2 Transcription

The kinetics and levels of mRNA specific for Hsp71 and Hsc73 were examined over development using Northern (Fig. 5) and dot blots. Embryos exposed to the teratogenic heat shock showed a rapid (20 min after exposure) increase in transcription of the *hsp71* gene. Accumulation of Hsp71 mRNA reached a peak (eight-fold above

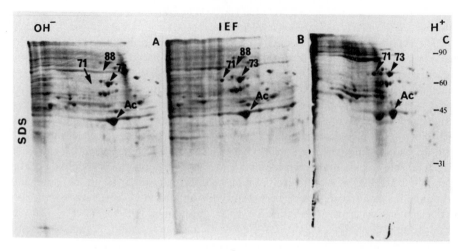

Fig. 4A-C. Fluorographs of two-dimensional gels comparing the Hsps synthesized in the neural plate of 9.5–day embryos exposed to various heat shocks. Embryos were labelled with [^{35}S] methionine for 60 min at 38.5 °C following heat shock. Proteins were extracted and separated by isoelectric focusing in the first dimension and SDS-PAGE in the second dimension. **A** Control embryos were cultured continuously at 38.5 °C. **B** Embryos exposed to a heat shock at 42 °C for 10 min. **C** Embryos heat shocked at 43 °C for 7.5 min. The positions of Hsp71 and Hsp88 along with Hsc71 are marked by *arrowheads*. Actin (*Ac*) serves as an internal marker; standard protein markers are indicated in kilodaltons

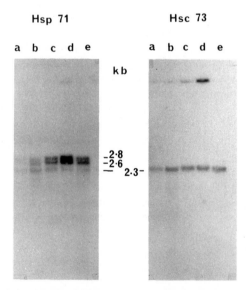

Fig. 5. Northern blot analysis of Hsp71 (2.8/2.6 kb) and the Hsc73 (2.3 kb) mRNAs of 9.5-day embryos. A total of 5 ug glyoxylated total RNA was subjected to electrophoresis in 1% agarose gels and transferred to nitrocellulose. The mRNA bands were identified by nick-translated *hsp71* and *hsc73* probes. Four temperature regimes were used: *a* control; *b* 42 °C for 10 min plus 15 min at 38.5 °C; *c* 42 °C for 10 min plus 60 min at 38.5 °C; *d* 43 °C for 7.5 min plus 60 min at 38.5 °C; *e* 42 °C for 10 min plus 60 min at 38.5 °C followed by 43 °C for 7.5 min

control) after 90 min and then quickly declined to control levels within 4 h. It is of interest to note that at this stage of development, 6–7 h is the average duration of a cell division cycle in the neuroectodermal cells in the neural plate. *Hsc73* transcription also increased but to a much lower level, being about two-fold above the control level. Comparison between the rate of transcription and the synthesis of Hsp71 suggests that the control of Hsp71 expression is occurring mainly at a transcriptional level.

Using *in situ* hybridization techniques, it was demonstrated that synthesis of Hsp71 mRNA was tissue specific (Walsh et al. 1989). Hsp71 expression after heat shock was observed, in the mesoderm and in specific areas of the neuroectoderm (Fig. 6). There was no transcription in the endoderm and yolk sac (mesoderm and endoderm) but high constitutive expression in the ectoplacental cone (paternally derived tissue). Within the neural plate, expression was particularly strong in the neuroectoderm layer corresponding to the neural ridge and forebrain placode. This region is important in the early differentiation and migration of the neural crest cells.

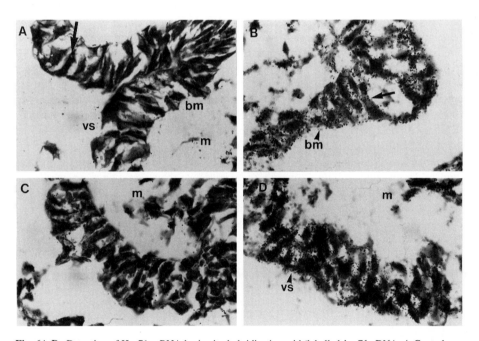

Fig. 6A-D. Detection of Hsp71 mRNA by in situ hybridization with labelled *hsp71* cDNA. **A** Control embryo with neuroectoderm (*arrow*) showing the ventricular surface (*vs*), basement membrane (*bm*) and mesoderm (*m*) at the midline of the neural plate. **B** Embryo heat shocked for 7.5 min 43 °C and probed with *hsp71* cDNA after 90 min. At these later time stages expression is now evident (*arrow*) throughout the neuroectoderm layer. Hsp71 mRNA is predominantly found in the cytoplasm and basement membrane (*bm*). **C** Neuroectoderm of control embryos with the anterior neuropore region. **D** Heat-shocked embryo (same as b) with *hsp71* gene expression in the anterior region of the neural plate. Magnification in all cases 1000-fold

4 Thermotolerance and Heat Shock Protein Synthesis

After a thermotolerant heat shock at 42 °C for 10 min followed by a further heat shock at 43 °C for 7.5 min, accumulation of Hsp71 mRNA was lower than that in embryos heat shocked at 43 °C alone. This observation suggests that *hsp71* transcription is not simply a passive response to the total amount of heat applied to the embryo. In addition, the temporal correlation between induction of hs mRNA and the onset and persistence of thermotolerance indicates that thermotolerance may be regulated at the transcriptional level. The suggestion that Hsps, particularly Hsp71 and Hsp27, play an important role in thermoprotection has been proposed for many systems (Lindquist and Craig 1988; Riabowol et al., 1989). For example, Rat-1 cells transfected with the *hsp71* gene and over-expressing the Hsp71 appear to have increased thermal resistance (Li et al. 1990). We propose that in 9.5-day rat embryos, acquisition of thermotolerance by the neuroectoderm cells may require the preexistence of hs gene products. Induction of Hsp71 thus serves a protective function if present before a damaging exposure. However, it does not repair cells already damaged by a severe heat shock. This proposed role for Hsp71 in thermotolerance is compatible with the suggested cellular function as a chaperone protein (Ellis 1987; Pelham 1989). Both the Hsp71 and another member of this gene family, the heat shock cognate (Hsc) 73 appear to have vital roles in protein stability within the cell. In addition, the Hsps appear to affect cellular protein degradation with an unfoldase-like activity, or to maintain protein conformations essential for intermembrane transport (Chiang et al. 1989). Thus, we assume that Hsp71 protects nascent proteins and prevents them from being denatured by the teratogenic heat shock. However, if the conformation of these proteins is already altered by heat, further association with Hsps may not be possible. Although the role of other Hsps in this respect was not studied in sufficient detail, there is evidence for their participation in thermoprotective effects as well. Table 1 summarizes our data on developmental and tissue-specific expression of hs genes in rats. In many species as in rat, the activation of *hsp* genes varies at different stages of development. For instance, mammalian embryos do not respond to heat shock by induction of Hsps until the blastula-morula stage. As with other stages, the ability to activate *hsp* genes and synthesize Hsps appears to correlate with the ability to acquire thermotolerance. In addition to their protective function, Hsps may play a role in normal development, hence their constitutive expression at the neural plate stage.

5 Heat Shock and Cell Cycle Changes

In vivo (Sect. 2) and in vitro studies have shown that heat causes marked changes to the cell cycle. Formation of the brain requires neuroectodermal and mesodermal induction of the neural folds, during which neuroectodermal nuclei migrate from the basement membrane through their cells to the ventricular surface where division occurs. Thus, the nuclei progress through the cell cycle Go, G1 and early S-phase near the basement membrane, migrate through the cell in mid-to-late S-phase and reach the ventricular surface in G2. Mitosis then occurs. Expansion of the surface of the neural plate is achieved by specific proliferation and orientation of these daughter cells to each other. The nuclei from the two daughter cells then travel back down the cell to the basement membrane to progress again through the cell cycle (Walsh and Morris 1989).

In the neural plate of 9.5-day rat embryos, a teratogenic exposure of 43 °C for 7.5 min inhibits progression of the cell cycle and migration of the nuclei through the neuroectoderm. Cell cycle inhibition results in the accumulation of nuclei at the basement membrane. Progression of the cell cycle at the G1-S interface is delayed for 1.2 h (Fig. 7). However, there appears to be no delay in S-phase but rather an enhanced progression of cells to mitosis. This results in a loss of cells in S-phase and a depletion of the middle layer of the neuroectoderm. Depletion of S-phase cells in the neural plate results in large irreversible morphological changes, and cell death follows, resulting in major developmental defects.

Thermotolerance is associated with an overall delay in the cell cycle. Progression through both the G1/S and S/G2 boundary is observed. Cell progression is delayed, resulting in an overall lengthening of the S-phase and the overall cell cycle time. Cells exposed to heat shock in the middle to late S-phase are more susceptible to thermal injury. Thermotolerance may simply be an inhibition of DNA synthesis at the most sensitive stages of the cell cycle (Walsh and Morris 1989). Lengthening of the cell cycle may also be associated with thermotolerance and with expression of Hsps. It has been shown that in non-stressed mammalian cells, Hsp71 synthesis is tightly regulated

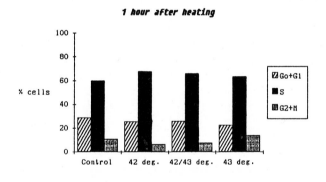

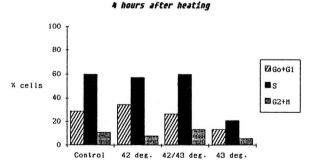

Fig. 7. DNA histograms of neuroectoderm cells at 1 h and 4 h after heat shock expressed as percentage of cells in Go+G1, S and G2+M phases. Embryos were cultivated at 38.5 °C (control) or heat shocked for 10 min at 42 °C (42), for 10 min at 42 °C followed by 7.5 min at 43 °C after a 60 min recovery period (42/43) and for 7.5 min at 43 °C (43) respectively

during the cell cycle (Milarski and Morimoto 1986; Milarski et al. 1989). In the control rat embryo, the level of Hsp71 increases upon entry into S-phase and reaches a maximum in G2 and M-phases. These levels of Hsp71 are dramatically affected and increased after a thermotolerance-inducing and teratogenic heat shock.

6 Conclusion

The Hsps27, 71, 88 and the Hsc73 appear to play a major role in both normal embryonic development and the recovery of the neural plate after heat shock. The inducibility of the Hsps are developmentally regulated through organogenesis. The Hsps27, 88 and the Hsc73 are constitutively expressed at neural tube closure but they all also appear to respond to heat shock. However, *hsp71* transcription is induced mainly by stress (Table 1).

The heat shock response is highly efficient in protecting neuroectoderm cells from heat shock. Temperature elevations of 4 to 4.5 °C are non-teratogenic, provided cells had a prior exposure to a mild temperature elevation (3.5 °C) that induces the heat shock response. Activation of this protective response, i.e. acquired thermotolerance, requires a short recovery period of approximately 15 min. However, the heat shock response cannot protect cells against a preceding teratogenic heat shock, i.e. it is not able to repair cells damaged in this way. Regulated recovery of the neural plate is associated with transcription and translation of the Hsps and also their degradation. The Hsps appear to act as protein "chaperones" required for protein stability and correct folding for their intracellular transport during development.

Table 1. Expression of hs genes during rat development. The amount of hs mRNA was estimated from tissue by comparative cell number and protein concentration using dot blot analyses and corresponding gene specific probes. The detection limit is in the range of 3×10^5 mRNA molecules per ug of total RNA

Hsp/Hsc[a]	2-Cell stage	Blastula	Neural plate			Foetus			Adult rat		
			Ectod.	Mesod.	Endod.	Head	Heart	Liver	Head	Heart	Liver
Hsp88 Induced	-	+	+++	++	+	-	-	-	+	++	+
Non-ind.	-	++	+	+	-	-	-	-	-	+	-
Hsc73 Induced	-	+	+	+	-	-	-	-	-	-	-
Non-ind.	-	-	+++	++	+	-	-	-	-	+	+
Hsp71 Induced	-	+	+++++	+++	-	+	++	+++	+	++	+++
Non-ind.	++	-	++	-	-	-	-	-	-	-	-
Hsp27 Induced	-	-	++	+	+	+	+	+	+	+	+
Non-ind.	-	-	+	-	-	-	+	-	+	+	+

[a] In vitro heat shock was achieved by immersing the culture bottle in a calibrated water bath at 42 °C for 9.5 min. Control bottles were exposed to a water bath at 38 °C for 9.5 min.

In vivo heat shock samples were collected from whole pregnant rats immersed in a water bath to achieve a core temperature elevation of 42 °C for 9.5 min. (This required heating in a water bath for 20-25 mins depending on body weight). The animals were killed and tissue removed 90 min after heat shock.

The significance of Hsps and the changes to the cell cycle are unknown. The cell cycle is delayed at the G1/S and S/G2 interface during thermotolerance, preventing progression into late S-phase and mitosis, which are particularly sensitive to heat shock and cellular damage.

References

Chiang H-L, Terlecky SR, Plant CP, Dice JF (1989) A role for the 70 kD heat shock protein in lysosomal degradation of intracellular proteins. Science 246:382–385

Edwards MJ (1986) Hyperthermia as a teratogen: a review of experimental studies and their clinical significance. Teratog Carcinog Mutagen 6:563–582

Edwards MJ (1989) Hyperthermia and the developing central nervous system. Ann Res Inst Environ Med, Nagoya Univ, 40:355–364

Edwards MJ, Penny RHC (1985) Effects of hyperthermia on the myelograms of adult and fetal guinea-pigs. Br J Haematol 59:93–108

Edwards MJ, Mulley R, Ring S, Wanner RA (1974) Mitotic cell death and delay of mitotic activity in guinea-pig embryos following brief maternal hyperthermia. J Embryol Exp Morphol 32:593–602

Edwards MJ, Wanner RA, Mulley RC (1976) Growth and development of the brain in normal and heat-retarded guinea-pigs. Neuropathol Appl Neurobiol 2:439–450

Ellis HM, Horvitz HR (1986) Genetic control of programmed cell death in the nematode *C.elegans*. Cell 44:817–829

Ellis J (1987) Proteins as molecular chaperones. Nature 328:378–379

Klein NW, Vouler MA, Chatot CL, Pierro LJ (1980) The use of cultured rat embryos to evaluate the teratogenic activity of serum cadmium and cyclophamide. Teratology 21:199–208

Landry J, Chretien P, Lambert H, Hickey E, Weber LA (1989) Heat shock resistance conferred by expression of the human HSP27 gene in rodent cells. J Cell Biol 109:7–15

Layde PM, Edmonds LD, Erikson JD (1980) Maternal fever and neural tube defects. Teratology 21:105–108

Li GC, Li L, Liu R, Mak JY, Lee W (1990) Stable expression of human *hsp70* gene in rodent cells confers thermal resistance. Heat Shock Conference Workshop, Ravello, Italy

Lindquist S, Craig E.A. (1988) The heat-shock proteins. Annu Rev Genet 22:631–677

Lipson A (1988) Hirschsprung disease in the offspring of mothers exposed to hyperthermia during pregnancy. Am J Med Genet 29:117–124

Milarski KL, Morimoto RI (1986) Expression of human Hsp 70 during the synthetic phase of the cell cycle. Proc Natl Acad Sci USA 83:9517–9521

Milarski KL, Welch WJ, Morimoto R (1989) Cell cycle-dependent association of HSP70 with specific cellular proteins. J Cell Biol 108:413–423

New DAT, Coppola PT, Terry S (1973) Culture of explanted embryos in rotating tubes. J Reprod Fertil 35:135–138

Nilsen NO (1985) Vascular abnormalities due to hyperthermia in chick embryos. Teratology 30:237–251

Petersen NS (1990) Effects of heat and chemical stress on development. In: Scandalios FG (ed) Advances Genetics, vol 28: Genomic responses to environmental stress. Academic Press, San Diego, CA

Pelham HRB (1989) The selectivity of secretion protein sorting in the endoplasmic reticulum. Biochem Soc Trans 17:795–802

Riabowol KT, Mizzen LA, Welch WJ (1988) Heat shock is lethal to fibroblasts microinjected with antibodies against hsp70. Science 242:433–436

Upfold JB, Smith MSR, Edwards MJ (1989) Quantitative study of the effects of maternal hyperthermia on cell death and proliferation in the guinea-pig brain on day 21 of pregnancy. Teratology 39:173–179

Walsh DA, Hightower LE, Klein NW, Edwards MJ (1985) The induction of the heat shock proteins during early mammalian development. Heat shock, Cold Spring Harbor Lab Symp 2:92

Walsh DA, Klein NW, Hightower LE, Edwards MJ (1987) Heat shock and thermotolerance during early rat embryo development. Teratology 36:181–191

Walsh DA, Li K, Speirs J, Crowther CE, Edwards MJ (1989) Regulation of the inducible heat shock 71 genes in early neural development of cultured rat embryos. Teratology 40:321–334

Walsh DA, Morris VB 1989 Heat shock affects cell cycling. Teratology 40:583–592

Wanner RA, Edwards MJ, Wright RG (1976) The effects of hyperthermia on the neuroepithelium of 21 day guinea-pig foetus: histopathologic and ultrastructural study. J Pathol 118:235–244

Yuan J, Horvitz HR (1990) The *Caenorhabditis elegans* genes ced-3 and ced-4 act cell autonomously to cause programmed cell death. Dev Biol 138:33–41

5 Strain Differences in Expression of the Murine Heat Shock Response: Implications for Abnormal Neural Development

Mark D. Englen[1] and Richard H. Finnell[2]

1 Introduction

Neural tube defects are common congenital anomalies affecting approximately 1–2 per 1000 liveborn infants (Nakano 1973; Richards et al. 1972). Empiric risk figures, along with numerous clinical studies, indicate that neural tube defects are of a multifactorial origin, having both a genetic and an environmental component to their development (Martin et al. 1983; Holmes et al. 1976; Campbell et al. 1986). Many environmental factors have been implicated as inducers of neural tube defects in humans, including maternal vitamin and folate deficiencies (Smithells 1982), hyperthermia (Edwards et al. 1981, 1986; Fisher and Smith 1981), valproic acid (Bjerkedal et al. 1982; Mastroiacovo et al. 1983; Lindhout and Schmidt 1986) and retinoids (Keitzmann et al. 1986). Although several genetic hypotheses have been put forth to explain the inheritance of neural defects, no single hypothesis fits all of the data currently available.

A great deal of experimental evidence supports the idea that agents capable of inducing congenital malformations, including neural tube defects, do so by delaying normal developmental events (Spyker and Smithberg 1972; Fuyuta et al. 1978; Theodosis and Fraser 1978; Webster and Messerle 1980; Chernoff et al. 1986; Walsh and Morris 1989). With this in mind, German (1984) hypothesized that the induction of heat shock proteins (Hsps), either by hyperthermia or other environmental agents, preempted the synthesis of essential proteins during development. The resulting failure of these essential proteins to function at critical times (developmental delay) would alter the temporal coordination of morphogenesis and lead to the development of congenital malformations. German's hypothesis suggested to us that the extent of Hsp synthesis may be correlated with an animal's genetically determined susceptibility to heat-induced neural defects. This idea has been extensively tested in our laboratory using highly inbred strains of mice differing from one another in terms of their sensitivity to environmentally induced neural tube defects.

[1] Department of Veterinary and Comparative Anatomy, Pharmacology and Physiology, College of Veterinary Medicine, Washington State University, Pullman, Washington 99164–6520, USA
[2] Department of Veterinary Anatomy and Public Health, College of Veterinary Medicine, Texas A & M University, College Station, Texas 77843, USA

Results and Problems in Cell Differentiation 17
Heat Shock and Development
Hightower and Nover (Eds.)
©Springer-Verlag Berlin Heidelberg 1991

2 The Heat Shock Proteins

Heat stress in all organisms induces the synthesis of a specific set of proteins collectively known as the heat shock proteins (Hsps). Other stresses, such as anoxia and glucose starvation, and certain chemicals also induce the production of these highly conserved proteins. The Hsps comprise several gene families encoding proteins of distinct molecular weights (see Introduction Table 1, Neidhardt et al. 1984). Proteins of 100–110 kD, 90 kD, 70 kD, and several lower molecular weight proteins between 16–40 kD are produced by most eukaryotic cells following exposure to selected noxious stimuli (Lindquist 1986; Subjeck and Shyy 1986). Hsp70 and 90 have been found in all organisms examined, and are the most well-characterized of the Hsps. These proteins were originally observed in cells which had been heat stressed, and it was hypothesized early on that Hsps served to protect the cell from the toxic effects of hyperthermic stress (Lindquist and Craig 1988). This hypothesis was bolstered by the observation that acquired thermotolerance was correlated with the induction of the Hsps (Mitchell et al. 1979; Li and Laszlo 1985). However, it has since been shown that both Hsp70 and 90 are also abundant in many cell types at normal temperatures, and their synthesis is only further enhanced by a sufficient heat treatment.

The synthesis of Hsps has been associated with neural tube defects induced by chemical agents such as cadmium and retinoic acid, in addition to hyperthermia (Layton and Ferm 1980; Courgeon et al. 1984; Anson et al. 1987). Several recent studies have suggested possible functions for Hsps during critical periods in morphogenesis in response to teratogenic insults. Using a monoclonal antibody specific for Hsp90, Sanchez et al. (1988) showed by indirect immunofluorescence that Hsp90 was associated with the microtubules in cultured mammalian epithelial cells. These authors speculated that Hsp90 may shuttle proteins along the microtubule architecture from sites of synthesis to the nucleus or plasma membrane. A primary function of Hsp70 appears to be the regulation of protein assembly and disassembly both during normal growth and after heat shock, a hypothesis originally suggested by Varshavsky and colleagues (Finley et al. 1984) and later extended by Pelham (1986). Minton et al. (1982) have hypothesized that the protective effect of Hsps in stressed cells may be a generalized, nonspecific one, whereby Hsps may protect sensitive proteins from denaturation. The Hsps would then help the embryo maintain its temporal coordination of developmental processes, by assuring that gene products that are developmentally regulated, such as growth factors, are expressed when they are required, even under adverse conditions. In terms of our own experimental models, we reasoned that strains of mice sensitive to heat-induced teratogenesis may show less induction of Hsps than resistant strains. Alternatively, sensitive strains might show prolonged synthesis of Hsps, and this synthesis at the expense of essential proteins would put the embryo at risk for a defect in neural tube closure. Therefore, the failure to protect sensitive regulatory proteins, or the failure to resume programmed developmental processes would account for the teratogenic effect of maternal hyperthermia.

Table 1. The effect of maternal hyperthermia treatment on the number of implants, resorptions, liveborn fetuses, exencephalic fetuses, and exencephalic litters

Strain	Treatment day[a]	No. litters	Implants (Mean + SEM)	Resorptions (%)	No. liveborn	No. exencephaly (%)	No. exencephalic litters (%)
DBA/2J	Control	10	6.7 + 0.56	26.9	49	0(0)	0(0)
	8.0	10	5.8 + 0.55	43.0	33	0(0)	0(0)
	8.5	10	5.8 + 0.61	53.4	27	0(0)	0(0)
	9.0	14	5.5 + 0.51	26.0	57	0(0)	0(0)
C57BL/6J	Control	8	5.0 + 0.78	10.0	36	0(0)	0(0)
	8.0	15	6.9 + 0.27	22.3	80	8(10.0)	8(53.3)
	8.5	14	6.7 + 0.38	38.3	58	1(1.7)	1(7.1)
	9.0	10	5.4 + 0.65	13.0	47	0(0)	0(0)
SWR/J	Control	10	9.4 + 0.92	4.3	90	0(0)	0(0)
	8.0	10	9.6 + 0.58	43.8	54	4(7.4)	1(10)
	8.5	14	10.1 + 0.50	7.9	139	19(13.7)	5(33)
	9.0	10	9.0 + 0.73	11.1	80	4(5.0)	2(20)
LM/Bc	Control	12	9.8 + 0.48	6.8	109	0(0)	0(0)
	8.0	13	8.1 + 0.50	44.8	58	1(1.7)	1(7.7)
	8.5	14	9.9 + 0.31	9.4	125	17(13.6)	7(50)
	9.0	14	9.4 + 0.44	6.1	124	1(0.8)	1(7.1)
SWV	Control	10	12.6 + 0.56	7.9	116	1(0.9)	1(10)
	8.0	11	11.8 + 1.18	36.2	88	1(1.1)	1(10)
	8.5	10	13.6 + 0.37	22.1	106	47(44.3)	9(90)
	9.0	10	12.5 + 0.43	3.2	121	1(0.8)	1(10)

[a] For explanations see text.

3 Strain Differences in Heat-Induced Neural Tube Defects

Previous work with mice had clearly demonstrated a multifactorial etiology for exencephaly, a neural tube defect analogous to early anencephaly in humans (Cole and Trasler 1980). It had also been shown that a single, 10-min waterbath exposure produced an elevated (13.6%) frequency of exencephaly in C_3 H mice (Chernoff et al. 1983, 1984; Webster and Edwards 1984). The results of these studies prompted an examination of the importance of both the maternal and embryonic genotype in determining susceptibility to heat-induced neural tube defects. A hierarchy of susceptibility to hyperthermia-induced neural tube defects was found among five inbred strains of mice (Finnel et al. 1986).

Female mice were exposed to a 10-min 43 °C hyperthermic treatment on days 8:0, 8:12 (8 days plus 12 h) or 9:0 of gestation. Control mice were exposed to a 10–min 38 °C treatment. As seen in Table 1, hyperthermic treatment on gestational day 8:12 produced the highest peak response frequency of exencephalic fetuses, and the following strain differences were noted: SWV, 44.3%; SWR/J, 13.7%; LM/BC, 13.6%; C57BL/6J, 1.7% and DBA/2J, 0%. Hyperthermic treatment on gestational day 8:0 or 9:0 produced less that 10% exencephalic fetuses in all five strains tested although the C57BL/6J strain embryos were most sensitive on gestational day 8.0 (10%). Control

Table 2. Percent exencephaly after hyperthermia treatment on day 8.5 in reciprocal crosses

Strain of dam	Strain of sire		
	SWV	LM/Bc	C57BL/6J
SWV	44.3	9.3	5.2
LM/Bc	11.0	13.6	2.2
C57BL.6J	11.0	7.0	1.7

mice for all five strains produced less than 1.0% spontaneous exencephalic fetuses. Reciprocal crosses between SWV, LM/Bc, and C57BL/6J were performed to determine whether several genes or only a few genes are involved in mediating hyperthermic teratogenicity. When SWV dams were crossed to either C57BL/6J or LM/Bc sires, the percent of exencephalic fetuses dropped from 44.3% to less than 10% (Table 2). This result indicates that it is the embryo's genotype and not the maternal genotype that is the critical factor in determining susceptibility to heat-induced neural tube defects. The crosses between the low (C57BL/6J) and moderate (LM/Bc) response strains resulted in 0-14% exencephalic fetuses (Table 2). This result, together with those observed for the crosses involving SWV dams, suggests that there are at least two levels of response to heat-induced neural tube defects. The first, in the range of 0-14%, probably involves several genes, and the second, in the area of 44% involves at most only one or two genes.

4 The Murine Heat Shock Response

4.1 The Heat Shock Response in the Murine Embryo and Lymphocyte

The experiments discussed in the preceding section revealed distinct differences in susceptibility to the teratogenic effects of a hyperthermic shock among inbred strains of mice. These results provided a particularly attractive model in which to test German's hypothesis that the heat shock response results in an overall embryonic developmental delay, which leads to abnormal morphogenesis and birth defects (German 1984). According to this hypothesis, a particular strain's susceptibility to heat-induced neural defects should be reflected in the diversion of normal protein synthesis by embryonic or maternal cells to the production of Hsps and/or a delay in the resumption of normal protein synthesis.

Pregnant dams from sensitive (SWV) and resistant (DBA/2J) mouse strains were exposed to a single, 10-min. 43 °C hyperthermic treatment on gestational day 8:12. Control dams were similarly treated in a 38 °C waterbath. At selected time points post-treatment, both embryos and maternal lymphocytes were collected and labeled with ^{35}S-methionine. Samples of both tissue types were prepared for SDS-PAGE and electrophoresed on 12-20% gradient gels. Autoradiograms were prepared from the dried gels and the bands of newly synthesized proteins were quantified by scanning densitometry. As shown in Fig. 1, both strains displayed an identical response to heat shock, in that immediately following treatment there was an initial inhibition of total protein

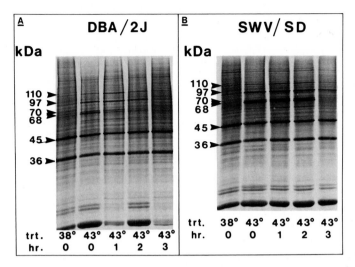

Fig. 1. Autoradiographs of SDS-PAGE gels containing lymphocyte samples collected 0-3 h following either a control (38 °C) or hyperthermia (43 °C) *in vivo* treatment. Lymphocytes were labeled for 1 h at 38 °C with ^{35}S-methionine and for each gel equal counts of radioactivity were loaded onto each gel lane. A DBA/2J lymphocytes. B SWV/SD lymphocytes

synthesis, accompanied by the appearance of Hsps migrating with molecular weights of 68, 70, 97 and 110 kD. In the DBA/2J strain (Fig. 1A), enhanced synthesis of these Hsps was no longer seen at 2 h post-treatment. In the SWV strain (Fig. 1B), Hsp synthesis was observed up to 3 h (gestational day 8:15) following the brief hyperthermic insult.

The patterns of protein synthesis observed in the embryos and lymphocytes suggested differences in the quantities of the individual proteins. Densitometric analyses were performed on representative autoradiograms to further evaluate the strain differences in protein synthesis. As seen in Fig. 2A, immediately following the hyperthermic insult, lymphocytes obtained from DBA/2J dams showed a decrease in total protein synthesis, compared to the 38 °C controls. There was, however, clear evidence of Hsp synthesis, which accounted for a great deal of the protein synthetic activity in the heat-shocked cells. In the SWV lymphocytes sampled immediately after the heat treatment (Fig. 2C), there was a far greater decrease in total protein synthesis compared to the controls and to the DBA/2J lymphocytes (Fig. 2A). By 2 h post-treatment, the densitometry tracings of the DBA/2J lymphocytes were nearly identical to that of the control cells (Fig. 2B), except for a few low-molecular weight bands where greater synthesis was observed in the heat-shocked cells. In the heat-sensitive SWV lymphocytes, protein synthesis did not begin to return to control levels until 2 h following the thermal shock (Fig. 2D), although at this time protein synthesis was still markedly depressed. While some differences were observed in the synthesis of non-Hsps between embryos and lymphocytes of both strains, it is apparent from these studies that isolated maternal lymphocytes are highly reflective of the embryonic heat shock response in both strains of mice examined. Furthermore, the results of these studies demonstrate that a hyperthermic insult sufficient to induce neural tube defects in mouse embryos is capable of altering the normal patterns of protein synthesis during gestational days 8:0-9:14, the period of neural tube closure.

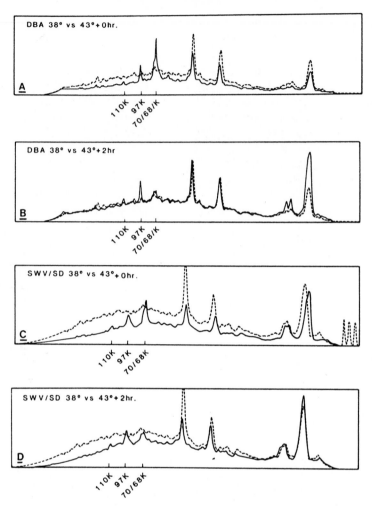

Fig. 2A-D. Representative densitometry tracings of autoradiographs obtained from SDS-PAGE gels of SWV and DBA/2J lymphocytes. Equal counts of radioactivity were loaded onto each gel lane. Each panel represents a composite overlay of densitometry tracings of protein synthesis by lymphocytes obtained immediately or at 2 h following treatment. *Dashed line* 38 °C (control) treated lymphocytes; *solid line,* 43 °C (hyperthermia) treated lymphocytes. **A** DBA/2J lymphocytes obtained immediately following control and hyperthermia treatment. **B** DBA/2J lymphocytes obtained 2 h following control and hyperthermia treatment. **C** SWV/SD lymphocytes obtained immediately following control and hyperthermia treatment. **D** SWV/SD lymphocytes obtained 2 h following control and hyperthermia treatment

The alterations in the patterns of protein synthesis displayed by both the embryos and the maternal lymphocytes following the teratogenic heat shock are closely paralleled by the delay in neural tube closure following maternal hyperthermia. In the heat-resistant DBA/2J embryos, neural tube closure under control conditions is completed by gestational day 9:11. When DBA/2J embryos are exposed to the hyperthermic shock, neural tube closure is delayed 1 h, until gestational day 9:12. In the heat-sensitive SWV strain, neural tube closure is normally completed by gestational day 9:14. However, when heat shocked, those embryos that successfully complete neural tube closure do so no earlier than gestational day 10:6, a delay of 16 h when compared to controls. Thus, a strong association exists between the embryo's ability to resume normal protein synthesis following a teratogenic heat treatment, the length of time that neural tube closure is delayed, and the embryo's susceptibility to heat-induced neural tube defects. In the DBA/2J embryos, protein synthesis rapidly returns to normal following a brief heat shock and neural tube closure is delayed only 1 h. This brief interruption in the embryo's normal timetable of development allows sufficient time for the embryo to recover before neural tube development is adversely affected. This is demonstrated by the resistance of the DBA/2J strain to heat-induced exencephaly (see Table 1). On the other hand, the SWV strain produces Hsps for an extended period of time following the heat treatment, and the Hsp synthesis is paralleled by a reduction in total protein synthesis. As a possible consequence of the delayed return to normal protein synthesis, a lengthy delay in neural tube closure has been observed. As SWV embryos were found to be highly sensitive to heat-induced neural tube defects, a possible connection between the Hsp synthesis and the delayed neural tube closure was suggested. Therefore, it is the embryo's genetically determined ability to minimize adverse effects on protein synthesis, thereby limiting the temporal disruption of embryogenesis and permitting normal events to proceed, that appears to be the key factor in determining susceptibility to heat-induced neural tube defects.

4.2 A Genetic Basis for Strain Differences in the Murine Heat Shock Response

These initial investigations demonstrated that the induction and duration of Hsp synthesis by both the embryo and the maternal lymphocytes reflects the susceptibility of an inbred strain of mice to neural tube defects following a brief thermal insult. The studies which used embryonic tissue were, however, limited by the small amounts of tissue that were readily obtainable and by the rapid differentiation of the embryo during the period of neural tube closure. Since the lymphocyte response to heat shock adequately parallels that of the embryo, the lymphocyte heat-shock assay was expanded to compare the kinetics of protein synthesis by lymphocytes from sensitive (SWV/SD) and resistant (DBA/2J) inbred strains, to that of lymphocytes taken from the F_1 offspring of reciprocal crosses between these two strains. The lymphocyte assay thus provided a convenient means to examine the pattern of transmission of the sensitivity trait as reflected by the heat shock response.

Lymphocytes from adult SWV/SD and DBA/2J mice were labeled with ^{35}S-methionine and autoradiograms were prepared as described in the preceding Section 4.1. Reciprocal crosses between SWV/SD and DBA/2J mice were performed and lymphocytes from the F_1 adults (90 days of age) were examined for Hsp induction and duration of synthesis as described above for the parental strains. As seen in previous experiments, DBA/2J lymphocytes synthesized Hsps68 and 70 up to 1 h post-treatment, while SWV/SD lymphocytes continued to synthesize Hsps68 and 70 up to 2 h post-treatment (see Fig. 1A and 1B). Synthesis of Hsp97 was evident at all times in both SWV/SD and DBA/2J lymphocytes; Hsp110 synthesis was highly variable in these experiments and showed no consistent pattern in either strain. When new protein synthesis by lymphocytes from both DBA/2J X SWV/SD crosses was examined, a pattern of Hsp induction similar to the DBA/2J parental strain was found (Fig. 3A, B). In particular, Hsps68 and 70 were observed immediately following and at one hour

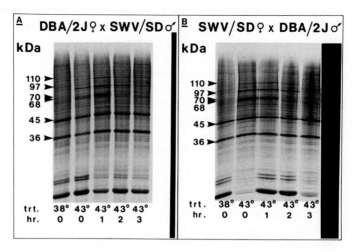

Fig. 3A,B. Autoradiograph of SDS-PAGE gel containing lymphocyte samples of the F_1 offspring of recip-rocal DBA/2J X SWV/SD crosses collected 0-3 h following either a control (38 °C) or hyperthermia (43 °C) *in vivo* treatment. Lymphocytes were labeled for 1 h at 38 °C with ^{35}S-methionine and for each gel equal counts of radioactivity were loaded onto each gel lane. **A** Lymphocytes from the F_1 offspring of matings between the DBA/2J females and SWV/SD males. **B** Lymphocytes from the F_1 offspring or matings between the SWV/ SD females and DBA/2J males

post-treatment, i.e., the same pattern of synthesis seen with DBA/2J lymphocytes. The changes in total protein synthesis by the F_1 offspring also mimicked that of the heat-resistant DBA/2J strain. Thus the patterns of both Hsp and normal protein synthesis, and the relative susceptibility to exencephaly following heat shock (Sect. 3) had reverted in the F_1 offspring of both reciprocal crosses to that of the DBA/2J parental strain. Furthermore, the results obtained using the reciprocal crosses indicate that the more rapid return to normal protein synthesis observed in the DBA/2J strain is the dominant trait, as is the resistance to heat-induced exencephaly (Sect. 3). This, in turn, suggests a genetically determined link between the pattern of heat-induced Hsp synthe-sis and susceptibility to heat-induced exencephaly. As discussed in previous sections, German (1984) theorized that the extent to which normal protein synthesis is displaced by Hsp synthesis following heat shock governs the extent of developmental delay and thereby the likelihood of inducing congenital malformations. The pattern of Hsp synthesis and the resistance to heat-induced neural malformations observed in the F_1 progeny is consistent with this idea. Moreover, these results further suggest that a general shutdown of protein synthesis and the induction of Hsp synthesis is controlled by only a few, perhaps regulatory, genes.

4.3 An In Vitro Model of the Murine Heat Shock Response

The physiologic and genetic factors which govern the susceptibility of an inbred strain to neural tube defects following heat shock are obviously complex. An in vitro model using short-term (0–6 h) cultures of mature, differentiated cells to study the murine heat shock response allows for a direct evaluation of the effects of hyperthermia on the cells, independent of any complicating physiological differences. In addition, an in vitro assay using human lymphocytes or other cell types offers the prospect of a simple means to screen the teratogenic potential of chemical or physical agents, based upon the induction of the Hsps. As previously reported (Bennett et al. 1990; Mohl et al. 1990) and discussed above, the lymphocytes from adult mice had proven to be a useful model of the murine heat shock response for both adult and embryonic cells. Accordingly, Hsp induction was examined in lymphocyte cultures from our two test strains of mice used previously (SWV/SD and DBA/2J), as well as in lymphocytes from the adult F_1 progeny of the reciprocal crosses. Immediately following isolation and separation of the lymphocytes, the cells were exposed to a 10-min 45 °C heat shock in a heat-controlled oven. The temperature for the in vitro experiments was increased 2 °C over the 43 °C temperature used in the previously discussed in vivo studies to ensure that a temperature of 43 °C was maintained in the culture media. Control lymphocytes were kept at 38 °C. Cell viability, as measured by trypan blue exclusion, was not significantly different in any of the heat-shocked lymphocyte cultures when compared to those of control lymphocytes ($p \leq 0.05$). Following heat exposure, aliquots of the lymphocyte cultures were removed at different time points, and labeled with ^{35}S-methionine. As previously described, autoradiograms were prepared from SDS-PAGE gels; individual bands on the autoradiograms were quantitated by scanning densitometry.

As seen in other experiments using adult lymphocytes (see Fig. 1A, B), the synthesis of Hsps68 and 70 was observed immediately and up to 1 h following heat treatment in DBA/2J lymphocytes, and up to 2 h after heat treatment in SWV/SD lymphocytes. The synthesis of Hsp97 was observed immediately after heat treatment in DBA/2J lymphocytes and up to 3 h later; in SWV/SD lymphocytes Hsp97 synthesis continued for an additional hour, until 4 h post-treatment. Lymphocytes from both DBA/2J X SWV/SD reciprocal crosses showed patterns of Hsp induction similar to that of the DBA/2J parental strain, a result seen in previous in vivo experiments employing lymphocytes from DBA/2J X SWV/SD reciprocal crosses (Sect. 4.2). In particular, in lymphocytes from both reciprocal crosses, the synthesis of Hsps68 and 70 was seen up to 1 h post-treatment and Hsp97 was observed up to 2 h post-treatment, a pattern of Hsp synthesis characteristic of the DBA/2J strain following heat shock. As before, the differences noted in the autoradiograms in the synthesis of Hsps68, 70, and 97 were verified by scanning densitometry. Thus, the results of these in vitro studies were consistent with the results of previous studies using DBA/2J and SWV/SD lymphocytes and embryos (Sects. 4.1, 4.2). In addition to establishing the potential value of the in vitro lymphocyte assay as a method of screening suspected teratogenic agents, the results of the in vitro studies reinforce the concept that developmental delay is the common thread linking the susceptibility to heat-induced neural tube defects to the induction of the Hsps. Additional studies examining the in vitro heat shock response of lymphocytes exposed to selected teratogenic and nonteratogenic agents are currently in progress.

5 Conclusions

Murine lymphocytes and embryos both respond to a hyperthermic insult by synthesizing Hsps. In fact, the results of the studies summarized in this Chapter provide evidence for the excellent correlation between the heat shock response of maternal lymphocytes and that of embryonic tissue. As the isolated lymphocytes did mimic the embryo's response, it opens the possibility of using genetically defined cells that are easily obtained to assess an individual's susceptibility to an adverse reproductive outcome when exposed to a potentially teratogenic agent.

It was of interest to note the differences in the induction and duration of Hsp synthesis between inbred strains sensitive (SWV) and resistant (DBA/2J) to heat-induced neural tube defects. These differences raise some very interesting issues. For example, these studies revealed that the relative synthesis of Hsps is prolonged, while normal protein synthesis is more depressed in the SWV strain, than in the DBA/2J strain. This is important, for it is the SWV strain that is the most susceptible to heat-induced exencephaly. This suggests that the induction and prolonged synthesis of the Hsps is directly related to an increased response frequency of exencephalic fetuses. However, both the SWV and DBA/2J strains produced Hsps, a clear indication that their presence alone is not sufficient to explain the observed differences between these two strains in susceptibility to heat-induced neural tube defects. The induction of the Hsps may, in fact, under some conditions protect the embryo from a later, more severe environmental stress (Mirkes 1987; Walsh et al. 1987, see Chap. 4). Yet the reversion by the F_1 hybrid progeny of SWV X DBA/2J to the pattern of protein synthesis characteristic of the resistant DBA/2J parental strain adds support to a cause-and-effect relationship between Hsp synthesis and neural tube defects, since the F_1 progeny are also significantly more resistant to heat-induced exencephaly.

In terms of German's (1984) stress hypothesis of teratogenesis, susceptibility to teratogenic effects of heat shock may be related to the embryo's inability to shorten the period of developmental arrest, a period characterized by Hsp synthesis. For example, if the embryonic stress hypothesis is correct, and normal protein synthesis is disrupted by the maternal hyperthermia treatment, then embryos with shortened periods of arrested protein synthesis should be resistant, while embryos with protracted periods of abnormal protein synthesis should be most susceptible to neural tube defects. Alternatively, the heat shock response may be homeostatic. In this case, Hsp induction by genetically resistant embryos would limit the period of arrested protein synthesis. This would minimize the temporal disruption of embryogenesis, permitting the normal events to proceed and result in normal development. Embryos incapable of inducing Hsp synthesis would experience a longer delay and be at an increased risk for congenital malformations.

It is also possible that the two inbred mouse strains used in these experiments differed in the actual amount of thermal damage that they sustained during the course of the hyperthermic treatment. If this were the case, then one would reasonably expect that the more severely damaged strain of embryos would express an extended duration of Hsp synthesis and accumulation, and that this would be positively correlated with sensitivity to heat-induced neural tube defects. A number of different groups have convincingly demonstrated in mammalian cells, rat embryo cultures, and in non-eukaryotic organisms (Walsh et al. 1987, see review by Lindquist and Craig 1988), that the amount of Hsp70 produced increases in direct proportion to the severity of the hyperthermic shock and consequently, to the amount of thermal damage. Although the actual amount of Hsps synthesized were not quantified in our studies, a qualitative assessment of the intensity of the bands on the autoradiograms (Fig. 1A, B), suggests that the neural tube defect sensitive SWV inbred mouse strain produced and accumulated more Hsp68/70 when compared to the DBA/2J strain. The prolonged synthesis of the Hsps and delayed return to normal non-Hsp synthesis placed the SWV embryos at greater risk for a neural tube defect.

At this point in time, the one conclusion that can be made is that it is the embryo's genetically determined ability to tolerate developmental delay that is the critical factor in determining teratogenic susceptibility or resistance. Further studies at the molecular level using genetic profiling of neuroepithelial cells will help to better elucidate the precise relationship between the induction and presence of Hsps in the developing embryo and the production of neural tube defects.

Acknowledgments. This work was supported in part by research grant number ES04326 from the National Institutes of Health to Dr R.H. Finnell. The authors thank Ms. Pat Ager for her technical assistance, and Ms. Catherine Smith and Ms. Paula Perron for their excellent editorial and secretarial assistance.

References

Anson JF, Hinson WG, Pipkin JL, Kwarta RF, Hansen DK, Young JF, Burns ER, Casciano DA (1987) Retinoic acid induction of stress proteins in fetal mouse limb buds. Dev Biol 121:542–547

Bennett GD, Mohl VK, Finnell RH (1990) Embryonic and maternal heat shock responses to a teratogenic hyperthermic insult. Reprod Toxicol 4:113–19

Bjerkedal TA, Czeixel A, Goujard J, Kallen B, Mastroiacova P, Nevin N, Oakley G, Robert E (1982) Valproic acid and spina bifida. Lancet ii:1096

Campbell LR, Dayton DH, Sohal GS (1986) Neural tube defects: a review of human and animal studies on the etiology of neural tube defects. Teratology 34:171–187

Chernoff GF, Golden JA, Edwards MJ (1983) Heat-induced neural tube defects in C3H/lg-ml mice. Proc Greenwood Genet Ctr 2:93–94

Chernoff GF, Golden JA, Finnel RH (1984) Heat-induced exencephaly in the mouse: evidence for induction of the defect prior to neural tube closure. Proc Greenwood Genet Ctr 3:108

Chernoff GF, Golden JA, Seymour MA (1986) Neural tube closure in mouse embryos following a hyperthermic exposure. Proc Greenwood Genet Ctr 5:102–103

Cole WA, Trasler DG (1980) Gene-teratogen interaction and insulin-induced mouse exencephaly. Teratology 22:125-139

Courgeon AM, Maisonhaute C, Best-Belpomme M (1984) Heat shock proteins are induced by cadmium in *Drosophila* cells. Exp Cell Res 153:515–521

Edwards MJ (1981) Clinical disorders in fetal brain development: defects due to hyperthermia. In: Hetzel BS, Smith RM (eds) Fetal brain disorders: recent approaches to the problem of mental deficiency. Elsevier/North Holland Biomedical Press, Amsterdam, pp 335–364

Edwards MJ (1986) Hyperthermia as a teratogen: a review of experimental studies and their clinical significance. Teratog Carcinog Mutagen 6:563–582

Finley D, Crechanover A, Varshavsky A (1984) Thermolability of ubiquitin-activating enzyme from the mammalian cell cycle mutant ts85, Cell 37:43–55

Finnell RH, Moon SP, Abbot LC, Golden JA. Chevnoff GF (1986) Strain differences in heat-induced neural tube defects in mice. Teratology 33:247–252

Fisher NL, Smith DW (1981) Occipital encephalocel and early gestational hyperthermia. Pediatrics 68:480–483

Fuyuta MT, Fujimoto F, Hirata S (1978) Embryotoxic effects of methylmercuric chloride administered to mice and rats during organogenesis. Teratology 18:353–366

German J (1984) Embryonic stress hypothesis of teratogenesis. Am J Med 76:293–301

Holmes LB, Driscoll SG, Atkins L (1976) Etiologic heterogeneity of neural tube defects. N Engl J Med 294:365–369

Keitzman H, Schwarze I, Grote W, Ravens U, Janig U, Harms D (1986) Embryonale Fehlbildung bei Etretinat-Therapie der Mutter wegen Morbus Darier. Dtsch Med Wochenschr 111:60-62

Layton WM, Ferm VH (1980) Protection against cadmium induced limb malformations by pretreatment with cadmium or mercury. Teratology 21:357–360

Li GC, Laszlo A (1985) Thermotolerance in mammalian cells: a possible role for the heat shock proteins. In: Atkinson BG, Walden DB (eds) Changes in eukaryotic gene expression in response to environmental stress. Academic Press, Orlando, pp 349–371

Lindhout D, Schmidt D (1986) In-utero exposure to valproate and neural tube defects. Lancet i:1392–1393

Lindquist S (1986) The heat-shock response. Annu Rev Biochem 55:1151-1191

Lindquist S, Craig EA (1988) The heat shock proteins. Annu Rev Genet 22:631-677

Martin RA, Fineman RM, Jorde LB (1983) Phenotypic heterogeneity in neural tube defects: a clue to causal heterogeneity. Am J Med Gen 16:519-525

Mastroiacova P, Bertollini R, Morandini S, Segni G (1983) Maternal epilepsy, valproate exposure, and birth defects. Lancet ii:499

Minton KW, Karmin P, Hahn GM, Minton AP (1982) Nonspecific stabilization of stress-susceptible proteins by stress-resistant proteins: a model for the biological role of heat shock proteins. Proc Natl Acad Sci USA 79:7107–7111

Mirkes PE (1987) Hyperthermia-induced heat shock response and thermotolerance in postimplantation rat embryos. Dev Biol 119:115–122

Mitchell HK, Moller G, Petersen NS, Lipps-Sarmiento L (1979) Specific protection from phenocopy induction by heat shock. Dev Genet 1:181–192

Mohl VK, Bennett GD, Finnell RH (1990) Genetic differences in the duration of the lymphocyte heat shock response in mice. Genetics 124:949-955

Nakano KK (1973) Anencephaly: a review. J Ment Defic Res 1:4–15

Neidhardt FC, VanBoglen RA, Vaughn V (1984) The genetics and regulation of heat shock proteins. Annu Rev Genet 18:295–329

Pelham HRB (1986) Speculations on the functions of the major heat shock and glucose-regulated proteins. Cell 46:959–961

Richards IDG, Roberts CJ, Lloyd S (1972) Area differences in prevalence of neural tube malformations in South Wales. Br J Prev Soc Med 26:89–93

Sanchez ER, Redmond T. Scherrer LC, Bresnick EH, Welsh MJ, Pratt WB (1988) Evidence that the 90-kDa heat shock protein is associated with tubulin-containing complexes in L cell cytosol and in intact PtK cells. Mol Endocrinol 2:756–760

Smithells RW (1982) Neural tube defects: prevention by vitamin supplements. Pediatrics 69:498–499

Spyker JM, Smithberg M (1972) Effects of methylmercury on prenatal development in mice. Teratology 5:181–190

Subjeck JR, Shyy TT (1986) Stress proteins of mammalian cells. Am J Physiol 250:C1–C17

Theodosis DT, Fraser FC (1978) Early changes in the mouse neuroepithelium preceding exencephaly induced by hypervitaminosis. Teratology 18:219-232

Walsh DA, Morris VB (1989) Heat shock affects cell cycling in the neural plate of cultured rat embryos: a flow cytometric study. Teratology 40:583–592

Walsh DA, Klein NW, Hightower LE, Edwars MJ (1987) Heat shock and thermotolerance during early rat embryo development. Teratology 36:181–191

Webster WS, Edwards MJ (1984) Hyperthermia and the induction of neural tube defects in mice. Teratology 29:417–425

Webster WS, Messerle K (1980) Changes in the mouse neuroepithelium associated with cadmium induced neural tube defects. Teratology 21:79–88

Part II

Cell-Specific and Developmental Control of Hsp Synthesis

6 The Expression of Heat Shock Protein and Cognate Genes During Plant Development

Jill Winter and Ralph Sinibaldi [1]

1 Introduction

"He had been eight years upon a project of extracting sunbeams out of cucumbers..." (Swift 1726). At one time it may have seemed equally ludicrous to look for heat shock proteins in the absence of heat. Nonetheless the pursuit of developmentally or metabolically regulated Hsps has begun and there is now a solid foundation for future research in this area. The wealth of this information has been generated in nonplant systems and thus we have divided this Chapter into two sections. The first section summarizes reports of the developmental expression of heat shock genes. This is not an exhaustive documentation, but a synopsis of reports that have implications for plants. Generalities drawn from yeast and *Drosophila* research have been applicable to many systems that utilize this highly conserved network of proteins. More detail on Hsp functions from yeast, *Drosophila*, and mammalian systems can be gleaned from other Chapters in this Volume, or the current review by Lindquist and Craig (1988). A useful review for a comparison of sequences between plant and animal heat stress proteins, as well as a comprehensive review of heat shock in plants, is that of Neumann et al. (1989). The last section of this Chapter will focus on the expression of plant Hsps when development and heat shock coincide. In some organs, specific developmental programs preclude a normal heat-shock response. A tally of such events and a review of heat shock protein levels in nonstressed tissues will hopefully give the reader some clues as to how plants utilize heat shock genes and their cognates during growth and development.

The first visual observation of heat-induced gene activation in animals was made by Ritossa (1962), who observed that several new chromosome puffs were present in the polytene chromosomes of *Drosophila busckii* after the insects experienced a temperature elevation from 20 °C to 37 °C. The same set of puffs could also be seen with dinitrophenol or sodium salicylate treatment of *Drosophila* salivary glands. This observation was extended to other species of *Drosophila* (Berendes and Holt 1964; Ritossa 1964), where it became apparent that some of the chromosomal loci that responded to heat shock also exhibited normal developmentally programmed puffing (see Ashburner and Richards 1976; Ashburner and Bonner 1979 for reviews). This was the first indication that some of the heat shock genes were also expressed in normal development.

The initial report that plants express Hsps during heat shock came from Barnett et al. in 1980. Plant scientists were so eager for solutions to the agronomic problem of

---

[1] Sandoz Crop Protection, 975 California Avenue, Palo Alto, California 94304, USA

Results and Problems in Cell Differentiation 17
Heat Shock and Development
Hightower and Nover (Eds.)
©Springer-Verlag Berlin Heidelberg 1991

heat stress that developmentally regulated Hsps were not investigated until the end of the decade.

After several of the yeast and *Drosophila* heat shock genes were cloned and Hsps were analyzed from many species, numerous reports of developmental expression began to appear. These reports were based on either mRNA cross hybridization to cloned genes or comigration on two-dimensional polyacrylamide gels of proteins. More precise observations have been recently made using gene specific probes, i.e., gene-specific 5' and 3' untranslated coding region sequences and monoclonal antibodies. Because gene specific probes have not always been available, there is some confusion in the literature about which family members are being expressed at any given time. The similarity between family members in each class of h*sp* is responsible for the technical difficulty in distinguishing between members. This homology is also responsible for the somewhat ambiguous nomenclature in the field and throughout this Chapter. In this chapter, we will use *hsc* for "heat shock cognate", a term originally coined to describe heat shock family members that are expressed in the absence of stress (Craig et al. 1983). This terminology is complicated by family members that are expressed both in the presence and the absence of stress. These later members are usually still referred to as *hsp* to be consistent with earlier characterizations. Hsp is also used for members expressed only during stress. It will take some time to correlate identity and function for each member of some of these large (Hsp70, Hsp20) families and until then we can only catalogue what is currently known and speculate on the rest.

2 Classes of Heat Stress Proteins and the Putative Functions of the Family Members

2.1 Hsp104

The gene for Hsp104 was recently cloned from yeast and it appears to be the principle component responsible for induced thermotolerance in yeast (Sanchez and Lindquist 1990). The gene has yet to be cloned from other organisms, though large Hsps have been documented in several systems. Mammalian cells produce proteins of 100 and 110 kD which do not appear to have counterparts in *Drosophila*. In plants, there are reports of heat-induced proteins of 108 kD in maize (Baszczynski et al. 1982), 95 kD in tomatos (Nover and Scharf 1984), 120 kD in tobacco (Meyer and Chartier 1983) and 110 kD in soybean (Vierling and Key 1985). The relationship of these heat-induced plant proteins of higher molecular weight to the larger mammalian Hsps or the yeast Hsp104 is unclear.

Immunological localizations of the mammalian Hsp100 show concentration in the Golgi (Lin et al. 1982). Antibodies raised against mouse Hsp110 detect protein localized in the nucleoli of both control and heat shocked cells (Subjeck et al. 1983). Immunolocalizations of Hsp110 show associations with the fibrillar component of the nucleolus, the site of rDNA chromatin. Ribonuclease treatment of fixed cells eliminates the staining (Subjeck et al. 1983), suggesting an RNA-Hsp110 association. Since ribosome production and assembly is sensitive to heat stress (Nover et al. 1986), it has been postulated that Hsp110 is induced to protect it, but direct evidence to support this hypothesis is lacking. Antibody studies with Hsp104 reveal that it too is a nuclear protein (Borkovich and Lindquist, cited in Lindquist and Craig 1988), however, any relationship between Hsp104 and mammalian Hsp110 remains to be determined.

In summary, there is no well-documented developmental expression of Hsps in this class of heat induced proteins. However, both the 100- and 110-kD proteins are found in normal mammalian cells and are glucose regulated. Their induction patterns are very complex under various conditions (Hightower and White 1982; Welch et al. 1983; Sciandra et al. 1984; Kasambalides and Lanks 1985; Whelan and Hightower 1985) and one cannot preclude some form of developmental expression of these large Hsps.

2.2 Hsp90

Genes in this class are also referred to as *hsp82-83* (*Drosophila*) and *hsp80-110* (plants and mammals). In most cells, proteins of the Hsp90 family are relatively abundant at control temperatures and the levels are further increased by heat. In *Drosophila melanogaster* there appears to be only one Hsp83 gene. Yeast has two genes in this family, encoding nearly identical proteins. One member is strongly heat induced and expressed constitutively at a low level; the other is only moderately heat induced and is constitutively expressed at a high level (Borkovich et al. 1989).

The carboxyl-termini of proteins encoded by these genes are generally divergent with the exception of the four terminal amino acids which are glu-glu-val-asp in all but one eukaryotic Hsp90 (Neumann et al. 1989). Vertebrates have an additional Hsp90 that has a signal sequence for endoplasmic reticulum (ER) transport (Kulomaa et al. 1986; Mazzarella and Green 1987; Moore et al. 1987; Sorger and Pelham 1987). Grp94 is larger than the cytosolic proteins with an estimated molecular weight of 94–108 kD versus 92–97 kD for the cytosolic forms. The carboxy terminus of Grp94 is extended 24 amino acids beyond the former terminus of glu-glu-val-asp. The four amino acids at the carboxy terminus of Grp94 are lys-asp-glu-leu, which are also found in corresponding Grp78, a glucose regulated protein belonging to the Hsp70 family. These four amino acids provide a signal for retention in the ER (Munro and Pelham 1987), which is also found in other ER proteins.

The cytosolic and ER Hsp90 proteins do not appear to be coordinately regulated. The ER form is induced upon glucose starvation (much the same as mammalian Hsp100 and 110) and is considered a glucose regulated protein or Grp (Sciandra and Subjeck 1983), while the cytosolic Hsp90 is induced upon glucose restoration. Heat, steroids, and other agents also induce the Grp94 ER proteins, but the responses vary with cell type. The entire Hsp90 family of proteins may not be abundantly expressed in all cells types. Suggestive evidence for this is based on some observations in *Drosophila*. RNA isolated from fruit flies that have been subjected to a heat shock has a 20-fold higher level of *hsp83* transcripts than RNA from control flies; whereas only a 20% increase in *hsp83* transcripts is observed in RNA isolated from heat treated *Drosophila* Kc tissue culture cells compared to control RNA (O'Conner and Lis 1981).

In addition to heat induction and tissue general expression, *Drosophila hsp83* is developmentally regulated during oogenesis (Zimmerman et al. 1983). Sequences necessary for ovarian expression have been shown to reside in the interval between positions −880 and −170 (mRNA start is position 1; Xiao and Lis 1989). Sequence

comparisons with other *hsps hsps26* and *hsp27*) expressed in ovaries uncover a conserved sequence box of seven nucleotides, CGTTTTG. However, direct evidence that
this sequence box or several shorter related sequences are responsible for the developmental expression of *hsp83* in ovaries is lacking (see Chap. 7).

Developmental regulation of the *hsp90s* has been described in detail in the murine
system. Barnier et al. (1987) demonstrated that mouse fibroblasts synthesize two
related but distinct Hsp90 proteins, one of which is highly induced by heat shock. They
also observed that in contrast to mouse fibroblasts, murine embryonic cells expressed
both *hsp90s* at high levels. Moore et al. (1989) subsequently cloned, analyzed and
compared the *hsp90s* from mouse. Using gene specific probes from the two murine
hsp90s, Lee (1990) further described the developmental expression of these two genes.
Lee detected *hsp84* expression in embryos, brain, thymus, lung, liver, kidney, adrenal
gland, uterus, ovary, and testes. Expression of *hsp86* was more limited to embryos,
placenta, brain and testes. In testes, the presence of Hsp86 was detected at all stages;
however, its highest level was in day 21 testes (during meiotic prophase). Lee (1990)
could not detect *hsp86* mRNA in the mutant strain *atrichosis* which lacks germ cells
but has all the somatic components of the testes.

The role Hsp86 plays in germ cells of mouse remains undetermined, as well as the
function of Hsp84 in somatic tissue (if there are different functions in the two types
of tissues). In *Drosophila,* one *hsp90* is apparently sufficient to serve the function
in both somatic and germline tissue. In yeast, the two Hsp90 proteins are highly related
to each other and one or the other is required for growth at higher temperatures
(Borkovich et al. 1989). These observations are consistent with an *hsp90* gene
duplication and subsequent sequence divergence without obvious functional diver-
sification.

In plants, there are reports of heat induced proteins in this size class (see Neumann
et al. 1989) but very little is known about the developmental expression of this family.
Moreover, only a few of the plant genes in this class have been cloned, sequenced and
characterized. In the maize genome, cross hybridizing sequences to the *Drosophila*
hsp83 were first detected by Dietrich and Sinibaldi (1983) and the suggestion of
an *hsp83* multigene family in plants was proposed. Expression of these *hsp83* homologous sequences into RNA was confirmed (Dietrich and Sinibaldi 1984) and
genomic clones for 2 of the *hsp83*-related genes were isolated (Dietrich et al. 1986).
The coding region for one of the *hsp83s* has been entirely sequenced, the second gene's
coding region partially sequenced, and the promoters from both analyzed in transient
assays. The promoters act in an analogous fashion to the yeast genes described earlier;
one is predominately heat induced and the other is constitutive in nature. Interestingly,
hybrid selected translations of mRNA reveal three distinct size proteins (of ≈ Mr 81, 82
and 85 kD) on SDS polyacrylamide gels and the two that have been cloned encode
proteins that terminate in glu-glu-val-asp. With the exception of the ER retained hsps,
this carboxy terminal sequence is common to most Hsp90 and Hsp70 proteins (see
Neumann et al. 1989). The two maize *hsp90* genes do have unique 5' untranslated
leader regions and thus provide some of the tools necessary to study the developmental
regulation of these highly homologous genes in maize. Another plant species where an
entire *hsp80* gene has been cloned and the coding region sequenced is *Brassica*
oleracea (Kalish et al. 1986). Analysis of the promoter from this gene in transient
assays indicates that it directs the constitutive expression of *hsp80* (Cannon et al.

1987). Expression of the *Brassica* gene was analyzed in several plant organs with no indication of differential expression. More detailed experiments need to be performed before any definitive statements on the developmental expression of Hsp90 can be made.

2.3 Hsp70

These proteins are encoded by a multigene family and are expressed under a variety of physiological conditions. The *hsp70* gene families are probably best characterized in *Drosophila* and yeast. The *Drosophila melanogaster* family includes five to six copies of *hsp70*, one copy of the heat inducible *hsp68* and at least seven other related genes that are expressed during normal growth. These related genes have been denoted as *hsc1-7* and map cytologically to multiple distinct chromosomal loci (Lindquist and Craig 1988), and At least one of these Hsc3, is analogous to the mammalian glucose regulated protein Grp78. In yeast, there are at least nine Hsp70 family members (Lindquist and Craig 1988). Eight of these have been classified on the basis of structural and functional similarities, and the ninth, *KAR2,* has been recently identified (Rose et al. 1989). On the nine yeast genes, the expression of six of them is increased by heat shock (*SSA1, SSA3, SSA4, SSC1, SSD1* and *KAR2*). The expression of two of them is decreased by heat shock and one gene is unaffected by heat stress, i.e. the level of expression remains the same. The four *hsp70s* in the *SSA* group of genes make up an essential family, at least one of which is required for viability. The *SSC1* and *KAR2* genes are also essential genes.

Results from many laboratories implicate the Hsp70 family members in a wide array of non-stress related cellular processes including DNA replication (Wu and Morimoto 1985), transport of proteins across membranes (Deshaies et al. 1988), and the binding of proteins in the endoplasmic reticulum (Kassenbrock et al. 1988). At least two previously characterized proteins, BIP (immunoglobulin heavy chain binding protein) (Kassenbrock et al. 1988), and clathrin uncoating ATPase (Chappell et al. 1986), have been shown to be Hsp70 family members. These various Hsp70 functions are seemingly unrelated. However, they all appear to involve some sort of protein-protein interaction and they appear to involve Hsp70 catalyzed ATP hydrolysis.

Hsp70 family members have also been associated with a variety of developmental events. In 8-day mouse ectoderm and in embryonal carcinoma cells, Bensaude and Morange (1983) detected a protein having the same electrophoretic mobility on two-dimensional gels as an Hsp70. The developmental expression of the *Drosophila hsc* genes has been detected by Craig and associates and is discussed below in comparison to plant studies (Ingolia and Craig 1982a; Craig et al. 1983; Palter et al. 1986). More detailed reports of developmental expression of *hsp70* and related genes in vertebrates has been recently summarized by Morimoto and Milarski (1990). Most notable is the expression of an *hsc70* during sperm cell development in the germline of mouse and humans (Zakeri et al. 1988; Zakeri and Wolgemuth 1987).

Hsp and Hsc70s have been identified in the presence and absence of stress in several plants (Neumann et al. 1989). Two *hsc/hsp70* cDNAs from nonstressed tomato ovaries have been sequenced by Duck et al. (submitted, 1990). One of these (referred to as *hsc1*) has been used to define transcript levels throughout the non-stressed tomato plant (Duck et al. 1989). These studies utilized part of the coding region of the *hsc*-cDNA (which is capable of detecting transcripts from several family members) as a source of RNA-probe. The conclusions from in situ hybridizations and Northern blot analyses are that *hsc*- homologous transcripts are present at some level in all but xylem, unfertilized ovules, and mature pollen. There are exaggerated transcript levels in ovaries, developing embryos and the meristematic sections of root tips. An intermediate level of transcript was detected in the transmitting tissue of the style and immature anthers, and a low to intermediate level in the inner and outer phloem throughout the plant. Surprisingly, there are similarities between the observations made in these studies and *in* situ studies of *Drosophila* tissue sections. Utilizing antibody probes, Palter et al. (1986) observed a low level of Hsc/Hsp70 in all tissues except ovules and a high level of Hsc/Hsp70 in ovaries and during embryogenesis. Given the obvious dissimilarities between *Drosophila* and tomato, three explanations for these similarities come to mind: (1) Hsc/Hsp70 is present in high levels in metabolically active cells; (2) some conserved developmental function requires the aid of Hsc/Hsp70 in ovaries and embryos; and (3) coincidence. Vierling and Sun (1987) have also detected *hsc*-*hsp70* transcripts in seeds from nonstressed pea plants and this is discussed in more detail in a later section.

As mentioned earlier, Hsc70s have been noted for late meiotic expression in the male germ line of rats and humans. Pollen is the plant equivalent of sperm and although meiotic stages of pollen have been examined for Hsp18 expression (Bouchard 1990; Bouchard and Walden 1990), similar studies for proteins of the Hsp70 family have not been performed.

2.4 Hsp60

This class of heat shock protein is also referred to as Hsp58 and appears to be localized in the mitochondria of eukaryotes. The expression of the genes that encode this Hsp appears to be constitutive, and the expression level increases two- to three-fold after heat shock. To date there are no reports of developmental expression of these genes. Sequence data has been obtained for the *hsp* genes from a number of species (Neumann et al. 1989). There is extensive sequence homology between the various genes and they all seem to have N-terminal targeting sequences for mitochondria. Additionally, there is an Hsp60 that is chloroplast-associated and binds ribulose-biphosphate carboxylase subunit protein in wheat (Hemmingsen et al. 1988).

2.5 Low Molecular Weight Hsps (Hsp20 Family)

This class represents the most diverse group of Hsps. The number of genes that encode these Hsps varies from organism to organism; with *S. cerevisiae* having a single gene (Petko and Lindquist 1986), to greater than 30 in some higher plants (Mansfield and Key 1987). A great deal of variance is also observed in the molecular weights, which range from 16 (Russnak and Candido 1985) to 40 kD (Nene et al. 1986) in size. Generally, the low-molecular-weight Hsps show greater homology within the species than outside species. They appear to have similar hydropathy profiles and limited regions of amino acid identity. One of the most conserved regions in virtually all the low molecular weight Hsps lies near the carboxy terminus and consists of the sequence gly-val-leu-thr-leu-val-ile-aa$_1$-aa$_2$-pro. The relative degree of sequence divergence outside of this conserved region could be the result of gene duplication and subsequent change or drift during evolution.

One feature that is highly conserved in the small Hsps is their structural properties. The small Hsps have the propensity to form large aggregates or polymeric structures during heat shock. These structures are referred to as heat shock granules (Nover et al. 1983, 1989; Arrigo et al. 1985; Schuldt and Kloetzel 1985). In addition to the aggregation phenomenon, the small Hsps share a second physical property, phosphorylation. There are reports of phosphorylation of small Hsps in mammals (Kim et al. 1984; Arrigo and Welch 1987) and *Drosophila* (Rollet and Best-Belpomme 1986) under various conditions. There is no indication that the phosphorylation is universal among all small Hsps or that it serves any purpose. One final physical property that the small Hsps may share is that they bind RNA. This is observed in *Drosophila* (Kloetzel and Bautz 1983), sea urchin (Rimland et al. 1988), and tomato (Nover et al. 1989), but not in yeast (Lindquist and Craig 1988). The cellular functions of any of the small Hsps remains highly speculative. However, due to their diverse nature it is likely that they serve multiple functions. They have been proposed to be involved in thermotolerance, protection of thylakoid membranes and protection of RNA molecules (see Neumann et al. 1989 for discussion), but convincing evidence in support of any the proposed functions is still lacking. It is hoped that the study of developmental expression of these genes may lead to more plausible and testable hypotheses on the functions of these Hsps. The reports of developmental expression of the small Hsps in many species are numerous and several are detailed below.

In *D. melanogaster* there are six or seven small Hsps (designated Hsp22, 23, 26, 27; and Gene 1) which clearly belong to the small Hsp family (Ingolia and Craig 1982b; Sirotkin and Davidson 1982; Southgate et al. 1983; Ayme and Tissieres 1985). All the small *hsp* genes appear to be expressed in the late larval to early pupal transition in *D. melanogaster* (Sirotkin and Davidson 1982; Cheny and Shearn 1983; Ayme and Tissieres 1985). Messenger RNAs complimentary to Hsp27, Hsp26 (along with Hsp83) are found in adult nurse cells (Zimmerman et al. 1983) and are apparently passed into developing oocytes. Furthermore, transcripts of Hsp23 and Gene1 are found in young adults shortly after eclosion (Ayme and Tissieres 1985). At least some of the developmental expression of the small *hsp* genes can be attributed to ecdysone stimulation (Ireland and Berger 1982; Berger and Woodward 1983; Thomas and Lengyel 1986). Temporal pulses of ecdysone have been shown to induce specific chromosomal puffs and thus regulate developmental expression of sets of genes (Ashburner and Richards 1976). Using specific *lacZ-hsp26* promoter fusions the expression of the *hsp26* gene has been thoroughly studied in *Drosophila,* and results indicate a complex assortment of expression in spermatocytes, nurse cells, imaginal discs, epithelial tissues, and neurocytes (Glaser et al. 1986). The regulatory elements for ovarian expression (Cohen and Meselson 1985) and spermatocyte expression (Glaser and Lis 1990) have been localized to define upstream regions in the *hsp26* promoter.

In yeast, there is only one small heat shock protein, Hsp26, and deleting it has no apparent effects on thermotolerance, growth at high temperature, spore development or germination (Petko and Lindquist 1986). In numerous experiments with *hsp26* deleted strains, no measurable effects on various cellular responses and processes could be detected (see Lindquist and Craig 1988). Therefore, Hsp26 appears to be a nonessential protein. However, it could be involved in some yet uninvestigated cellular process (i.e., recombination) or an unrelated protein could compensate for its absence. The developmental expression of *hsp26* in yeast coincides with meiosis and sporulation (Kurtz et al. 1986). Moreover, the timing of expression of RNA for *hsp26* and *27* in *Drosophila* ovaries is also coincident with meiosis.

In plants, the small *hsp* genes have been categorized into at least four different classes based on sequence similarities and are designated class I, III, IV and VI (Schöffl et al. 1986). Two of these classes are closely related, I and VI, each of which encodes 9-12 cytoplasmic proteins in the 17-19 kD range (Schöffl and Key 1982, 1983). Class IV is comprised of genes encoding 20–23 kD proteins that are localized in chloroplasts (Vierling et al. 1988, 1989). This class of small Hsps has only a very limited region of amino acid identity (gly-val-leu-thr-leu-val-ile-aa$_1$-aa$_2$-pro; discussed previously) with the cytoplasmic small Hsps. More recently, a novel type of small *hsp* gene (which contains an intron), encoding a protein of Mr 26 kD, has been isolated and characterized (Czarnecka et al. 1988; Hagen et al. 1988). There is no sequence homology with the previously mentioned cytoplasmic or chloroplast localized small Hsps, but the synthesis of its product is definitely stress induced. This classification of the low molecular weight Hsps is somewhat confusing, i.e., classes II, III and V are no longer mentioned in current Hsp literature. Perhaps a more reasonable classification of the low molecular weight Hsps in plants would include only three distinct categories: (1) cytoplasmic 15-18 kD (combining class I and VI; as they are related); (2) chloroplast localized 20-23 kD (previously class IV); and (3) the intron containing 26 kD (in a class by itself).

The majority of the plant small hsp genes have been isolated from soybean, but others have been cloned from pea (Vierling et al. 1988), wheat (McElwain and Spiker 1989), lily (Bouchard 1990), and maize (Dietrich et al. 1987; Nieto-Sotelo et al. 1990). With the exception of the lily and maize *hsp18* genes, there is little information about the developmental expression of the small *hsp* genes. Bouchard (1990) has recently characterized a set of meiotic prophase repeat transcript clones of *Lilium* (EMPRs) that have sequence homology to small cytoplasmic Hsps. These cDNA clones detect transcripts in lily microsporocytes that are specific to the meiotic prophase interval. One of the lily cDNA clones was used as a probe to isolate a maize genomic clone (Dietrich et al. 1986). The analysis of the promoter from the maize gene revealed that it could direct the heat induced synthesis of a fused marker gene (*β* -glucuronidase) in transient assays in maize protoplasts. Therefore, the maize gene represents a heat shock gene that was isolated on the basis of its sequence homology to the lily meiotically expressed gene. Subsequently, Bouchard and Walden (1990) have shown that RNAs complimentary to this gene are found in developing maize anthers. Bouchard (personal communication) has recently discovered that only a subset of the cytoplasmic small *hsp* genes are expressed in maize anthers during an interval encompassing meiotic prophase. These observations are consistant with the ones previously reported in yeast and fruit flies and raises points of discussion that will be addressed in the pollen development section.

2.6 Other Heat Shock Proteins

There have been many reports of the heat-induced expression of proteins not originally considered Hsps (Lindquist and Craig 1988; Neumann et al. 1989). One of the more interesting of these proteins is ubiquitin, which is found in elevated amounts after heat shock of many organisms (Bond and Schlesinger 1986; Parag et al. 1987; Bond et al. 1988). In yeast one of the four polyubiquitin genes is essential for stress tolerance, although it is nonessential for normal vegetative growth (Finley et al. 1987).

Heat induction of ubiquitin has also been described in maize (Christensen and Quail 1989), *Arabidopsis* (Burke et al. 1988), barley (Gausing and Barkardottin 1986), and oat (Vierstra et al. 1989). In *Arabidopsis* (Burke et al. 1988) there are indications that at least one of the genes may be heat regulated in a tissue specific manner. In maize, Christensen and Quail have recently described transcriptional and post transcriptional regulation of polyubiquitin during heat shock. Maize cells contain three different size classes of mRNA homologous to the polyubiquitin cDNA. The level of only one of these classes increases three- to four-fold during the first 3h of heat shock. Because of ubiquitin's role as a tag for the proteolysis of denatured proteins, there is speculation that it may help cells recover from damage incurred during stress.

One common denominator among *hsp* genes of all classes is the heat shock element (HSE) of the promoter. Years ago, Pelham and Bienz (1982) showed that this simple sequence was sufficient to confer heat inducibility upon a thymidine kinase gene in monkey COS cells. The spacing of this element is commonly within 35 bases upstream of the TATA box, although some variation has been observed. There are cases of altered spacing as well as multiple HSEs in both plant and animal systems (Neumann et al. 1989). Recent work by Xiao and Lis (1990) demonstrates that minor changes in a *Drosophila* HSE can direct organ specific expression in the absence of heat stress. Although Czarnecka et al. (1989) and Schöffl et al. (1989) have appraised several plant *hsp* promoters for sequences required for heat induction, no attempts have been made to integrate developmental expression data into these promoter maps. In general, the study of developmental expression of plant *hsp/hsc* genes is still in its infancy.

3 Hsps and Hscs Expressed During Plant Development

Past studies of Hsp (or Hsc) synthesis during plant development have concentrated on the ability of plants to induce Hsps in response to thermal stress. One purpose of this research has been to identify potential breeding tools (such as superior Hsp profiles) for crop improvement. This practical emphasis, as well as the desire to manipulate the regulation and function of Hsps during stress, has resulted in a very limited number of publications on plant Hsp (Hsc) profiles in the absence of stress.

We have organized this section into four parts: seeds and seedlings, leaves and roots, flowers, and pollen. This discussion covers experiments that evaluate the induced and endogenous levels of certain Hsps in a variety of plants. Thus, the reader must persevere through comparisons between species, between Hsps and sometimes between mRNA and protein levels.

3.1 Seeds and Seedlings

Very young seedlings of sorghum (Ougham and Stoddart 1986) and maize (Riley 1984) are heat sensitive and unable to make Hsps during the first 16 h of germination. Ougham and Stoddart (1986) have shown that the more heat-tolerant sorghum geno- type produces Hsps at an earlier developmental time that the more sensitive line. These data are only correlative, but suggest that the survival of the germinating seedling may depend on its ability to produce Hsps in response to early heat shocks. If this conclu- sion is correct, one might expect to find plants that store Hsps or their mRNAs in their seeds to show superior heat tolerance during germination. This hypothesis can be addressed indirectly by first reviewing the literature that documents the presence of Hsps and their mRNAs in developing, mature and germinating seeds, and secondly by focusing on reports that attempt to correlate stored hsps to thermal tolerance during germination.

3.1.1 Hsps and *hsp* mRNAs During Seed Development

Duck et al. (1989) report *hsc/hsp70* transcript accumulation in non-heat-shocked developing seed of tomato, both in the inner integument and in embryos. They did not investigate protein content or discuss germination. Vierling and Sun (1987) report that *hsp* 19, 20, and 70 mRNAs accumulate in non-heat-shocked pea seed (embryonic axes and cotyledons) and these transcripts rapidly decrease in abundance within 7 h of imbibition. Whether these seed transcripts are ever translated, either in the pea seed or during imbibition is unknown. The levels of endogenous and inducible Hsc/Hsps and mRNAs in wheat embryos has been investigated rather thoroughly by Helm and Abernethy (1990). They conclude that the developing embryo contains family mem- bers from several classes of Hsp, both as mRNAs and proteins. Whether the wheat embryo was heat shocked or not seems to have little impact on the *hsp/hsc* mRNAs and protein levels that are observed. Additionally, a subset of the wheat Hsps are detectable by a coomassie blue stain of proteins from non-heat-shocked dry seed and in vitro translations indicate that *hsp* transcripts are also present in the mature wheat seed. The Hsps that are present during wheat embryo development appear to be induced by stresses other than heat, possibly dehydration, the metabolic status of the cell, or "developmental stress". Alternatively, they may respond to a purely developmental signal that has nothing to do with stress, as will be discussed later in this Chapter.

The endogenous *hsp* transcripts of wheat embryos are absent after 3 h of imbi- bition at 25 °C, and *hsp* transcripts can be induced within 3h of imbibition. This rapid disappearance of endogenous heat shock mRNAs during early germination in wheat is consistent with Vierling's findings in pea. Wheat differs from maize and sorghum in that Hsps cannot be induced quite so early after imbibition in the later plants.

Zimmerman's analysis of carrot somatic embryos gives yet a different view of *hsp* expression during development (Zimmerman et al. 1989). Zimmerman finds that *hsp17.5* transcripts are almost undetectable in non-shocked and heat-shocked develop- ing embryos from the callus to the plantlet stage. Transcription of these genes can only readily be detected after heat shock induction of the callus and again during and after the heart-torpedo stage of development. This is just opposite to the findings of Helm and Abernethy that are discussed above, where hs mRNAs and proteins are present whether the developing embryos have been shocked or not. There are two possible explanations for these differences: (1) The *hsp17.5* cDNA probe used by Zimmerman et al. is from a family member that is silent during development and other *hsp* cDNA probes would have indicated that transcripts are present or inducible during development; or (2) carrot somatic embryos and wheat may simply differ in their developmental expression of Hsps.

Despite Zimmerman's inability to detect embryonic transcription of *hsp17.5*, even after a heat shock, she does demonstrate a dramatic accumulation of Hsps after the heat shock. This apparent contradiction can be explained by a translational bias for rare *hsp* transcripts that allow the cell to turn undetectable amounts of transcript into copious amounts of protein. Clearly there is documentation in the literature (Bienz and Gurdon 1982) for translational bias of *hsp* mRNAs during stress. Zimmerman further supports this argument by giving evidence that this translational bias is at the expense of normal translation, resulting in phenocopies rather than normally developed embryos. This could be another strategy, though seemingly a costly one, for getting around a block in transcriptional activation. This finding would imply that: (1) Hsps are necessary for survival, for it is mainly the phenocopies that have presumably performed this feat of translational bias, which survive the shock; and (2) that certain developmental stages may result in a temporary trancriptional block for *hsp* genes (and perhaps many other genes) and that plant embryos unable to overcome this block are heat sensitive during this stage.

3.1.2 Heat Tolerance During Germination and Endogenous Hsps

Wheat seedlings show their greatest heat tolerance during the first 8 h of imbibition, though there is nothing remarkably different in the Hsp profile that is induced before and after the 8-h point. Perhaps developmentally induced proteins (unrelated to Hsps) exist that have protective functions. Abernethy et al. (1989) make reference to "developmentally dependant proteins" in wheat seeds and these may warrant further study.

Unexpectedly, in the first 8 h of imbibition, the wheat seedlings cannot exhibit "induced thermotolerance " (the acquisition of tolerance to a severe shock due to a preceding mild shock), yet after 12 h they can. Helm and Abernethy's finding (1990) that induced thermotolerance may be independent of Hsp induction has also been observed in the germinating pollen of *Tradescantia* (Xiao and Mascarenhas 1985), as will be discussed below, and in cultured cowpea cells (Heuss-La Rosa et al. 1987). These observations require further analysis.

It would be interesting to know if pea seedlings, like wheat, have their greatest heat tolerance within the first hours of imbibition, since pea has also been shown (Vierling and Sun 1987) to store *hsp* mRNAs in the embryo. Although the stored *hsp* mRNAs rapidly vanish after imbibition, and we thus speculate that they have developmental as opposed to stress related functions, the possibility of a very early stress related function that affects the next several hours of germination cannot be eliminated.

Although some plants produce *hsp* mRNAs and Hsps during embryo development, their rapid disappearance during imbibition is curious. Early germination is a vulnerable time for a plant, for the plant must endure temperature extremes without leaves and roots to moderate the effects. It is surprising that there are plants unable to launch an Hsp defense during this period; one can only speculate that for these plants there may be a selective disadvantage in doing so, or that other developmental programs may have higher priority.

3.2 Roots and Leaves

Lobing in chrysanthemum leaves is a temperature-induced phenotype. The temperature prevailing during initiation and early differentiation determines the leaf shape such that every plant carries a permanent record of the temperatures to which it has been subjected in the shapes of its successive leaves (Schwabe 1969). The molecular basis for these apparent alterations in cell division rates is unknown. Heat, like light, is a key environmental signal for the development of many plants, and very little, if anything, is known about Hsps playing any role in this signal transduction. Most of the Hsps implicated in development are thought to be responding to developmental or metabolic signals and not directly to the environment. For plants in particular this may be an oversimplification. Given that there is evidence that some parasitic protozoans (Hunter et al. 1984; Lawrence and Robert-Gero 1985; Van der Ploeg et al. 1985) cannot differentiate into their parasitic form until they receive a heat shock (which is a signal that they are in the bloodstream of a warm-blooded host), it is not unreasonable to speculate that plants may use their heat sensory system to direct development also.

Much of the molecular biology that is known about heat shock in plant leaves has been acquired in the last decade. The mRNA and protein profiles for many species of plants (Burke and Orzech 1988; Neumann et al. 1989) readily show a dramatic induction of Hsps in leaf tissues when temperatures rise 10 °C or more above the ambient growth temperature. Neumann et al. (1989) have compared tomato leaf, root, ovary, and sepal and shown that they each have slightly different Hsp profiles under the same stress conditions. This may be an indication that some of the heat-inducible family members of each Hsp class are also "tissue specific" but this will require more work for clarification. Early work by Baszczynski et al. (1983) indicates that the root and shoot of maize seedlings show identical Hsp patterns after heat shock, however, this comparison was made on 1 D gels and thus the conclusion may be premature. Even on 1 D gels, however, Necchi et al. (1987) have been able to detect differences in the heat shock profile of wheat roots and coleoptiles. As in every organ so far discussed, the tissue specific response may be characteristic of one species and not another, thus generalizations are difficult to make. Duck et al. (1989) have shown a concentration of *hsc/hsp70* transcript in the lateral root tips of tomato. They hypothesize that one of the Hsp70 family members may be involved in the active cell division that is ongoing in these meristems.

3.3 Flowering

As emphasized above, many plants appear to use heat, in conjunction with light, to regulate their developmental program. Specifically, several plants (*Xanthium*, Hammer and Bonner 1938: *Kalanchoe blossfeldiana*, Spear and Thimann 1954; *Ipomea nil*, Imamura 1953) which require a short day for floral induction, will fail to flower if exposed to a high temperature during the inductive dark period (Nakayama 1958). The role of the Hsps in this reversal of floral induction, if any, has not been defined. It is possible that the arrest in expression that accompanies severe heat stress could be responsible for preventing the developmental transition from the vegetative to the floral meristem. (This theory could easily be tested by using induced thermotolerance to avoid the severity of translational arrest that often accompanies severe shocks).

The specific organs of several flowers have been analyzed for their ability to induce or constitutively synthesize Hscs/Hsps. Duck et al. (1989) have shown a high level of *hsc/hsp70* mRNA in the ovaries and transmitting tissue of non-heat shocked developing tomato flowers, and Neumann et al. (1989) have shown that heat shocked tomato sepals and ovaries produce a full spectrum of Hsps.

It is interesting to note that in some plants self-incompatibility can be overcome by a mild heat shock (De Nettencourt 1977). Whether this represents a change in the style (transmitting tissue) or the germinating pollen, is unknown. Although several laboratories have investigated heat damage to flowers [often observing flower drop (El Ahmadi and Stevens 1979) and male sterility (Yoshida et al. 1981)], very little work has been done to document corresponding Hsp levels. Pollen is the best characterized part of the flower with respect to endogenous and induced Hsp levels.

3.4 Pollen

3.4.1 Heat Stress During Pollen Development

Few studies are available on developing pollen due to the technical difficulties inherent to such studies. Frova et al. (1989) have studied the synthesis of Hsps throughout maize pollen development. They have analyzed post-meiotic midvacuolate microspores (as their earliest time point), early and advanced mitosis, and early and late "advanced maturation" stage; as well as dehydrated and germinating pollen. They report that only two sporophytic Hsps, of 84 and 72 kD, are synthesized consistently in the heat shocked gametophyte, regardless of genotype or developmental stage. They state that several more Hsps can be induced early in pollen development (the strongest response being in uninucleate microspores that synthesize the sporophytic 102, 84, 72, and 18 kD Hsps) but not later in development, and they make analogies to yeast in which similar observations have been made for Hsp70.

As mentioned, Bouchard's work from lily (1990) and maize (Bouchard and Walden 1990) indicate that *hsp18* family member mRNAs seem to be abundantly expressed during meiosis (in the absence of heat stress) in developing pollen. It is unknown whether heat stress induces *hsp18* during this developmental phase in maize and lily. The finding of an *hsp18* that is transcribed during late prophase of meiosis is consistent with the expression of the equivalent gene during meiosis in sporulating *S. cerevisiae* (Kurtz et al. 1986), and during the meiotic period of *Drosophila* oogenesis

(Zimmerman et al. 1983). This similarity in the developmental *hsp/hsc* expression is intriguing, since, except for the process of meiosis, the developmental events leading to the yeast ascospore, the fruit fly oocyte, and the plant microspore are highly dissimilar. The finding of small *hsp/hsc* expression correlated with meiosis in organisms from three eukaryotic kingdoms raises the possibility that small Hsps/Hscs have a specific function in the meiotic cells of organisms that is part normal development. Furthermore, the specific expression of *hsc70* and *hsp86* genes during a period of mammalian testes development (see Chapt. 9) that coincides with meiosis raises the possibility that a subset of the Hsps are serving some stress or development related function in meiotic cells.

Various developmental programs may be stressful to cells. Theodorakis et al. (1989) suggest that the "stress" activated heat shock transcription factor (HSF) is responsible for induction of *hsps* during avian erythrocyte differentiation (see Chap. 11). This would imply that development is being perceived as just another stress, such as heat or heavy metals, in this system. In other systems (reviewed in Lindquist and Craig 1988) this is not so apparent, since some of the developmentally induced *hsps* have sequences aside from the heat shock element that are necessary for induction.

3.4.2 Heat Shock During Pollen Germination

Cooper et al. (1984) surveyed several maize organs for Hsp synthesis in response to heat shock. Although some of their conclusions are limited by the resolution of 1-D gels, they state that 10 Hsps can be synthesized in leaf, root, scutella, coleoptiles and suspension cells, but not in germinating pollen. Frova et al. (1989) have also reported that germinating maize pollen does not make sporophytic Hsps in response to a heat shock. Similar conclusions have been drawn from analyses of heat shocked *Tradescantia* (Xiao and Mascarenhas 1985).

Both Cooper and Frova note two proteins around 70 kD that are synthesized in the absence of stress by germinating pollen. Since neither group used 2-D gel analysis or Hsp70 antibodies to identify this protein, its identity remains speculative. Cooper et al. (1984) notice two germinating-pollen specific Hsps of 56 and 92 kD, whereas Frova sees only a 54 kD inducible Hsp. The different genotypes, or germination protocols used in these studies, may account for this difference. (Frova in fact tested several genotypes and demonstrates that there are genotypic differences between both sporophytic and developing pollen Hsp profiles).

Van Herpen et al. (1989) have reported that lily and tobacco both make some sporophytic hsps during pollen germination. Since the authors present the Hsp profiles on 2-D gels and the sporophytic Hsp profiles on 1-D gels, it is difficult to conclude from this paper whether the pollen Hsps are identical to the sporophytic Hsps or not. Although a previous publication (Schrauwen et al. 1986) from this laboratory concluded that heat-shocked lily and petunia germinating pollen did *not* make sporophytic Hsps, this conclusion was drawn by comparing both the petunia and the lily heat-shocked pollen profiles with only the heat-shocked petunia root profile. Schrauwen et al. may have been misled due to differences between lily and petunia sporophytic profiles. Nonetheless, Van Herpen et al. (1989) find that germinating pollen Hsp synthesis is dependent on active transcription (cordycepin sensitive), which is at least consistent with Xiao and Mascarenhas's findings (1985) that Hsps are not made early in development and stored for utilization during germination. Consistent with this theme are the results of Duck et al. (1989) that show that *hsc70* homologous transcripts are not detectable as stored mRNA in mature tomato pollen.

Grasping for generalizations about the heat shock response of pollen is complicated by the apparent variability both within and between species. Furthermore, it may be artificial to evaluate heat shock responses via laboratory inflicted stress and in vitro germination assays. The effects of whole plant stress in pollen that is shielded by its surrounding tissues may be quite different from laboratory simulations thereof; however, it is difficult to incorporate radiolabeled amino acids under the more natural conditions.

One point worthy of consideration, when comparing the heat shock response of different tissues, is that different organs may heat shock at different temperatures. Certainly several reports exist which indicate that different sets of Hsps are produced from different temperatures (Zivy 1987; Necchi et al. 1987) and what is "optimal" for induction in one tissue may not be so for another. It is a tedious point to make, since it implies that each organ would have to be examined at several temperatures and then these several profiles compared between organs, but it nonetheless may warrant some investigation. The relatively low heat-shock temperature of A*rabidopsis* (Coldiron-Rachlin et al. 1986; Binelli and Mascarenhas 1990) came as a surprise to many and thus it may also be prudent to evaluate the heat shock response of specific organs over a range of temperatures.

If we take the above-mentioned pollen data at face value, then the following conclusions can be drawn: under the conditions used, some plants do not make sporophytic Hsps during pollen germination (maize, *Petunia, Tradescantia*), other plants do lily, tobacco); some plants make pollen germination-specific hsps (maize, lily, tobacco), and other plants do not. How these results bear upon thermotolerance is hard to evaluate.

At least in *Tradescantia*, germinating pollen can become thermotolerant in the absence of obvious sporophytic Hsp synthesis, provided that the raise in temperature is a gradual one (Xiao and Mascarenhas 1985). Van Herpen et al. (1989) who observed Hsc and Hsp synthesis in the germinating pollen of lily and tobacco, claim that the germinating pollen survives heat shocks well, even in the absence of the conditioning preshocks that are necessary to obtain thermotolerance in *Tradescantia* pollen. Whether this survival is Hsp-mediated or not has not been directly proven. This is an important issue, for, if Hsps truly protect pollen from heat induced germination failure, then this very significant agronomic problem could be addressed via biotechnology. Since several traits seem to be involved in thermotolerance (Schoper et al. 1987) it has always seemed naive to think that any significant improvements could be made by altering Hsp expression.

As for Hscs during pollen development, it would be surprising not to see any since so many of the Hscs seem to serve metabolic roles and so many of these roles are played out as the pollen mother cell goes through meiosis and mitosis, differentiates into spores and dehydrates. In addition to the *hsp18* mRNA found in prophase of maize and lily pollen, Duck et al. (1989) reported hybridization of a tomato *hsc70* RNA probe to tapetal tissue in sections of immature tomato anthers. Now that plant nucleic acid probes and antibodies are available for almost every class of Hsp, it will be easier to evaluate *hsc* and *hsp* expression during pollen development.

4 Conclusions

Tomato, maize, pea, lily and wheat obviously produce Hsps-Hscs in the absence of heat. Sometimes these Hsps-Hscs occur in tissue (seed) that is unable to induce Hsps during stress and one might speculate that they are stored to compensate for this shortcoming. Alternatively, they may function as part of a programmed arrest in gene expression when seeds dehydrate and prepare to enter into a suspended animation. Hsps and-or their cognates may function during these periods of selective expression, possibly protecting the cells from the consequences of dramatic but transient changes in protein synthesis.

In other cases, such as the ovaries of the developing tomato, there is no obvious reason to infer a stress-related function (since ovaries can induce Hsps after stress) and these Hscs most probably have a developmental role. We should keep in mind that it may be somewhat of a misnomer to refer to cognates that are found during development as having developmental roles. It is possible that some of the "developmental" Hscs will have quite mundane metabolic roles for which there is an exaggerated need during development. Thus, an Hsc that is involved in glycosylation reactions may have high concentrations in tissues that actively synthesize or secrete glycoproteins during development. This "developmental" activity may also occur at lower and more difficult to detect levels in tissues that are not differentiating. Other Hsps, such as the Hsp18 of lily and maize meiosis, are expressed during events that are quite specific to development and which are known to utilize several classes of Hsps in other organisms as well (Lindquist and Craig 1988).

Hsps are not the only stress-inducible plant proteins found during a presumably stress free development. A recent report by Lotan et al. (1989) indicates that some of the pathogenesis-related proteins have high endogenous levels during florigenesis. It is quite possible that all of the stress-related proteins are evolutionary derivatives of proteins that have non-stress-related functions during development. Ironically, we may only discover the nature of these proteins by making tools from the more abundant stress induced proteins and then using these tools to seek out the alleged progenitors.

Heat shock protein/cognate developmental analysis in plants has just begun. Hopefully, future analyses will reveal data that enlighten us about the function of Hsp family members both in the presence and absence of stress.

References

Abernethy RH, Thiel DS, Petersen NS, Helm K (1989) Thermotolerance is developmentally dependent in germinating wheat seed. Plant Physiol 89:569–576

Arrigo AP, Darlix JL, Khandjian EW, Simon M, Sphar PF (1985) Characterization of the prosome from *Drosophila* and its similarity to the cytoplasmic structures formed by the low molecular weight heat–shock proteins. EMBO J 4:399–406

Arrigo AP, Welch WJ (1987) Characterization and purification of small 28 000–dalton mammalian heat shock protein. J Biol Chem 262:15359–15369

Ashburner M, Richards G (1976) The role of ecdysone in the control of gene activity in the polytene chromosomes of *Drosophila*. In: Lawrence PA (ed) Insect development. Wiley, New York, pp 203–225

Ashburner J, Bonner J (1979) The induction of gene activity in *Drosophila* by heat shock. Cell 17:241–254

Ayme A. Tissieres A (1985) Locus 67B of *Drosophila melanogaster* contains seven, not four, closely related heat shock genes. EMBO J 4:2949–2954

Barnett T, Altschuler M, McDaniel CN, Mascarenhas JP (1980) Heat shock induced proteins in plant cells. Dev Genet 1:331–340

Barnier JV, Bensaude O, Morange O, Babinet C (1987) Mouse 89 kD heat shock protein. Two polypeptides with distinct developmental expression regulation. Exp Cell Res 170:186–194

Baszczynski CL, Walden DB, Atkinson BG (1982) Regulation of gene expression in corn (*Zea mays L.*) by heat shock, Can J Biochem 60:569–579

Baszczynski CL, Walden DB, Atkinson BG (1983) Regulation of gene expression in corn (*Zea mays L.*) by heat shock. II. In vitro analysis of RNAs from heat shocked seedlings. Can J Biochem 61:395–403

Bensaude O, Morange M (1983) Spontaneous high expression of heat shock protein in mouse embryonal carcinoma cells and ectoderm from day 8 mouse embryo. EMBO J 2:173–177

Berendes HD, Holt Th KH (1964) The induction of chromosomal activities by temperature shocks. Genen Phaenen 9:1–7

Berger EM, Woodward MP (1983) Small heat shock proteins in *Drosophila* may confer thermal tolerance. Exp Cell Res 147:437–442

Bienz M, Gurdon JB (1982) The heat-shock response in *Xenopus* oocytes is controlled at the translational level. Cell 29:811–819

Binelli G, Mascarenhas JP (1990) *Arabidopsis:* sensitivity of growth to high temperature. Dev Genet (in press)

Bond U, Schlesinger MJ (1986) The chicken ubiquitin gene contains a heat shock promoter and expresses an unstable mRNA in heat shocked cells. Mol Cell Biol 6:4601–4610

Bond U, Agnell N, Hass AL, Redman K, Schlesinger MJ (1988) Ubiquitin in stressed chicken embryo fibroblasts. J Biol Chem 263:2384–2388

Borkovich KA, Farrelly FW, Finkelstein DB, Tavlien J, Lindquist S (1989) Hsp82 is an essential protein that is required in higher concentrations for growth of cells at higher temperatures. Mol Cell Biol 9:3919–3930

Bouchard RA (1990) Characterization of expressed meiotic prophase repeat transcript clones of *Lilium*: meiosis–specific expression, relatedness, and affinities to small heat shock protein genes. Genome 33:68–79

Bouchard RA, Walden DB (1990) Stage–specific expression of a small heat shock gene in anthers. Maize Gene Coop News Lett 64:122

Burke JJ, Orzech KA (1988) The heat shock response in higher plants: a biochemical model. Plant Cell Envir 11:441–444

Burke TJ, Callis J, Viertstra RD (1988) Characterization of a polyubiquitin gene from *Arabidopsis thaliana*. Mol Gen Genet 213:435–443

Cannon C, Wood S, Kalish F, Brunke K (1987) Sequence of a genomic clone for an hsp81 gene from *Brassica oleracea*. J Cell Biol 105:246a

Chappell TG, Welch WJ, Schlossman DM, Palter KB, Schlesinger MJ, Rothman JE (1986) Uncoating ATPase is a member of the 70 kilodalton family of stress proteins. Cell 45:3–13

Cheny C, Shearn MA (1983) Developmental regulation of *Drosophila* disc proteins; synthesis of a heat shock protein under non–heat shock conditions. Dev Biol 95:325–330

Christensen AH, Quail PH (1989) Sequence analysis and transcriptional regulation by heat shock of polyubiquitin transcripts from maize. Plant Mol Biol 12:619–632

Cohen RS, Meselson M (1985) Separate regulatory elements for the heat-inducible and ovarian expression of the *Drosophila* hsp26 gene. Cell 43:737–746

Coldiron–Raichlin P, Dietrich PS, Sinibaldi RM (1986) The heat shock response of *Arabidopsis thaliana*. J Cell Biol 103:176a

Cooper P, Ho TD, Hauptmann RM (1984) Tissue specificity of the heat-shock response in maize. Plant Physiol 75:431–441

Craig EA, Ingolia TD, Manseau LJ (1983) Expression of heat shock cognates during heat shock and development. Dev Biol 99:418–426

Czarnecka E, Nagao RT, Key JL, Gurley WB (1988) Characterization of Gmhsp26-A, a stress gene encoding a divergent heat shock protein of soybean: heavy metal induced inhibition of intron processing. Mol Cell Biol 8:1113–1122

Czarnecka E, Key JL. Gurley WB (1989) Regulatory domains of the Gmhsp 17.5-E heat shock promoter of soybean. Mol Cell Biol 9:3457–3463

De Nettencourt D (1977) In: Frankel R, Gall GAE, Linskens HF, (eds) Monographs on theoretical and applied genetics. Springer, Berlin Heidelberg New York, pp 106–107

Deshaies RJ, Koch BD, Werner-Washburne M, Craig EA, Schekman R (1988) A subfamily of stress proteins facilitates translocation of secretory and mitochondrial precursor polypeptides. Nature 332:800–805

Dietrich PS, Sinibaldi RM (1983) Sequence homology between *Zea mays* DNA and *Drosphila* heat shock genes. J Cell Biol 97:150a

Dietrich PS, Sinibaldi RM (1984) Expression of *Zea mays* DNA sequences homologous to *Drosophila* heat shock genes, J Cell Biol 99:451a

Dietrich PS, Bouchard RA, Sinibaldi RM (1986) Isolation of Maize 83, 70 and 18 Kd heat shock genes. J Cell Biol 103:311a

Dietrich Ps, Bouchard RA, Silva EM, Sinibaldi RM (1987) The complete sequence of a maize 18 Kd heat shock gene. J Cell Biol 105:245a

Duck J, McCormick S, Winter J (1989) Heat Shock protein hsp70 cognate gene expression in vegetative and reproductive organs of *Lycopersicon esculentum*. Proc Natl Acad Sci USA 86:3674–3678

El Ahmadi AB, Stevens MA (1979) Reproductive responses of heat-tolerant tomatoes to high temperatures. J Am Soc Hort Sci 104:686–691

Finley D, Ozkaynak E., Varshavsky A (1987) The yeast polyubiquitin gene is essential for resistance at high temperatures, starvations and other stresses. Cell 48:1035–1046

Frova C, Taramino G, Binelli G (1989) Heat shock proteins during pollen development in maize. Dev Genet 10:324–332

Gausing K, Barkardottin R (1986) Structure and expression of ubiquitin in higher plants. Eur J Biochem 158:57–62

Glaser RL, Lis JT (1990) Multiple, compensatory regulatory elements specify spermatocyte–specific expression of the *Drosophila melanogaster* hsp26 gene. Mol Cell Biol 10:131–137

Glaser RL, Wolfner MF, Lis JT (1986) Spatial and temporal pattern of HSP26 expression during normal development. EMBO J 5:747–754

Hagen G, Uhrhammer N, Guilfoyle TJ (1988) Regulation of expression of an auxin-induced soybean sequence by cadmium. J Biol Chem 263:6442–6443

Hamner KC, Bonner J (1983) Photoperiodism in relation to hormones as factors in floral initiation and development. Bot Gaz 115:361–364

Helm KW, Abernethy RH (1990) Heat shock protein and their mRNAs in dry and early inbibing embryos of wheat. Plant Physiol 93:1626–1633

Hemmingsen SM, Woolford C, Van Der Vies SM, Tilly K, Dennis DT, Georgopoulos CP, Hendrix RW, Ellis RJ (1988) Homologous plant and bacterial proteins chaperone oligomeric protein assembly. Nature 333:330–335

Heuss–LaRosa K, Mayer RR, Chery JH (1987) Synthesis of only two heat shock proteins is required for thermoadaptation in cultured cowpea cells. Plant Physiol 85:4–7

Hightower LE, White FD (1982) Preferential synthesis of rat heat–shock and glucose-regulated proteins in stressed cardiovascular cells. In: Schlesinger MJ, Ashburner M, Tissiere A (eds) Heat shock from bacteria to man. Cold Spring Harbor Lab, New York, pp 369–378

Hunter KW, Cook CL, Hayunga EG (1984) Leishmanial differentiation in vitro: induction of heat shock proteins. Biochem Biophys Res Commun 125:755–760

Imamura A (1953) Photoperiodic initiation of flower primordia in Japanese morning glory *Pharbitis nil* chois. Proc Jpn Acad 29:368–373

Ingolia TD, Craig EA (1982a) *Drosophila* gene related to the major heat shock induced gene is transcribed at normal temperatures and not induced by heat. Proc Natl Acad Sci USA 79:525–529

Ingolia TD, Craig EA (1982b) Four small *Drosophila* heat shock proteins are related to each other and to mammalian α–crystallin. Proc Natl Acad Sci USA 79:2360–2364

Ireland RC, Berger EM (1982) Synthesis of low molecular weight heat shock peptides stimulated by ecdysterone in a cultured *Drosophila* cell line. Proc Natl Acad Sci USA 79:855–859

Kalish F, Cannon C, Brunke K (1986) Characterization of a putative Hsp81 gene in *Brassica oleracea* which is both constitutive and inducible. J Cell Biol 103:176a

Kasambalides EJ. Lanks KW (1985) Anatagonistic effects of insulin and dexamethasone on glucose–regulated and heat shock protein synthesis. J Cell Physiol 123:283–287

Kassenbrock CK, Garcia PD, Walter P, Kelly R (1988) Heavy–chain binding protein recognizes aberrant polypeptides translocated in vitro. Nature 333:90–93

Kim Y–J, Shuman J, Sette M, Przybyla A)1984) Nuclear localization and phosphorylation of three 25-kilodalton rat stress proteins. Mol Cell Biol 4:468–474

Kloetzel P, Bautz EKF (1983) Heat shock proteins are associate with hnRNP in *Drosophila melanogaster* tissue culture cells. EMBO J 2:705–710

Kulomaa MS, Weigel NL, Klein-Sek DA, Beattie SA, Connelly OM (1986) Amino acid sequence of a chicken heat shock protein derived from the complementary DNA nucleotide sequence. Biochemistry 25:6244–6251

Kurtz S, Rossi J, Petko L, Lindquist S (1986) An ancient development induction: heat-shock proteins induced in sporulation and oogenesis. Science 231:1154–1157

Lawrence F, Robert–Gero M (1985) Induction of heat shock stress proteins in promastigotes of three *Leishmania* species. Proc Natl Acad Sci USA 82:4414–4417

Lee S (1990) Expression of hsp86 in mail germ cells. Mol Cell Biol 10:3239–3242

Lin JJC, Welch WJ, Garrels JI, Feramisco J (1982) The association of the 100 Kd heat shock protein with the Golgi apparatus. In:Schlesinger MJ, Ashburner M, Tissieres A (eds) Heat shock from bacteria to man. Cold Spring Harbor Lab, New York, pp 267–273

Lindquist S, Craig EA (1988) The heat shock proteins. Annu Rev Genet 22:631–677

Lotan T, Ori N, Fluhr R (1989) Pathogenesis-related proteins are developmentally regulated in tobacco flowers. Plant Cell 1:881–867

Mansfield MA, Key JL (1987) Synthesis of the low molecular weight heat shock proteins in plants. Plant Physiol 84:1007–1017

Mazzarella RA, Green M (1987) Grp99, an abundant, conserved glycoprotein of the endoplasmic reticulum, is homogous to the 90–KD heat shock protein (hsp90) and the 94–Kd glucose regulated protein (Grp94). J Biol Chem 262:8875–8883

McElwain EF, Spiker S (1989) A wheat cDNA clone which is homologous to the 17 Kd heat shock protein gene family of soybean. Nuel Acids Res 17:1764

Meyer J, Chartier Y (1983) Long lived and short lived heat shock protein in tobacco mesophyll protoplasts. Plant Physiol 72:26–32

Moore SK, Kozak C, Robinson EA, Ulrich SJ, Appella E (1987) Cloning and nucleotide sequence of the murine hsp84 cDNA and chromosome assignment of related sequences. Gene 56:29–40

Moore SK, Kozak C, Robinson EA, Ulrich SJ, Appella E (1989) Murine 86– and 84 Kd heat shock proteins, cDNA sequences, chromosome assignments and evolutionary origins. J Biol Chem 264:5343–5351

Morimoto Rl, Milarski KL (1990) Expression and function of vertebrate hsp70 genes. In Georgopoulos C, Tissieres A, Morimoto RI (eds). Stress proteins in biology and medicine. Cold Spring Harbor Laboratory Press, New York, pp 323–359

Munro S, Pelham HRB (1987) A C–terminal signal prevents secretion of luminal ER proteins. Cell 48:899–907

Nakayama S (1958) Studies on the dark process in the photoperiodic response of *Pharbitis* seedlings. Sci Rep Tohok Univ S4 Biol 24:137

Necchi A, Pogna NE, Mapelli S (1987) Early and late heat shock proteins in wheats and other cereal species. Plant Physiol 84:1378–1384

Nene V, Dunne DW, Johnson KW, Taylor DW, Cordingley JS (1986) Sequence and expression of a major egg antigen from *Schistosoma mansoni*. Homologies to heat shock proteins and alpha – crystallins. Mol Biochem Parasitol 21:179–188

Neumann D, Nover L, Parthier B, Rieger R, Scharf K-D, Wollgiehn R, Nieden U (1989) Heat shock and other response systems of plants. Biol Zentralbl, pp 1–156

Nieto–Sotelo J, Vierling E, Ho T-HD (1990) Cloning, sequence analysis and expression of a cDNA encoding a plastid localized heat shock protein in maize. Plant Physiol (in press)

Nover L, Scharf KD, Neumann D (1983) Formation of cytoplasmic heat shock granules in tomato cell cultures and leaves. Mol Cell Biol 3:1648–1655

Nover L, Scharf KD (1984) Synthesis, modification and structural binding of heat shock proteins in tomato cell cultures. Eur J Biochem 139:303–313

Nover L, Munsche F, Neumann D, Ohme K, Scharf KD (1986) Control of ribosome biosynthesis in plant cell cultures under heat shock conditions ribosomal RNA. Eur J Biochem 160:297–304

Nover L, Scharf KD, Neumann D (1989) Cytoplasmic heat shock granules are formed from precursor particles and contain a set of mRNAs. Mol Cell Biol 9:1298–1308

O'Conner D, Lis J (1981) Two closely linked transcription units within 63B heat shock puff locus of *D. melanogaster* display strikingly different regulation. Nucl Acids Res 9:5075–5092

Ougham HJ, Stoddart JL (1986) Synthesis of heat shock protein and acquisition of thermotolerance in high–temperature tolerant and high–temperature susceptible lines of sorghum. Plant Sci 44:163–167

Palter KB. Watanabe M, Stinson L, Mahowald AP, Craig EA (1986) Expression and localization of *Drosophila melanogaster* hsp70 cognate proteins. Mol Cell Biol 6:1187–1203

Parag HA, Raboy B, Kulka RG (1987) Effect of heat shock on protein degradation in mammalian cells: involvement of the ubiquitin system. EMBO J 6:55–61

Pauli D, Tonka C–H, Ayme-Southgate A (1988) An unusual split *Drosophila* heat shock gene expressed during embryogenesis, pupation and in testis. J Mol Biol 200:47–53

Pelham H, Bienz M (1982) DNA sequences required for transcriptional regulation of the *Drosophila* hsp70 heat-shock gene in monkey cells and *Xenopus* oocytes. In: Schlesinger MJ, Ashburner M, Tissieres A (eds) Heat shock from bacteria to man. Cold Spring Harbor Lab, New York, pp 43–48

Petko L, Lindquist S (1986) Hsp26 is not required for growth at high temperatures, nor for thermotolerance, spore development, or germination. Cell 45:885–8

Riley GJ (1984) Effects of high temperature on RNA–synthesis during germination of maize *(Zea mays L.)* Plant Sci Lett 35:201–206

Rimland J, Akhayat O, Infante D, Infante AA (1988) Developmental regulation and biochemical analysis of 21 Kd heat shock protein in sea urchins. J Cell Biochem (Suppl) 12D:271

Ritossa FM (1962) A new puffing pattern induced by heat shock and DNP in *Drosophila*.Experientia 18:571–573

Ritossa FM (1964) Experimental activation of specific loci in polytene chromosomes of Drosophila. Exp Cell Res 36:515–523

Rollet E. Best–Belpomme H (1986) HSP26 and 27 are phosphorylated in response to heat shock and ecdysterone in *Drosophila melanogaster* cells. Biochem Biophys Res Commun 14:426–433

Rose MD, Misraond L, Vogel J (1989) KAR2, a karyogamy gene, is the yeast homologue of mammalian BIP/GRP78. Cell 57:1211–1216

Russnack RH, Candido PM (1985) Locus encoding a family small heat shock genes in *Caenorhabditis elegans*: two genes duplicated to form a 34.8 kilobase inverted repeat. Mol Cell Biol 5:1268–1278

Sanchez Y, Lindquist SL (1990) HSP104 required for induced thermotolerance. Science 248:1112–1115

Schöffl F, Key JL (1982) An analysis of mRNAs for a group of heat shock proteins of soybean using cloned cDNAs. J Appl Genet 1:301–314

Schöffl F, Key JL (1983) Indentification of a multigene family for small heat shock proteins in soybean and physical characterization of one individual gene coding region. Plant Mol Biol 2:269–278

Schöffl F, Baumann G, Raschke E, Beavan M (1986) The expression of heat-shock genes in higher plants Philos Trans R Soc Lond 314:453–468

Schöffl F, Rieping M, Baumann G, Bevan M, Angermueller S (1989) The function of plant heat shock promoter elements in the regulated expression of chimaeric genes in transgenic tobacco. Mol Gen Genet 217:246–253

Schoper JB, Lambert RJ, Vasilas BL, Westgate ME (1987) Plant factors controlling seed set in maize. Plant Physiol 83:121–125

Schrauwen JAM ,Reijnen WH, De Leeuw HCGM, van Herpen MMA (1986) Response of pollen to heat stress. Acta Bot Neerl 35:312–327

Schuldt C, Kloetzel BM (1985) Analysis of cytoplasmic 19S ring-type particles in *Drosophila* which contain hsp23 at normal growth temperature. Dev Biol 11:65–74

Schwabe WW (1969) Morphogenic responses to climate. In: Evans LT (ed) The induction of flowering. Cornell University Press, lthaca, New York, pp 318–333

Sciandra JJ, Subjeck JR (1983) The effects of glucose on protein synthesis and thermosensitivity in Chinese hamster ovary cell. J Biol Chem 258:12091–12093

Sciandra JJ, Subject J, Hughes CS (1984) Induction of glucose-regulated proteins during anaerobic exposure and of heat–shock proteins after reoxygenation. Proc Natl Acad Sci USA 81:4843–4847

Sirotkin K, Davidson N (1982) Developmentally regulated transcription from *Drosophila melangaster* chromosomal site 67B. Dev Biol 89:196–210

Sorger PK, Pelham HRB (1987) The glucose–regulated protein grp94 is related to heat shock protein hsp90. J Mol Biol 194:341–344

Southgate R, Ayme A, Voelmy R (1983) Nucleotide sequence analysis of the *Drosophila* small heat shock gene cluster at locus 67B, J Mol Biol 165:35–37

Spear I, Thimann KV (1954) The interrelation between CO_2 metabolism and photoperiodism in *Kalanchoe*. ll. Effect of prolonged darkness and high temperature. Plant Physiol 29:414–417

Subjeck J, Shyy TT, Shen J, Johnson J (1983) Association between the mammalian 110 000 dalton heat–shock protein and nucleoli. J Cell Biol 97:139–1395

Swift J (1726) Voyage to la Puta. In: Gulliver's travels. Benjamin Motte Chapter 5,.

Theodorakis NG, Zand DJ, Kotzbauer PT, Williams GT, Morimoto RI (1989) Hemin–induced transcriptional activation of the HSP70 gene during erythroid maturation in K562 cells is due to a heat shock factor mediated stress response. Mol Cell Biol 9:3166–3173

Thomas SR, Lengyel JA (1986) Ecdysteroid-regulated heat shock gene expression during *Drosophila melanogaster* development. J Cell Physiol 127:451–456

Van der Ploeg LH, Giannini SH, Cantor CR (1985) Heat shock genes: regulatory role for differentiation in parasitic protozoa. Science 228:1443–1436

Van Herpen MMA, Reijnen WH, Schrauwen JAM, De Groot, PFM, Jager JWJ, Wullems GJ (1989) Heat shock proteins and survival of germinating pollen of *Lilium longiflorum* and *Nicotiana tabacum*. J Plant Physiol 134:345–351

Vierling E, Key JL (1985) Ribulose 1.5–bisphosphate carboxylase synthesis during heat shock. Plant Physiol 78:155–162

Vierling E, Sun A (1987) Developmental expression of heat shock protein. In: Cherry J (ed) Environmental stress in plants: biochemical and biophysical mechanisms. Springer, Berlin Heidelberg New York Tokyo, pp 343–354

Vierling E, Nagao RT, DeRocher AE, Harris LM (1988) A heat shock protein localized to chloroplasts is a member of a eukaryotic superfamily of heat shock proteins. EMBO 7:575–581

Vierling E, Nagao RT, DeRocher AE, Harris LM (1989) The major low molecular weight heat shock protein in chloroplasts shows antigenic conservation among diverse higher plant species. Mol Cell Biol 9:461–468

Vierstra RD, Langon SM, Schaller GE (1989) Complete amino acid sequence of ubiquitin from the higher plant *Avena sativa*. Biochemistry 25:3105–3108

Welch WJ, Garrels Jl, Thomas GP, Lin JJC, Feramisco J (1983) Biochemical characterization of the mammalian stress proteins and idenity of 2 stress proteins as glucose and Ca^{+2-} inonophore regulated proteins. J Biol Chem 258:7102–7111

Whelan SA, Hightower LE (1985) Differential induction of glucose–regulated and heat shock proteins. J Cell Physiol 125:251–258

Wu BJ, Morimoto RI (1985) Transcription of the human hsp70 gene is induced by serum stimulation. Proc Natl Acad Sci 82:6070–6074

Xiao C-M, Mascarenhas JP (1985) High temperature-induced thermotolerance in pollen tubes of *Tradescantia* and heat shock proteins. Plant Physiol 78:887–890

Xiao H, Lis JT (1989) Heat shock and developmental regulation of the *Drosophila melanogaster* hsp83 gene. Mol Cell Biol 9:1746–1753

Xiao H, Lis JT (1990) Closely related DNA sequences specify distinct patterns of developmental expression in *Drosophila melanogaster*. Mol Cell Biol 10:3272–3276

Yoshida S, Satake T, Mackill DS (1981) High temperature stress in rice. Int Rice Res Inst 67:1–15

Zakeri Z, Wolgemuth DJ (1987) Developmental-stage-specific expression of the HSP70 gene family during differentiation of the mammalian male germ line. Mol Cell Biol 7:1791–1796

Zakeri Z, Wolgemuth DJ, Hunt CR (1988) Identification and sequence analysis of a new member of the mouse HSP70 gene family and characterization of its unique cellular and developmental pattern of expression in the male germ line. Mol Cell Biol 8:2925–2932

Zimmerman JL, Petri WL, Meselson M (1983) Accumulation of specific subsets of *D. melanogaster* heat shock mRNAs in normal development without heat shock. Cell 32:1161–1170

Zimmerman JL, Apuya N, Darwish K, O'Carroll C (1989) Novel regulation of heat shock genes during carrot somatic embryo development. Plant Cell 1:1137–1146

Zivy M (1987) Genetic variability for heat shock proteins in common wheat. Theor Appl Genet 74:209–213

7 Expression of Heat Shock Proteins During Development in *Drosophila*

André Patrick Arrigo[1] and Robert M. Tanguay[2]

1 Introduction

Historically, heat shock proteins (Hsp) were defined to be proteins induced by heat or by various types of chemical stresses. However, numerous early observations suggested that heat shock genes could be expressed without any stress either during normal development and/or during cellular differentiation. Mirault et al. (1978) first suggested that some of the Hsps in *Drosophila melanogaster*, namely Hsp83 and Hsp70, were made at a low level under normal conditions in the absence of cellular stress. This observation which was made in cultured cells was also found to be applicable to animals as shown by the early demonstration of Chomyn and Mitchell (1982) that Hsp83 was expressed in various tissues of larvae and pupae. Using gene probes, Mason et al. (1984) showed the presence of heat shock transcripts at various stages of development of *D. melanogaster*. More recently, studies using specific antibodies and/or P-element-mediated transformation have shed some new light on the cell-, tissue-and development-specific expression of the various Hsps in the absence of stress. It is becoming evident that most heat shock genes are expressed at some time or in specific tissues during development and cellular differentiation. The inducibility of the heat shock response in *Drosophila* has been known to be stage-dependent as shown by Dura in 1981. Thus in terms of inducibility, the fruit fly has a behavior similar to that observed in many other organisms (for reviews see Bond and Schlesinger 1987 and this Vol.). Here, we review studies on the expression of the Hsps in the absence of stress with special emphasis on the small heat shock proteins of *D. melanogaster*.

2 Expression of Hsp83

In *D. melanogaster*, there is a single gene encoding Hsp83. As mentioned above, this Hsp is expressed in the absence of stress both in cultured cells (Mirault et al. 1978; Carbajal et al. 1986) as well as in animals (Chomyn and Mitchell 1982). However, as shown in Fig. 1, its level of expression can vary during development and aging. In flies, *hsp83* transcripts have been shown to be developmentally induced in ovaries

[1] Université Claude Bernard Lyon 1, Centre de Génétique Moléculaire et Cellulaire CNRS-UMR-106, F-69622 Villeurbanne, France
[2] Ontogénèse et Génétique Moléculaire, Centre de Recherches du CHUL, Ste-Foy, Québec, Canada G1V 4G2

Results and Problems in Cell Differentiation 17
Heat Shock and Development
Hightower and Nover (Eds.)
© Springer-Verlag Berlin Heidelberg 1991

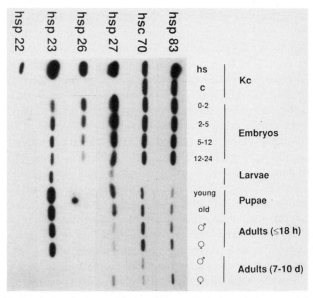

Fig. 1. Immunoblot analysis of Hsps during embryogenesis and development of *D. melanogaster*. *D. melanogaster* Kc III-10 cells (*hs*) were heat shocked for 1 h at 35 °C and allowed to recover overnight at 23 °C. Control cells (*c*) were kept at 23 °C. Embryos were staged as indicated. Larvae from third instar and pupae were separated into two groups, young (white) and old (close to hatching). Young (<18 h) and old (7–10 days) adults were sexed. Cells, embryos and animals were directly homogenized in SDS buffer. Equal amounts of protein were separated on SDS gels as described previously (Vincent and Tanguay 1982), transferred to nitrocellulose sheets and blotted (Carbajal et al. 1986). The anti-Hsp83 and anti-Hsc70 antibodies are described respectively in Carbajal at al. (1990) and Beaulieu and Tanguay (1988). The other antibodies against the small Hsps are from Tanguay (unpubl.)

where they can be found in high abundance in nurse cells but not in epithelial cells (Zimmerman et al. 1983). Biochemical fractionation and immunofluorescence studies indicate that Hsp83 is a cytosolic protein (Carbajal et al. 1986,1990). Recent observations by immunoelectron microscopy show that some Hsp83 is also found in the nucleus. This Hsp is present as a dimeric structure *in situ* (Carbajal et al. 1990). Both of these properties are consistent with cellular function(s) similar to those suggested for its homologue in mammals, Hsp90 (reviewed in Lindquist and Craig 1988). These properties and the observation that the expression of Hsp83 is correlated with peaks of ecdysterone during embryogenesis (Thomas and Lengyel 1986) warrant further studies on the quantitative analysis of expression of this Hsp during organogenesis and development.

3 Expression of Hsp70 and Its Cognates

The *hsp70* gene family of *Drosophila* comprises a large number of related members some of which are stress-inducible *(hsp)* while the others, referred to as the cognates *(hsc)*, are constitutively expressed. Five copies of *hsp70* genes and seven of the related *hsc70* (labelled *hsc1* to *7*) have been described so far (Lindquist and Craig 1988). Since these proteins share extensive homologies, the question of their individual expression during development has been more difficult to tackle. Using cDNA extension, Craig et al. (1983) reported that *hsc4* transcripts were abundant during all phases of late development. *Hsc1* and *hsc2* transcripts were not detected in embryos nor in larvae but were found in adults. The expression of *hsc4* has been measured at the protein level using antibodies specific for Hsc4 (Palter et al. 1986; Beaulieu and Tanguay 1988). As shown in Fig. 1, *hsc4* is expressed at all stages of development but is particularly abundant during embryogenesis as previously reported by Palter et al. (1986). Using in situ hybridization, Perkins et al. (1990) recently reported that *hsc4* transcripts were particularly enriched in cells undergoing rapid growth as well as in cells of tissues active in endocytosis like the garland gland. Whole-mount immunocytochemistry using an antibody specific for this Hsc also shows a higher level of Hsc4 in dividing cells (Tanguay, unpubl.). These results are consistent with those presented in Fig. 1 which show a reduced level of Hsc4 in aged flies.

Immunoelectron microscopic localization data indicate that Hsp70 and Hsc4 (= Hsc70) are not localized in the same intracellular compartments: in the nucleus, Hsc4 is localized on perichromatin RNP fibrils while Hsp70 is mainly enriched in the nucleolus (Carbajal et al. submitted). *Hsp68* and *hsp70* transcripts have been reported at a very low level at all stages of development (Mason et al. 1984). The specific expression of the different members of this family during development remains to be examined in detail with specific probes and antibodies. Such studies will be of interest as it is becoming clear that the different members of this family of closely related genes seem to perform basal functions in a number of different cellular processes (Lindquist and Craig 1988).

4 Expression of the Small Hsps

4.1 Gene Structure and Control of Expression in the Absence of Stress

Drosophila melanogaster cells contain four main small heat shock proteins with apparent molecular weights of 22, 23, 26, and 27 kD (Mirault et al. 1978). The genes encoding these Hsps are clustered within 12 kb of DNA at locus 67B (Corces et al. 1980; Craig and McCarthy 1980; Wadsworth et al. 1980; Voellmy et al. 1981). Three other developmentally regulated genes (genes 1,2 and 3) are found within that locus (Sirotkin and Davidson 1982) and their transcripts are also weakly inducible by heat shock (Ayme and Tissières 1985). However, no protein products have yet been detected for these genes. With the exception of gene 2, the six other genes present at this locus are related. Thus *hsp22,23,26* and *27* and genes 1 and 3 share sequence homology in two domains of their coding region. One conserved region comprises the first 15 amino-terminal amino acids. This region is very hydrophobic and resembles signal peptides regulating interactions with membranes (Arrigo and Pauli 1988). A second domain consists of 80 residues in the second half of the molecule. This region is highly conserved and shows homology with the carboxy-terminal region of alpha-A,B crystallins, a group of small proteins (19-20 kD) of lens cells (Ingolia and Craig 1982;

Southgate and Voellmy 1983). Immediately after the crystallin-like domain, Hsp22, 23, 26 and 27 share a conserved region of 25 residues. This last domain is not observed in gene 1 or 3 nor in alpha-crystallins.

In a study designed to isolate genes preferentially transcribed at the beginning of the pupal period, Sirotkin and Davidson (1982) were the first to report transcription of the small *hsp* genes in absence of heat shock during development. This observation was confirmed and extended by several other investigators. Using in situ hybridization, Zimmerman et al. (1983) showed abundant transcripts of *hsp27* and *26* in the female germ-lines and in early embryos. Mason et al. (1984) also reported the accumulation of mRNA for the small Hsps at various post-embryonic stages of normal development. Low levels of *hsp23* mRNA were also detected during early embryogenesis (Pauli et al. 1989; Tanguay 1989). mRNAs for *hsp 23, 26* and *27* are also abundant in late third instar larvae and young pupae. At this stage of development, all seven genes of locus 67B are expressed (Ayme and Tissières 1985; Pauli and Tonka 1987; Pauli et al. 1988, 1989).

In 1983, Cheney and Shearn were the first to report the expression of the small Hsps at the protein level. These investigators showed that Hsp23 is synthesized in imaginal discs isolated from late (but not from early) third instar larvae. Quantitative analysis of the expression of Hsp23 and Hsp27 was performed using specific antibodies recognizing these polypeptides. These studies showed that the maximal accumulation of Hsp23 and 27 occurred several hours (in case of embryos) or several days (in case of pupae) after the maximal level reached by their corresponding mRNAs (Arrigo 1987; Arrigo and Pauli 1988; Pauli et al. 1989). Consequently, stability is probably an important factor which regulates the level of these Hsps during development. Figure 1 shows new data concerning the pattern of accumulation of Hsp26 and Hsp22 during embryogenesis and development. Hsp26 is expressed in early embryos and its concentration declines as embryos age. It is also expressed in aged female flies. No expression of Hsp22 was observed at any stage. A major conclusion from these observations is that each individual Hsp has its own characteristic pattern of expression during embryogenesis and metamorphosis. This contrasts with the situation in response to heat shock when the major small Hsps show a somewhat similar behavior.

A second important finding was that the expression of the small Hsps shows tissue specificity during development (Glaser et al. 1986; Glaser and Lis 1990; Pauli et al. 1990; Tanguay 1989; see below). This phenomenon contrasts with the stress-induced expression of these polypeptides which occurs in almost every tissue, with the exception of differentiated gametes and preblastoderm embryos (Dura 1981; Zimmerman et al. 1983; Bonner et al. 1984). Several laboratories have investigated the signals involved in the expression of the small Hsps during development. Except perhaps in the case of *hsp22* (Klemenz and Gehring 1986), the heat shock regulatory element (HSE) does not seem to play a role in the developmental induction of the other *hsp* genes. This suggests that other regulatory sequences are involved. For example, steroid receptor binding sequences were found upstream of *hsp23* and *27* genes (Mestril et al. 1986; Riddihough and Pelham 1986, 1987). This was consistent with a regulatory role for the molting hormone ecdysterone as suggested by the observation that it triggered the expression of the small Hsps in tissue culture cells and in isolated imaginal discs (Ireland and Berger 1982; Ireland et al. 1982). In addition, the expression of these polypeptides in late third instar larvae-young pupae corresponded to a peak of accumulation of ecdysterone (Handler 1982; Thomas and Lengyel 1986; Dubrovsky and Zhimulev 1988). However, it is not yet clear whether this hormone regulates the expression of the small Hsps in tissues other than the imaginal discs. This is rather difficult to assess as ecdysterone may play a role in regulating mRNA stability and the half-life of *hsp* mRNAs may vary between tissues (Vitek and Berger 1984; Pauli et al. 1989).

The mechanisms specifying the tissue-specific expression of the small *hsps* appear to be complex. Indeed, analysis of the coarse regulatory sequences of several *hsp* genes shows that they differ from one tissue or one stage to another (Cohen and Meselson 1985; Glaser et al. 1986; Klemenz and Gehring 1986; Hofman et al. 1987; Glaser and Lis 1990). For example, in the case of the *hsp26* gene, sequences 350–500 bp upstream of the cap site appear to be important for the expression in the female germ line (Cohen and Meselson 1985). In contrast, the expression in the male germ line is controlled by an element residing in the first 278 bp in front of the gene (Glaser et al. 1986) and which appears to be made of multiple elements with redundant functions (Glaser and Lis 1990). Future studies using P-element-mediated transformations should provide a better description of these various regulatory sequences and of their functions during development.

4.2 Tissue-Specific Expression of Hsp27

Recently, the pattern of expression of Hsp27 in tissues was analyzed using immunological detection with an affinity-purified antiserum on thin sections of the developing insect (Pauli et al. 1990). These results were compared to the expression of *hsp27* mRNA using in situ hybridization. The uniform Hsp27 staining in preblastoderm and blastoderm embryos is consistent with the immunoblot data (Arrigo and Pauli 1988; Fig. 1). In addition to the polypeptide already accumulated during oogenesis, translation of *hsp27* mRNA had to take place during this preblastoderm period, since RNA was not detectable after this period (Zimmerman et al. 1983). Hsp27 was then slowly degraded over the entire embryonic period (Arrigo and Pauli 1988). In first, second and early third instar larvae, Hsp27 was only detected in the central nervous system and in the gonads (Fig. 2A). In addition to these tissues, all imaginal discs accumulated Hsp27 at the end of the third instar larval stage. Transcription and translation of *hsp27* in imaginal discs coincided with the beginning of their differentiation. In these tissues, *hsp27* mRNA disappeared before the middle of the pupal period while the protein was more progressively degraded over the pupal stage (Pauli et al. 1989). The disc-derived adult tissues were devoid of Hsp27. Indeed, in late pupae, Hsp27 was only detected in the gonads, the central nervous system and in the eyes at the top of the ommatidia. In the newborn adult fly, Hsp27 was still in the gonads but neural expression was confined to a few clusters of cells (Fig. 2B,C). In male gonads, Hsp27 was expressed from the first instar larval stage and accumulated in mature sperm. In females, this Hsp was detected in nurse cells during the vitellogenic stages of oogenesis as well as in mature gametes of the adult. *Hsp27* transcription was very strong in both the adult ovaries and testes. Contrasting with this observation, no *hsp27* transcripts were detected in the adult central nervous system although this gene was highly active in this tissue during the larval stages. This suggested that the Hsp27 found in the central

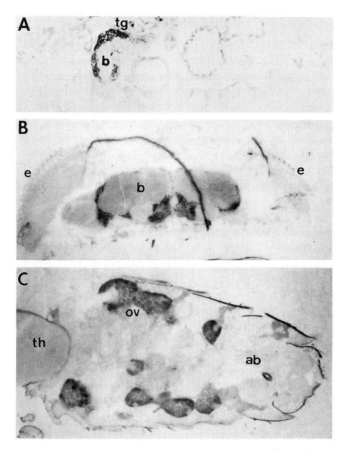

Fig. 2A-C. Immunolocalization of Hsps27 during *Drosophila* development. Thin sections of developing larvae and pupae were processed for immuno-histology using affinity-purified hsp27 antiserum as described in Pauli et al. (1990). **A** Sagittal section of early third instar larvae: a strong staining was found in the brain and the thoracic ganglion. The gonads are not visible in this section. **B-C** Sections of a newborn adult fly. **B** Longitudinal section of the head showing the presence of Hsp27 in the brain; **C** sagittal section of the abdomen: note the ovaries which are positive for Hsp27. Tissues of interest are indicated by the following abbreviations: *b* brain; *tg* thoracic ganglion; *e* eye; *ov* ovaries, *ab* abdomen; *th* thorax. The *dark coloration* reveals the presence of Hsp27

nervous system of pupae and adult flies was probably synthesized during the larval stages. As mentioned before, there is a lag of several days between the maximal expression of *hsp27* mRNA and the maximal expression of the corresponding polypeptide in pupae (Pauli et al. 1989). Consequently, modulations in the stability of Hsp27 may also regulate the tissue-specific level of this Hsp during development. It is tempting to classify the tissues described above in two groups: (1) the imaginal discs and the embryonic tissues which express Hsp27 transiently; (2) the central nervous system and the germ lines which express Hsp27 during a rather long period of development.

4.3 Tissue-Specific Expression of Hsp26

As shown in Fig. 1, Hsp26 is also expressed in early embryos and its titer declines as the embryo ages. While the specificity of Hsp26 expression at this stage remains to be ascertained by whole-mount immunostaining, the tissue distribution of *hsp26* mRNA has been investigated at later stages by in situ hybrization and by analyzing the expression of a *hsp26-lacZ* fusion gene product (Glaser et al. 1986; Glaser and Lis 1990). *Hsp26* mRNA is expressed in numerous tissues during development, including spermatocytes, nurse cells, epithelium, imaginal discs, proventriculus and neurocytes. However, since no studies using a specific antibody have been performed yet on these tissues, the kinetics of accumulation of this protein during development are unknown. During embryogenesis, Hsp26 accumulates to high levels (see Fig. 1) and this is consistent with the presence of large amounts of the corresponding mRNA at this stage of the development (Zimmerman et al. 1983; Pauli et al. 1989). At later stages, the immunoblotting assay did not detect a high level of expression (Fig. 1 and Tanguay, unpubl.). Thus, while the general developmental pattern of expression of Hsp26 resembles that described for Hsp27 during embryogenesis, it seems to differ from that of Hsp27 and Hsp23 at later stages.

4.4 Tissue-Specific Expression of Hsp23

In an immunoblot assay, Hsp23 was found to accumulate maximally during the pupal phase of the insect with an increase of about 30-fold between third instar larvae and the late pupal stages (Arrigo 1987). Similar results were obtained using a different antibody (Fig. 1 and Tanguay, in prep.). As shown in Fig. 1, Hsp23 is also expressed in young adults where it is found to be expressed in the brain and in gonads but in a pattern different from that of Hsp27 (Tanguay, in prep.). Low levels of Hsp23 were observed during embryogenesis (Arrigo 1987; Arrigo and Pauli 1988; Pauli et al. 1990; Fig. 1). However, as seen on the blot, its pattern of expression during embryogenesis differs from that of Hsp26 and Hsp27. Hsp23 expression peaks at about 5 h and remains high while Hsp26 and Hsp27 decline. Recently, an analysis of the distribution of this Hsp during early embryogenesis has been performed using whole-mount immunocytochemistry and in situ hybridization (Tanguay 1989). As shown in Fig. 3, the anti-Hsp23 stains a restricted set of cells located at regular intervals along the CNS. These cells were tentatively identified as midline precursor cells (MPC). Expression of Hsp23 starts at early germ band elongation and persists until hatching of the embryo. At later embryonic stages (stages 16–17), Hsp23 is also expressed in a very restricted number of cells of the brain. The specificity of the immuno-staining was confirmed by whole-mount in situ hybridization using an *hsp23*-specific probe (Tanguay, in prep.). Under these conditions, no labelling of the MPC cells was observed with an *hsp27* probe in contrast to a recent report (Haass et al. 1990). Notwithstanding this discrepancy, it is clear that the small Hsps are expressed in a specific pattern during neurogenesis in *Drosophila*. Finally, as shown in Fig. 1, no expression of Hsp22 (at least to the level observed during heat shock) has been detected yet by immunoblotting during embryogenesis. This does not preclude expression at a low level or in specific tissues which may be undetected in this assay.

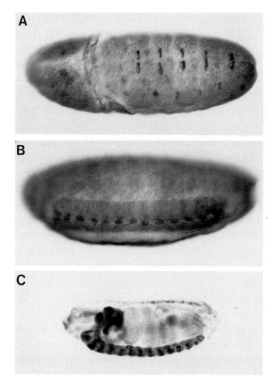

Fig. 3A-C. Expression of Hsp23 in *Drosophila* embryos. Whole-mount staining of *Drosophila* embryos was performed as described by Macdonald and Struhl (1986). An antibody prepared against a fusion Hsp23 protein construct was affinity-purified on nitrocellulose strips as described by Carbajal et al. (1986). Hsp23 is shown to be expressed in a small subset of cells tentatively identified as midline precursor cells disposed along the central nervous system (A-C). At a later stage, staining is also expressed in specific cells of the brain (**C**). Whole-mount in situ hybridization with a probe specific for *hsp23* shows localization of mRNA in the same cells (Tanguay 1989 and in prep.)

5 Cellular Localization and Function(s) of the Small Hsps During Development

At normal temperatures, the different Hsps can be either cytoplasmic or associated with the nuclear pellet (reviewed in Tanguay 1985). For example in S3 cells recovering from a heat stress or after ecdysterone stimulation, Hsp27 is nuclear (Beaulieu et al. 1989). Hsp23 is in the form of cytoplasmic granules in cultured cells (Duband et al. 1986) but shows a uniform cytoplasmic distribution in salivary gland cells (Arrigo and Ahmad-Zadeh 1981). Therefore, each of the small Hsps may play a role in a definite locale within the cell. During heat shock, most of these Hsps concentrate in or near the nuclear fraction (Arrigo et al. 1980,1985; Vincent and Tanguay 1982; Tanguay et al. 1985; Duband et al. 1986; Arrigo 1987; Beaulieu et al. 1989) This behavior is similar to that observed in vertebrate cells (Arrigo and Welch 1987; Collier et al. 1988; Arrigo et al. 1988; Arrigo 1990a), in plants (Nover et al. 1989) and in yeast (Rossi and Lindquist 1989). In this particular fraction, the small Hsps form very high molecular weight insoluble structures resembling the large aggregates formed by alpha-crystallin in cataract lens cells (Arrigo et al. 1988). However, in some cases such as in thermotolerant cells or tissues (Arrigo 1987; Arrigo et al. 1988) or in yeast cells active in respiration (Rossi and Lindquist 1989) this heat-induced phenomenon is not observed.

Whether induced by ecdysterone or during development, the small Hsps appear to behave similarly to those induced by heat shock. Indeed, in S3 cells exposed to the molting hormone ecdysterone, Hsp27 shows a nuclear localization. However, Hsp27 can be easily extracted from the nucleus of these cells with non-ionic detergents while after a heat shock this Hsp becomes resistant to this extraction (Beaulieu et al. 1989). During development, Hsp23 from late third instar larvae and from pupae behaves as a soluble cytoplasmic polypeptide, while after heat shock it is insoluble (Arrigo 1987). A similar observation was made for Hsp27, except that a fraction of this protein was insoluble at normal temperature (Arrigo and Pauli 1988). Both Hsp23 and 27, in their soluble form, sedimented as large molecular weight aggregates similarly to alpha-crystallin from bovine lens cells (Arrigo 1987; Arrigo and Pauli 1988). An intriguing possibility is that the stability of these proteins is modulated by their state of aggregation.

Cytological analysis of the localization of Hsp27 during development has recently been presented (Pauli et al. 1990). During embryogenesis, Hsp27 was cytoplasmic as well as nuclear, while after heat shock this Hsp was strictly nuclear. During the larval stages, Hsp27 seemed to accumulate in the nucleus of neuronal cells. Following heat treatment of the larvae, a predominant nuclear localization of Hsp27 was observed in most of the tissues. In adults, high levels of Hsp27 were detected in the nucleus of vasa deferents, nurse and neuronal cells. However, in most of these cells, detectable levels of Hsp27 were also observed in the cytoplasm. These observations suggest that during development, Hsp27 may play a role inside or at the periphery of the nucleus of cells of different tissues.

Two of *Drosophila* small Hsps (Hsp26 and 27) are phosphorylated in tissue culture cells exposed to either heat shock or ecdysterone (Rollet and Best-Belpomme 1986). Hsp27 is also probably phosphorylated during development (Arrigo and Pauli 1988). It is also interesting to note that alpha-crystallin is also a phosphoprotein (Voorter et al. 1989). The regulation of the phosphorylation of these Hsps appears to be complex. For example, in addition to heat shock, agents which either modulate cell division, cell differentiation (Welch 1985; Regazzi et al. 1988; Arrigo 1990b), or the immune response (Arrigo 1990b) increase the level of phosphorylation of the corresponding mammalian Hsp. Protein phosphorylation appears to play crucial roles in cell regulations and it may also regulate the function(s) of Hsp26 and Hsp27 during development.

The main unsolved question dealing with the small Hsps relates to their function(s). Unfortunately, no *Drosophila* mutants of these Hsps have been isolated, in spite of extensive mutagenesis of the 67 AD region (Leicht and Bonner 1988). Eissenberg and Elgin (1987) have constructed a mutant using a P-element insertion upstream of the Hsp27 coding sequence which partially disrupts the transcription of the gene. This mutant is nevertheless viable and Hsp27 is still detectable (Pauli and Arrigo, unpubl.). In vitro, specific inhibition of expression of individual small Hsps can be achieved using antisense RNA constructs (Nicole and Tanguay 1987). However, no distinct phenotype has been observed in cells where Hsp26 synthesis was inhibited by this approach (McGarry and Lindquist 1986). Berger and Woodward

(1983) provided early evidence for the involvement of the small Hsps in the development of thermotolerance in *Drosophila*. In other species, the analysis of mutants of
small Hsps is quite confusing. For example, a *Dictyostelium* mutant with an altered
pattern of expression of small Hsps is not viable at elevated temperature (Loomis and
Wheeler 1982). In yeast, deletion of the unique *hsp26* gene does not result in any altered phenotype (Petko and Lindquist 1986). Recently, Landry et al.(1989) reported
that elevated levels of human Hsp28 appear sufficient to confer a partial thermal
resistance to rodent cells. Expression of *Drosophila* Hsp27 under the control of the
human *hsp27* promotor in the same cells induces thermotolerance, suggesting that this
small Hsp can substitute for the human Hsp27 (Rollet et al., submitted).

Another hint for the function(s) of these polypeptides is related to their homology
with lens crystallins. Indeed, the small Hsps are believed to be ancestors of alpha-AB-
crystallins (Ingolia and Craig 1982; Wistow 1985; de Jong et al. 1989). Recently, some
lens crystallins have been shown to be aggregated metabolic enzymes (reviewed in
Wistow and Piatigorsky 1987, 1988). However, no enzymatic activity has yet been
found for alpha-AB-crystallin, despite that the fact that the mRNA encoding the B
form is detected not only in lens but also in heart, skeletal muscle, kidney, lung and
brain (Bhat and Nagineni 1989; Dubin et al. 1989). Alpha-B-crystallin also accumulates in the Rosenthal fibers of astrocytes from patients suffering of Alexander's
disease (neurological disorder) (Iwaki et al. 1989). Therefore, alpha-B-crystallin may
play a role other than that of a structural component of the lens. Further investigations
should determine whether this lens protein shares, in addition to structural homology,
functional activities with the small Hsps expressed during development.

Analysis of the particular pattern of expression of small Hsps during development
may also provide clues to their function(s). For example, it is interesting to note that
during embryonic and larval development, Hsp27 expression occurred predominantly
in tissues which contained highly proliferating cells, e.g. early embryo, ventral cord of
stage 14 embryos, central nervous system and gonads of larvae (Pauli et al. 1990; see
also Campos-Ortega and Hartenstein 1985; Lehner and O'Farell 1989 for a description
of the mitotic index in these tissues). Another type of accumulation of Hsp27 occurred
when imaginal disc cells stopped dividing and began to differentiate. Interestingly,
after complete differentiation of the discs, Hsp27 was no longer detectable (Pauli et al.
1990). This typical pattern of expression of Hsp27 is reminiscent of the work described
by Bielka et al. (1988) and Gaestel et al. (1989) on the mammalian small Hsp. Indeed,
following transplantation in mice, Ehrlich ascites tumor cells stop dividing concomi-
tantly with the accumulation of Hsp28. In both *Drosophila* imaginal discs and Ehrlich
ascites tumor cells, these proteins accumulate in cells arrested in late S/G2 phase
(Bryant 1987; Gaestel et al. 1989). An important finding was that the developmental
pattern of expression of Hsp27 resembled that of *Dras*, the *Drosophila* homologue of
the *ras* oncogene (Segal and Shila 1986). In addition, Hsp23 accumulates in the brain
during embryogenesis (Tanguay 1989 and in prep.), a situation reminiscent of the
pattern of expression of the *Drosophila* homologue of the *src* oncogene (Simon et al.
1985). Taken together, these observations suggest that the putative function(s) of the
small Hsps be related to both the proliferative and differentiated states of the tissues of
the developing insect.

6 Summary

Studies on the expression of heat shock proteins during development in *Drosophila* clearly show that individual Hsps accumulate in a tissue- and developmental stage-specific manner. This is in contrast to their coordinate expression in response to stress. Therefore, the Hsps may play at least two roles, one as housekeeping proteins during development and/or differentiation and the second one in restoring cellular functions after environmental stress. Research in the first two decades following the discovery of the heat shock response have focused on a search for functions in stressed cells. The next few years should bring us further understanding on the role of these fascinating proteins during development in *Drosophila* as well as in other eukaryotes.

Acknowledgements. This work was supported by the Sandoz Foundation (APA), the Centre National de la Recherche Scientifique (APA), and the Medical Research Council of Canada (RMT).

References

Arrigo AP (1987) Cellular localization of hsp23 during *Drosophila* development and subsequent heat shock. Dev Biol 122:39–48

Arrigo AP (1990a) The monovalent ionophore monensin maintains the nuclear localization of the human stress protein hsp28 during heat shock recovery. J Cell Science 96:419–427

Arrigo AP (1990b) Tumor necrosis factor induces the rapid phosphorylation of the mammalian heat shock protein hsp28. Mol Cell Biol 10:1276–1280

Arrigo AP, Ahmad-Zadeh C (1981) Immunofluorescence localization of a small heat shock protein hsp23) in *Drosophila melanogaster*. Mol Gen Genet 184:73–79

Arrigo AP, Pauli D (1988) Characterization of hsp27 and of three immunologically related polypeptides during *Drosophila* development. Exp Cell Res 175:169–183

Arrigo AP, Welch WJ (1987) Characterization and purification of the mammalian 28,000 dalton heat shock protein. J Biol Chem 262:15359–15369

Arrigo AP, Fakan S, Tissières A (1980) Localization of the heat shock induced proteins in *Drosophila melanogaster* tissue culture cells. Dev Biol 78:86–103

Arrigo AP, Darlix JD, Khandjian EW, Simon M, Spahr PF (1985) Characterization of the prosome from *Drosophila* and its similarity to the cytoplasmic structure formed by the low molecular weight heat shock proteins. EMBO J 4:2942–2954

Arrigo AP, Suhan J, Welch WJ (1988) Dynamic changes in the structure and locale of the mammalian low molecular weight heat shock protein. Mol Cell Biol 8:505–5071

Ayme A, Tissières A (1985) Locus 67B of *Drosophila melanogaster* contains seven, not four, closely related heat shock genes. EMBO J 4:2949–2954

Beaulieu JF, Tanguay RM (1988) Members of the *Drosophila* HSP 70 family share ATP-binding properties. Eur J Biochem 172:341–347

Beaulieu JF, Arrigo AP, Tanguay RM (1989) Interaction of *Drosophila* 27Kd heat shock protein with the nucleus of heat-shocked and ecdysterone-stimulated cultured cells. J Cell Science 92:29–36

Berger EM, Woodward MP (1983) Small heat shock proteins of *Drosophila* may confer thermal tolerance. Exp Cell Res 147:437–442

Bhat SP, Nagineni CN (1989) AlphaB-subunit of lens-specific alphaB-crystallin is present in other ocular and non-ocular tissues. Biochem Biophys Res Comm 158:319–325

Bielka H, Benndorf R, Jungham I (1988) Growth related changes in protein synthesis and in a 25 kDa protein of Ehrlich ascites tumor cells. Biomed Biochim Acta 47:557–563

Bond U, Schlesinger MJ (1987) Heat-shock proteins and development. Adv Genet 24:1–29

Bonner JJ, Parker-Thomburg C, Mortin MA, Pelham HRB (1984) The use of promoter fusions in *Drosophila* genetics: isolation of mutations affecting the heat shock response. Cell 37:979–991

Bryant PJ (1987) Experimental and genetic analysis of growth and cell proliferation in *Drosophila* imaginal discs. In: Bryant P (ed) Genetic regulation of development. Alan R. Liss, New York, pp 339–372

Campos-Ortega J A, Hartenstein V (1985) The embryonic development of *Drosophila* melanogaster. Springer, Berlin Heidelberg New York

Carbajal ME, Duband JL, Lettre F, Valet JP, Tanguay RM (1986) Cellular localization of *Drosophila* 83–kilodaltons heat shock protein in normal, heat-shocked and recovering cultured cells with a specific antibody. Biochem Cell Biol 64:816–825

Carbajal ME, Valet JP, Charest PM, Tanguay RM (1990). Purification of Drosophila hsp83 and immunoelectron microscopic localization. Eur J Cell Biol 52:147–156

Cheney CM, Shearn A (1983) Developmental regulation of *Drosophila* imaginal discs proteins: synthesis of a heat-shock protein under non-heat shock conditions. Dev Biol 95:325–330

Chomyn A, Mitchell HK (1982) Synthesis of the 84,000 dalton protein in normal and heat shocked *Drosophila melanogaster* cells as detected by specific antibody. Insect Biochem 12: 105–114

Cohen RS, Meselson M (1985) Separate regulatory element for the heat-inducible and ovarian expression of the *Drosophila* hsp26 gene. Cell 43:737–746

Collier NC, Heuser J, Aach-Levy M, Schlesinger M (1988) Ultrastructural and biochemical analysis of the stress granules in chicken embryo fibroblasts. J Cell Biol 106:1131–1139

Corces V, Holmgren R, Freund R, Morimoto R, Meselson M (1980) Four heat shock proteins of *Drosophila melanogaster* coded within a 12-kilobase region in chromosome subdivision 67B. Proc Natl Acad Sci USA 77:5390–5393

Craig E A, McCarthy BJ (1980) Four *Drosophila* heat shock genes at 67B: characterization of recombinant plasmids. Nucl Acids Res 8:4441–4457

Craig EA, Ingolia TD, Manseau LJ (1983) Expression of *Drosophila* cognate genes during heat shock and development. Dev Biol 99:418–426

de Jong WW, Hendricks W, Mulders JWM, Bloemendal H (1989) Evolution of eye lens crystallins: the stress connection. Trends Biochem Sci 14:365–368

Duband JL, Lettre F, Arrigo AP, Tanguay R (1986) Expression and cellular localization of HSP23 in unstressed and heat shocked *Drosophila* culture cells. Can J Genet Cytol 28:1088–1092

Dubin R A, Wawrousek EF, Piatigorsky J (1989) Expression of the murine alphaB-crystallin gene is not restricted to the lens. Mol Cell Biol 9:1083–1091

Dubrovsky EB, Zhimulev IF (1988) Trans-regulation of ecdysone-induced protein synthesis in *Drosophila melanogaster* salivary glands. Dev Biol 127:33–34

Dura JM (1981) Stage dependent synthesis of heat shock induced proteins in early embryos of *Drosophila melanogaster*. Mol Gen Genet 184:381–385

Eissenberg J C, Elgin SCR (1987) HSP28stl: a P-element insertion mutation that alters the expression of a heat shock gene in *Drosophila melanogaster*. Genetics 115:333–340

Gaestel M, Gross B, Benndorf R, Strauss M, Schunk W-H, Kraft R, Otto A, Böhm H, Stahl J, Drabsch H, Bielka H (1989) Molecular cloning, sequencing and expression in *Escherichia coli* of the 25-kDa growth-related protein of Ehrlich ascites tumor and its homology to mammalian stress proteins. Eur J Biochem 179:209–213

Glaser RL, Wolfner MF, Lis JT (1986) Spatial and temporal pattern of HSP26 expression during normal development. EMBO J 5:747–754

Glaser RL, Lis JT (1990) Multiple, compensatory regulatory elements specify spermatocyte-specific expression of the *Drosophila melanogaster* hsp26 gene. Mol Cell Biol 10:131–137

Haass C, Klein U, Kloetzel PM (1990) Developmental expression of *Drosophila melanogaster* small heat-shock proteins. J Cell Sci 96:413–418

Handler AM (1982) Ecdysteroid titers during pupal and adult development in *Drosophila melanogaster*. Dev Biol 93:73–82

Hofman E P, Gerring SL, Corces VG (1987) The ovarian, ecdysterone and heat-shock-responsive promoters of *Drosophila melanogaster* hsp27 gene react differently to perturbation of DNA sequence. Mol Cell Biol 7:973–981

Ingolia TD, Craig EA (1982) Four small *Drosophila* heat shock proteins are related to each other and to mammalian alpha-crystallin. Proc Natl Acad Sci USA 79:2360–2364

Ireland R C, Berger EM (1982) Synthesis of the low molecular weight heat shock proteins stimulated by ecdysterone in a cultured *Drosophila* cell line. Proc Natl Acad Sci USA 79:855–859

Ireland RC, Berger EM, Sirotkin K, Yund MA, Osterburg D, Fristom J (1982) Ecdysterone induces the transcription of four heat shock genes in *Drosophila* S3 cells and imaginal discs. Dev Biol 93:498-507

Iwaki T, Kume-Iwaki A, Leim RKH, Goldman J (1989) Alpha-crystallin is expressed in non-lenticular tissues and accumulates in Alexander's disease brain. Cell 57:71–78

Klemenz R, Gehring WJ (1986) Sequence requirement for expression of the *Drosophila melanogaster* heat shock protein hsp22 gene during heat shock and normal development. Mol Cell Biol 6:2011–2019

Landry J, Chretien P, Lambert H, Hickey E, Weber LA (1989) Heat shock resistance conferred by expression of the human HSP27 gene in rodent cells. J Cell Biol. 109:7–15

Lehner J, O'Farell PH (1989) Expression and function of *Drosophila* cylin A during embryonic cell cycle progression. Cell 56:957–968

Leicht BG, Bonner JJ (1988) Genetic analysis of chromosomal region 67 A-D of *Drosophila melanogaster*. Genetics 119:579–593

Lindquist S, Craig EA (1988) The heat-shock proteins. Annu Rev Genet 22:631–677

Loomis WF, Wheeler S (1982) Chromatin-associated heat shock proteins in *Dictyostelium*. Dev Biol 79:399–408

Macdonald PM, Struhl G (1986) A molecular gradient in early *Drosophila* embryos and its role in specifying the body pattern. Nature 324:537–545

Mason PJ, Hall LMC, Gausz J (1984) The expression of heat shock genes during normal development in *Drosophila melanogaster*. Mol Gen Genet 194:73–78

McGarry TJ, Lindquist S (1986) Inhibition of heat shock protein synthesis by heat inducible antisense RNA. Proc Natl Acad Sci USA 83: 399–403

Mestril R, Shiller P, Amin J, Klapper H, Jayakumar A, Voellmy R (1986) Heat shock and ecdysterone activation of *Drosophila melanogaster* hsp23 gene; a sequence element implied in developmental regulation. EMBO J 5:1667–1673

Mirault ME, Goldschmidt-Clermont M, Moran L, Arrigo AP, Tissières A (1978) The effect of heat shock on gene expression in *Drosophila melanogaster*. Cold Spring Harbor Symp Quant Biol 48:819–829

Nicole LM, Tanguay, RM (1987) On the specificity of antisense RNA to arrest in vitro translation of mRNA coding for *Drosophila* hsp23. Biosci Rep 7:239–246

Nover L, Scharf KD, Neumann D (1989) Cytoplasmic heat shock granules are formed from precursor particles and are associated with a specific set of mRNAs. Mol Cell Biol 9:1298–1308

Palter KB, Watanabe M, Stinson L, Mahowald AP, Craig EA (1986) Expression and localization of *Drosophila melanogaster* hsp70 cognates proteins. Mol Cell Biol 6:1187–1203

Pauli D, Tonka CH (1987) A *Drosophila* heat shock gene from locus 67B is expressed during embryogenesis and pupation. J Mol Biol 198:235–240

Pauli D, Tonka CH, Ayme-Southgate A (1988) An unusual split *Drosophila* heat shock gene expressed during embryogenesis, pupation and testis. J Mol Biol 200: 47–53

Pauli D, Arrigo A P, Vasquez J, Tonka CH, Tissières A (1989) Expression of the small heat shock genes during *Drosophila* development:comparison of the accumulation of hsp 23 and hsp27 mRNAs and polypeptides. Genome 31:671–676

Pauli D, Tonka CH, Tissières A, Arrigo AP (1990) Tissue specific expression of the heat shock protein hsp27 during *Drosophila melanogaster* development. J Cell Biol 111:817–828

Perkins LA, Doctor J S, Zhang K, Stinson L, Perrimon N, Craig EA (1990) Molecular and developmental characterization of the heat shock cognate 4 gene of *Drosophila melanogaster*. Mol Cell Biol 10:3232–3238

Petko L, Lindquist S (1986) Hsp26 is not required for growth at high temperature, not for thermotolerance, spore development or germination. Cell 45:885–894

Regazzi R, Eppenberger U, Fabbro D (1988) The 27,000 daltons stress proteins are phosphorylated by protein kinase C during the tumor promoter mediated growth inhibition of human mammary carcinoma cells. Biochem Biophys Res Comm 152:62–68

Riddihough G, Pelham HRB (1986) Activation of the *Drosophila* hsp27 promoter by heat shock and by ecdysone involves independent and remote regulatory sequences. EMBO J 5:1653–1658

Riddihough G, Pelham HRB (1987) An ecdysone response element in the *Drosophila* hsp27 promoter. EMBO J 6:3729–3734

Rollet E, Best-Belpomme M (1986) HSP 26 and 27 are phosphorylated in response to heat and ecdysterone in *Drosophila melanogaster* cells. Biochem Biophys Res Comm 141:426–433

Rossi J, Lindquist S (1989) The intracellular location of yeast heat shock protein 26 varies with metabolism. J Cell Biol 108:425–439

Segal D, Shila BZ (1986) Tissue locaiization of *Drosophila melanogaster* ras transcripts during development. Mol Cell Biol 6:2241–2248

Simon MA, Drees B, Kornberg T, Bishop M (1985) The nucleotide sequence and the tissue specific expression of *Drosophila* c-src. Cell 42:831–840

Sirotkin K, Davidson N (1982) Developmentally regulated transcription from *Drosophila melanogaster* site 67B. Dev Biol 89:196–210

Southgate R, Voellmy R (1983) Nucleotide sequence analysis of *Drosophila* small heat shock gene cluster at locus 67B. J Mol Biol 165:35–57

Southgate R, Mirault ME, Ayme A, Tissières A (1985) Organization, sequences and induction of heat shock genes. In: Atkinson BG, Walden DB (eds) Changes in gene expression in response to environmental stress. Academic Press, Orlando, pp 1–30

Tanguay RM (1985) Intracellular localization and possible functions of heat shock proteins. In: Atkinson BG, Walden DB (eds) Changes in gene expression in response to environmental stress. Academic Press, Orlando, pp 91–113

Tanguay RM (1989) Localized expression of a small heat shock protein, hsp23, in specific cells of the central nervous system during early embryogenesis in *Drosophila*. J Cell Biol 109:155a

Tanguay RM, Duband JL, Lettre F, Valet JP, Arrigo AP, Nicole L (1985) Biochemical and immunocytochemical localization of heat shock proteins in *Drosophila* culture cells. Intermediate filaments. Ann N Y Acad Sciences 455:712–714

Thomas SR, Lengyel JA (1986) Ecdysteroid-regulated heat shock gene expression during *Drosophila melanogaster* development. Dev Biol 115:434–438

Vincent M, Tanguay RM (1982) Different intracellular distribution of heat shock and arsenite induced proteins in *Drosophila* KC cells. J Mol Biol 162:365–378

Vitek M, Berger EM (1984) Steroid and high temperature induction of the small heat shock protein genes in *Drosophila*. J Mol Biol 178:173–189

Voellmy R, Goldschmidt-Clermont M, Southgate R, Tissières A, Levis R, Gehring WJ (1981) A DNA segment isolated from chromosomal site 67B in *D. melanogaster* contains four closely linked heat shock genes. Cell 45:185–193

Voorter CEM, Haard-Hoekman WA, Roersma ES, Meyer HE, Bloemendal H, de Jong, WW (1989) The in vivo phosphorylation sites of bovine alphaB-crystallin. FEBS Lett 259:1, 50–52

Wadsworth S, Craig EA, McCarthy BJ (1980) Genes for three *Drosophila* heat shock induced proteins at a single locus. Proc Natl Acad Sci USA 77:2134–2137

Welch WJ (1985) Phorbol ester, calcium ionophore, or serum added to quiescent rat embryo fibroblast cells result in the elevated phophorylation of two 28,000 dalton mammalian stress proteins. J Biol Chem 260:3058–3062

Wistow G (1985) Domain structure and evolution in alpha-crystallins and small heat shock proteins. FEBS Lett 181:1-6

Wistow G, Piatigorsky J (1987) Recruitment of enzymes as lens structural proteins. Science 236:1554–1556

Wistow G, Piatigorsky J (1988) Lens crystallins:the evolution and expression of proteins for a highly specialized tissue. Annu Rev Biochem 57:479–504

Zimmerman JL, Petri W, Meselson M (1983) Accumulation of a specific subset of *Drosophila melanogaster* mRNAs in normal development without heat shock. Cell 32:1161–1170

8 Regulation of Heat Shock Gene Expression During *Xenopus* Development

John J. Heikkila[1], Patrick H. Krone[1,2] and Nick Ovsenek[1,3]

1 Introduction

Heat shock-induced activation of the expression of a set of heat shock protein (*hsp*) genes giving rise to the accumulation of *hsp* mRNA and the synthesis of Hsps is characteristic of essentially all prokaryotic and eukaryotic cells (reviewed by Nover 1984: Atkinson and Walden 1985; Craig 1985; Lindquist 1986). However, in early stages of animal embryogenesis a heat shock response is not detectable. For example, during *Xenopus laevis* development heat shock-induced synthesis of Hsps does not occur in early cleavage stage embryos (Heikkila et al. 1985a ; Nickells and Browder 1985; Browder et al. 1989). Only immediately after the midblastula transition (MBT) of development, do embryos respond to heat shock by synthesizing Hsps with molecular weights of 87 (Hsp87) and 70 kd (Hsp70). Interestingly, the acquisition of the ability of the *Xenopus* embryos to synthesize hsps coincides with a dramatic increase in the acquisition of thermoresistance. This last finding is consistent with numerous studies suggesting that Hsp synthesis may play a role in the development of thermo-tolerance (Lindquist 1986). Also during *Drosophila* and sea-urchin development, heat-induced Hsp synthesis is not observed until embryos reach the blastoderm and blastula stage respectively (Dura 1981; Roccheri et al.1981; Heikkila et al. 1985a,b, 1986). Mouse and rabbit embryos are incompetent for heat-induced Hsp synthesis during the early cleavage stages of development but do synthesize Hsp70 at the blastocyst stage (Heikkila and Schultz 1984; Heikkila et al. 1985b). Given the similarity in the timing of the acquisition of the heat shock response in insect, echinoderm, amphibian and mammalian development, it is likely that this phenomenon has been conserved through evolution.

While a great deal of information is known about the activation of the *hsp* genes in somatic cells, relatively little is known about the mechanisms involved in the developmental regulation of these genes. In our studies of this phenomenon, we have chosen to work with *Xenopus* embryos since the eggs can be easily obtained in large quantities and fertilized in vitro. The large size of the eggs and oocytes (1 mm in diameter) also makes them ideal for microinjection studies (Heikkila 1990). Furthermore, a great deal of information is available about *Xenopus* embryonic development at the cell and molecular level. The following chapter will examine the results from our laboratory and others regarding the developmental regulation of *Xenopus hsp* genes with particular

[1] Department of Biology, University of Waterloo, Waterloo, Ontario N2L 3G1 Canada
[2] Fred Hutchinson Cancer Research Center, 1124 Columbia St., Seattle, WA 98104, USA
[3] Center for Developmental Biology, Department of Zoology, University of Texas, Austin, TX 78712, USA

Results and Problems in Cell Differentiation 17
Heat Shock and Development
Hightower and Nover (Eds.)
©Springer-Verlag Berlin Heidelberg 1991

emphasis on *hsp70* and a group of small molecular weight stress protein genes termed the *hsp30* family, which in contrast to *hsp70*, are not heat-inducible until the early tailbud stage of development.

2 Heat Shock-Induced Accumulation of Hsp and *Ubiquitin* mRNA in *Xenopus* Embryos is Developmentally Regulated

The acquisition of the ability to synthesize Hsp87 and 70 after the midblastula stage of *Xenopus* development reflects the heat shock-induced accumulation of *hsp* mRNAs (Fig. 1B; see Fig. 2B) (Bienz 1984a; Heikkila et al. 1985a; Heikkila et al. 1987b; Horell et al. 1987; Krone and Heikkila 1989). The levels of some preexisting messages such as actin mRNA are not appreciably affected under these conditions (Fig. 1C). As will be discussed below, the ability to synthesize *hsp* mRNA coincides with the activation of transcription of selected genes within the embryonic genome (Newport and Kirschner 1982a,b). Furthermore, exposure of *Xenopus* embryos to other stressors, such as sodium arsenite or ethanol, also induced a developmental stage-dependent accumulation of *hsp70* mRNA (Heikkila et al. 1987b). Thus, in addition to heat other stresses can also regulate *hsp* gene expression during development. It should be noted that low-level *hsp87* and *70* mRNAs are detectable in fertilized eggs and cleavage stage embryos and are probably of maternal origin (Heikkila et al. 1987b; Horell et al. 1987; Krone and Heikkila 1989).

Studies in chicken and yeast have shown that *ubiquitin* gene expression can be activated by heat shock (Bond and Schlesinger 1985; Ozkaynak et al. 1987). During *Xenopus* embryogenesis, heat shock also induced the accumulation of *ubiquitin* mRNA in a developmental stage-dependent pattern (Fig. 1A; Ovsenek and Heikkila 1988). While constitutive levels of *ubiquitin* mRNA were present throughout early develop-

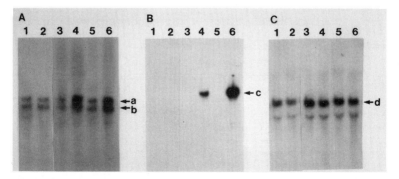

Fig. 1A-C. Developmental stage-dependent heat-induced accumulation of *hsp70* and *ubiquitin* mRNA during *Xenopus* development. Embryos were maintained at 22 °C or 33 °C for 1 h. **A** Total RNA (10 μg) was electrophoresed on formaldehyde-agarose gels, transferred to nitrocellulose, and then hybridized with the *ubiquitin* cDNA probe. **B** The blot was allowed to decay to background levels and then rehybridized with the *hsp70* probe. **C** This procedure was repeated and the blot was rehybridized with cytoskeletal *actin* cDNA. Transcript sizes: *a* 2.3 kb; *b* 2.6 kb; *c* 2.7 kb; *d* 2.3 kb. *Lane 1* 32-cell, 22 °C; *lane 2* 32-cell, 33 °C; *lane 3* late blastula, 22 °C; *lane 4* late blastula, 33 °C; *lane 5* neurula, 22 °C; *lane 6*, neurula, 33 °C (Ovsenek and Heikkila 1988)

ment, heat shock-induced accumulation of *ubiquitin* mRNA was not detectable until the late blastula stage (Fig. 1B). The biological significance of increased levels of ubiquitin in heat shocked embryos is not known. If heat shock results in an increased synthesis of abnormal or denatured protein in the embryo, increased synthesis of ubiquitin could facilitate its role in ATP-dependent proteolysis as suggested by Ciechanover et al. (1984)

In addition to *hsp87*, *hsp70* and *ubiquitin*, heat shocked *Xenopus* somatic and tissue culture cells transcribe a number of other heat shock genes, including a set of low molecular weight protein genes called the *hsp30* family (Bienz 1984a; Heikkila et al. 1987a; Darasch et al. 1988). This family of genes appears to be related to other low molecular-weight *hsp* genes found in other systems such as in *Drosophila* (Bienz 1984b). Interestingly, during *Xenopus* embryogenesis, we have found by RNase protection analysis that the *hsp30* genes are not heat-inducible until the early tailbud stage (stage 23–24) of development (Fig. 2A; Krone and Heikkila 1988;1989). Therefore, the regulation of *hsp30* gene expression during development differs dramatically from the expression of *hsp87*, *hsp70* and *ubiquitin* genes.

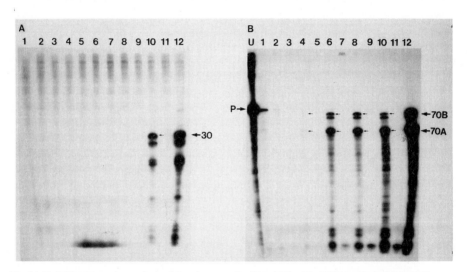

Fig. 2A,B. RNase protection analysis of A endogenous *hsp30* and B *hsp70* mRNA levels during early *Xenopus* development. Total RNA was isolated from embryos that had been maintained at either 22 °C or 33 °C for 1.5 h and 10 μg of this RNA was subjected to RNase protection analysis employing an *hsp30/cat* antisense RNA probe. The riboprobe that was used for detection of *hsp70* transcripts was obtained by SP6 RNA polymerase transcription of pG1 70H (Horrell et al., 1987). This 506 nucleotide antisense probe (*P*) protects a 476 nucleotide region of transcripts derived from the *hsp70B* gene as well as a 368 nucleotide region of mRNA transcribed from the *hsp70A* gene. Protected fragments were separated on either a 6% (*hsp30*) or 4% (*hsp70*) polyacrylamide/urea denaturing gels, *U* undigested *hsp70* probe; *lane 1* cleavage, 22 °C; *lane 2* cleavage, 33 °C; *lane 3* midblastula, 22 °C; *lane 4* midblastula, 33 °C; *lane 5* gastrula, 22 °C; *lane 6* gastrula, 33 °C; *lane 7* neurula, 22 °C; *lane 8* neurula, 33 °C;*lane 9* tailbud (stage 30–32), 22 °C; *lane 10*, tailbud (stage 30–32), 33 °C; *lane 11*, tadpole (stage 42), 22 °C; *lane 12*, tadpole (stage 42), 33 °C (After Krone and Heikkila 1989)

3 Pattern of Hsp and *Ubiquitin* mRNA Accumulation in Heat Shocked Embryos

The effect of temperature on *hsp70, hsp30* and *ubiquitin* mRNA accumulation in *Xenopus* embryos has been studied in detail (Heikkila et al. 1987b; Krone and Heikkila 1988). For example, the exposure of neurulae to temperatures ranging from 22–37 °C for 1 h revealed that *hsp70* mRNA accumulation was first detectable at 27 °C, with relatively greater levels at 30–35 °C and lower levels at 37 °C. A more complex effect of temperature on *hsp70* mRNA accumulation was observed in a series of time-course experiments (Fig. 3). While continuous exposure of embryos to heat shock induced a transient accumulation of *hsp70* mRNA, the temporal pattern of *hsp70* mRNA accumulation was temperature dependent. Exposure of embryos to 33–35 °C induced peak levels to Hsp70 mRNA within 1–1.5 h, whereas at 30 °C and 27 °C, maximum levels of *hsp70* mRNA accumulation occurred at 3 and 12 h, respectively. In recovery experiments following a 1–h heat shock at 33 °C the levels of *hsp70* mRNA decreased within 7 h (Heikkila et al.1987b). In a similar set of time-course and recovery experiments it was found that *ubiquitin* mRNA levels mimic the accumulation pattern of *hsp70* mRNA (Ovsenek and Heikkila 1988).

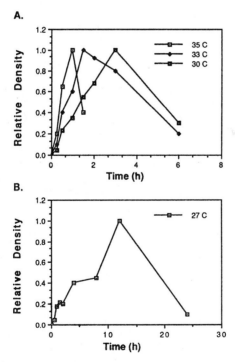

Fig. 3A,B. Time course of *hsp70* mRNA accumulation in *Xenopus* embryos at different temperatures. Neurula embryos were maintained at **A** 35, 33, and 30 °C; and **B** 27 °C for various periods of time. Total RNA (10 µg) was subjected to Northern hybridization analysis employing the labeled subclone of the *Xenopus hsp70* gene as a probe. The amount of *hsp70* mRNA in each lane was measured by densitometry of the autoradiograms and are expressed relative to the value (1.0) assigned to the most intense band in each experiment (After Heikkila et al. 1987b)

Since *hsp30* gene expression was found to be regulated quite differently during development compared to *hsp70* and *ubiquitin,* it was of interest to examine the pattern of expression at the tadpole stage when all these genes are heat inducible. Prolonged exposure of tadpoles to heat shock (33 °C) resulted in a coordinate accumulation of *hsp30, hsp70* and *ubiquitin* mRNA (Heikkila and Krone 1988). A coordinate temporal pattern of the three types of transcripts was also observed in embryos recovering from brief heat shock. Similar, coordinate patterns of *hsp30* and *hsp70* mRNA accumulation and decay during continuous temperature stress and recovery experiments have also been observed in *Xenopus* cultured kidney epithelial cells (Darasch et al. 1988). Thus although *hsp70, hsp30* and *ubiquitin* genes are inducible at different developmental stages, the genes in the tadpole are expressed coordinately during heat shock. In these studies the levels of certain preexisting mRNAs such as cytoskeletal and cardiac *actin* were unaffected by heat shock. Heat shock-induced stabilization has also been reported for control messages other than *actin* in *Drosophila* cultured cells (Storti et al. 1980), *Drosophila* pupae (Petersen and Mitchell 1982), and cultured tomato cells (Nover and Scharf 1984; Nover et al. 1989).

The transient pattern of *hsp30, hsp70* and *ubiquitin* mRNA accumulation observed in heat-shocked tadpoles may be mediated at the transcriptional and/or posttranscriptional level. A decrease in *hsp* mRNA stability could act in unison with decreased *hsp* gene transcription since it has been proposed that Hsp70 synthesis in *Drosophila* is autoregulated by a decrease in *hsp70* gene transcription as well as a destabilization of *hsp70* mRNA (Didomenico et al. 1982a,b). More recently, Peterson and Lindquist (1988,1989) have shown that regulation of *hsp70* mRNA degradation during recovery from heat shock in *Drosophila* is mediated by sequences present in the 3' noncoding region. This region of the mRNA is 75% AU-rich and contains two close matches to the UAUUUA consensus sequences which has been shown to confer instability to a number of mammalian messages such as *c-myc, c-fos,* and *interferon* (Shaw and Kamen 1986; Brawerman 1987; Vriz and Mechali 1989). Two of these mRNAs, *c-myc* and *c-fos,* have also been shown to be stabilized by heat shock (Andrews et al. 1987; Sadis et al. 1988). Peterson and Lindquist (1989) have proposed that *hsp70* mRNA degradation may be mediated by the same mechanism which is responsible for destabilizing these mRNAs and that the mechanism is nonfunctional at heat shock temperatures. Following heat shock, the mechanism would be reactivated and could recognize signals present in the 3' end of *hsp70* mRNA which target it for degradation. As will be discussed later, the 3' end of the *hsp30C* gene which we have recently isolated is also very AT-rich (76%) and the mRNA contains a perfect match to the UAUUUA consensus sequence. Thus degradation of *hsp* mRNAs during recovery from heat shock in *Xenopus* may be regulated in a similar fashion.

4 Involvement of Cis- and Trans-Acting Factors in the Developmental Regulation of Hsp70 Gene Expression

An effective technique for the examination of cis-acting DNA sequences involved in developmentally regulated transcription is the introduction of reporter gene constructions into developing embryos. This approach has proven effective in the identification of an enhancer element (Krieg and Melton 1985, 1987) as well as putative

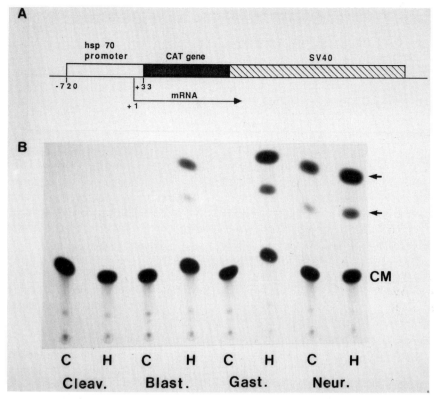

Fig. 4A,B. Expression of a microinjected *Xenopus hsp70/cat* fusion gene in *Xenopus* embryos. A *Xenopus laevis hsp70/cat* (pXL70CB) fusion gene containing 720 bp of upstream promoter sequence as well as 33 bp of DNA downstream of the *hsp70B* transcription initiation site. **B** CAT activity in *Xenopus* embryos microinjected with the *Xenopus hsp70/cat* fusion gene. CAT enzyme assays were performed using extracts from embryos maintained at either 22 °C (*C*) or 33 °C (*H*) for 1.5 h at the cleavage (*Cleav.*), blastula (*Blast.*), gastrula (*Gast.*) or neurula (*Neur.*) stages. Positive control was 0.05 units CAT. Acetylated forms of chloramphenicol (*CM*) are indicated by *arrows* (After Krone and Heikkila 1989)

inhibitory sequences (Steinbeisser et al. 1988) implicated in the developmental regulation of several *Xenopus* genes. In order to examine the mechanisms associated with the developmental stage-dependent expression of *hsp70* genes, a microinjected chimeric gene containing the *Xenopus hsp70B* promoter fused to the bacterial *chloramphenicol acetyl transferase (cat)* gene was microinjected into fertilized *Xenopus* eggs (Fig. 4A) and found to behave in a manner similar to the endogenous *hsp70* genes (Krone and Heikkila 1989). CAT enzyme assays revealed that heat-induced transcription was activated only after embryos reached the midblastula stage of development (Fig. 4B). Also, RNase protection analysis employing antisense riboprobes demonstrated that the *Xenopus hsp70/Cat* construct was correctly initiated (Fig. 5). These results indicate that the cis-acting sequences required for both heat-inducibility and activation of expression after the MBT are present on the *Xenopus hsp70* promoter. In a related study employing a human *hsp70/cat*, fusion gene which is constitutively expressed after the MBT, we found that a number of DNA sequence elements such as the CAAT,

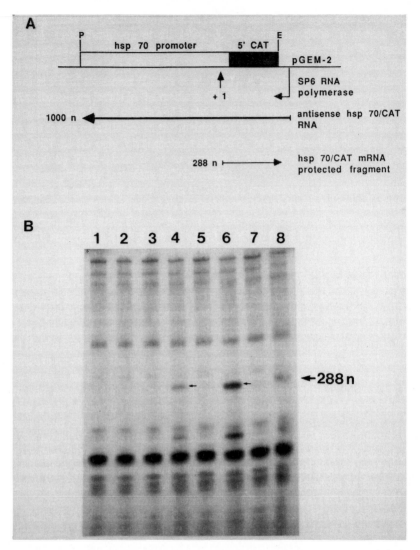

Fig. 5A,B. Transcription of the *hsp70-cat* fusion gene is correctly initiated in *Xenopus* embryos. **A** Antisense RNA probe utilized for RNase protection analysis of *Xenopus hsp70-cat* transcripts. The 1 kb Pst I-Eco RI fragment from *Xenopus hsp70-cat* which spans the transcriptional start size of the *hsp70* promoter was cloned into pGEM-2 to give the plasmid pG270CB. The 1000 nucleotide antisense RNA probe that is generated by SP6 RNA polymerase in vitro transcription of this construct protects 288 nucleotides of a correctly intiated *hsp70-cat* transcript. **B** RNase protection analysis of RNA isolated from control (22 °C for 1.5 h or heat shocked (33 °C for 1.5 h) embryos microinjected with the *hsp70-cat* fusion gene. Ten μg of total RNA was hybridized against the *hsp70-cat* antisense RNA probe, digested with RNase A and RNase T_1, and the digestion products separated on a 4% polyacrylamide-urea denaturing gel. The 288 nucleotide protected fragment of *hsp70-cat* mRNA is indicated by an *arrow*. *Lane 1* cleavage, control; *Lane 2* cleavage, heat shock; *Lane 3* late blastula, control; *Lane 4* late blastula, heat shock; *Lane 5* gastrula, control; *Lane 6* gastrula, heat shock; *Lane 7* neurula, control; *Lane 8* neurula, heat shock (After Krone and Heikkila 1989)

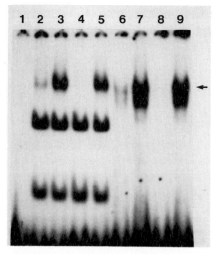

Fig. 6. Formation of an HSE-HSF complex is heat inducible in *Xenopus* embryos. DNA mobility shift assays were performed using 1 ng of end-labeled oligonucleotide corresponding to the proximal HSE of the *Xenopus hsp70B* gene promoter mixed with extracts from control or heat-shocked neurula stage embryos. Whole cell extracts were used in *lanes 2–5*, and nuclear extracts were used in *lanes 6–9*. *Lane 1* unbound probe. *Lanes 2* and *6* contain extracts from control embryos. *Lanes 3* and *7* contain extracts from heat shocked embryos (33 °C, 30 min). *Lanes 4* and *8*, contain heat shocked extracts with 50 ng of unlabeled HSE oligonucleotide. *Lanes 5* and *9* contain heat shocked extracts and 50 ng of CCAAT-PB competitor oligonucleotide. Specific complexes are indicated by the *arrow* (Ovsenek and Heikkila 1990)

purine-rich element, G-C element, ATF/API and TATA boxes were required for transcription (Ovsenek et al. 1990). Comparable deletion and linker-scanner mutant analysis employing the *Xenopus hsp70* promoter will undoubtedly reveal the sequence elements necessary for post-MBT heat-inducibility.

Heat-inducible synthesis of *hsp* mRNA is mediated in part by an interaction between the heat shock element (HSE; Pelham 1982; Pelham and Bienz 1982) and a transcription activating protein known as the heat shock factor (HSF). The *Xenopus hsp70B* promoter contains three HSEs which confer heat inducibility to a covalently linked β-globin gene in mammalian cells and *Xenopus* oocytes (Bienz and Pelham 1982; Bienz 1984b 1986). Therefore, a synthetic oligonucleotide corresponding to the proximal HSE of the *Xenopus* **hsp70B** gene was used in DNA mobility shift assays in an attempt to examine the basis for the stage-dependent expression of heat shock genes in *Xenopus* embryos. Is the developmental stage-independent expression of *hsp* genes related to HSF availability and/or inducibility? Initial studies revealed that sequence-specific HSE-binding activity was heat inducible in *Xenopus* embryonic extracts (Fig. 6). The heat inducibility of HSF binding in *Xenopus* embryos is in agreement with previous studies done with human, mouse, and *Drosophila* cells (Kingston et al. 1987; Sorger et al. 1987; Goldenberg et al. 1988; Mosser et al. 1988; Mezger et al. 1989), and more recently with *Xenopus* XTC cells (Zimarino et al. 1990). Interestingly, the presence of HSF binding was detected in heat shocked unfertilized eggs and cleavage stage embryos in which *hsp* gene expression is inhibited (Fig. 7).

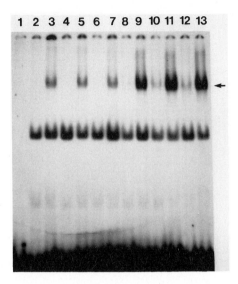

Fig. 7. Profile of HSF binding activity during early embryogenesis. A DNA mobility shift assay was performed with 1 ng of labeled HSE oligonucleotide mixed with whole cell extracts from control (22 °C) and heat shocked (33 °C), 30 min) embryos at different developmental stages. *Lane 1* unbound probe; *lane 2* control unfertilized eggs; *lane 3* heat shocked unfertilized eggs; *lane 4* control cleavage (stage 6); *lane 5* heat shocked cleavage; *lane 6* control early blastula (stage 8); *lane 7* heat shocked early blastula; *lane 8* control late blastula (stage 9); *lane 9* heat shocked late blastula; *lane 10* control gastrula (stage 10); *lane 11* heat shocked gastrula; *lane 12* neurula (stage 19); *lane 13* heat shocked neurula. The HSE-HSF complex is indicated by the *arrow* (Ovsenek and Heikkila 1990)

To date, the majority of experimental evidence indicates that there is essentially no mRNA transcription in pre-MBT embryos. Since a general inhibition or deactivation of maternally-derived transcription factors has not been excluded as possible mechanism for early transcriptional inhibition, the properties of HSF in cleavage stage and neurula stage embryos were compared. In time course experiments, in which both cleavage and neurula stage embryos were continuously exposed to heat shock temperatures, the activation of HSF was rapid and transient, returning to control levels by 2-3 h (Fig. 8A). Also, during recovery from a brief heat shock, HSF binding declined rapidly in both cleavage and neurula stage embryos (Fig. 8B). Thus the kinetics of HSF activation and deactivation appear to be similar in both pre- and post-MBT embryos. It is important to note that the effect of temperature on HSF activation as well as the temporal pattern of HSF activation and deactivation during continuous heat shock in *Xenopus* neurulae closely correlates with the studies examining *hsp70* mRNA levels (see Fig. 3; Heikkila et al. 1987b). This finding further supports the possibility that the pattern of *hsp70* mRNA accumulation in heat shocked *Xenopus* neurulae is regulated primarily at the transcriptional level.

The electrophoretic comigration of HSE-HSF complexes formed with extracts obtained before and after the MBT provided indirect evidence that the maternal form of HSF was similar in size to the zygotic form. This was confirmed by using a UV-crosslinking technique which demonstrates that the relative molecular mass of

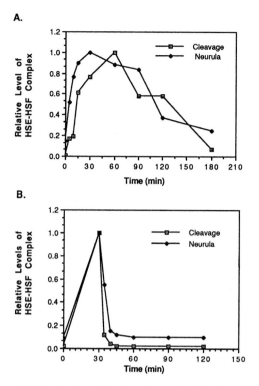

Fig. 8A,B. Time course of HSF activation during **A** continuous heat shock and **B** recovery in cleavage and neurula stage embryos. **A** Whole cell extracts were made from cleavage and neurula stage embryos incubated at the control temperature of 22 °C (time 0) or continually heated at 33 °C for the indicated time. **B** Cleavage and neurula stage embryos were treated at 22 °C (time 0) or at 33 °C for 30 min, and then allowed to recover at 22 °C. Relative HSE-HSF levels were measured by densitometry performed on autoradiograms of DNA mobility shift gels, and are expressed relative to the value (1.0) assigned to the most intensive band in each experiment (Ovsenek and Heikkila 1990)

Xenopus HSF is 88 kD in both cleavage and neurula stages (Fig. 9). The molecular mass of *Xenopus* HSF is similar to that reported for HSF in HeLa cells (93kDa; Larson et al. 1988), (83kD; Goldenberg et al. 1988), but smaller than in *Drosophila* SL-2 cells (110 kD; Wu et al. 1987).

The findings that both pre- and post MBT forms of HSF are similar in relative molecular mass and in the patterns of activation during time course and recovery experiments, indicate that maternal HSF may be similar or identical to the embryonic form of HSF. Thus, it appears that the mechanism involved in the stage-dependent transcription of heat shock genes, such as *hsp70* is not due to the absence of activatable HSF binding activity prior to the MBT. However, it is tenable that additional regulatory factors or a post-translational modification of HSF which may be necessary for transcriptional activation subsequent to HSE binding, are absent or blocked prior to the MBT.

It is possible that the mechanisms associated with the activation of selected genes at MBT also applies to the heat shock genes. In the model proposed by Newport and

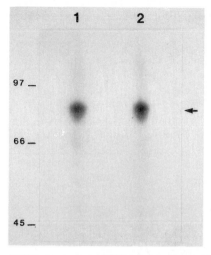

Fig. 9. Estimation of the size of HSF in pre- and post-MBT embryos. Comparative sodium dodecyl sulphate-polyacrylamide gel electrophoresis analysis of photoaffinity labeled HSF was performed using extracts from heat shocked cleavage and neurula stage embryos. Whole cell extracts from heat shocked cleavage stage (*lane 1*) and neurula stage (*lane 2*) embryos were UV-crosslinked to a [alpha-^{32}P]–labeled BrdU–substituted HSE oligonucleotide. The molecular size of *Xenopus* HSF was determined by comparison to the relative mobility of Coomassie blue-stained molecular weight markers (indicated on the *left*) which were run on the same gel (Ovsenek and Heikkila 1990)

Kirschner (1982a,b) and Kimelman et al. (1987), it was suggested that pre-MBT embryos are transcriptionally competent (i.e., contain functional transcription factors and RNA polymerase II) but are unable to synthesize RNA due to the rapid cell cycle. In addition to HSF, we have also identified the presence of RNA polymerase II transcription factors which bind to CCAAT, GC and ATF/AP-1-like sequences in pre-MBT embryos (Ovsenek et al. 1990). Assuming that HSF found in pre-MBT embryos is functional, it is likely that maternal HSF is involved in a heat shock response at the MBT. Given the role of Hsps in the acquisition of thermotolerance (Riabowol et al. 1988; Johnston and Kucey 1988), the obvious advantage of such a mechanism is that embryos would acquire the ability to transcribe *hsp* mRNA at the earliest possible time after the MBT.

Since the pool of cellular mRNAs which are stored in the oocytes are gradually replaced as *Xenopus* embryos become transcriptionally active, an intriguing question arises as to the time during development that zygotic HSF replaces the maternally derived protein. Because of the limited quantity of maternal factors, it is probable that HSF derived from transcription of the zygotic genome becomes active at some time shortly after the MBT. An increase in the level of activated HSF was detected by late blastula stage. However, it is unknown if these increased levels are due to (1) a more efficient activation of maternal HSF; (2) an increase in the translation from maternal mRNA encoding HSF; or (3) the activation of the zygotic HSF gene. Undoubtedly, this question and others will be answered with the eventual cloning of the *Xenopus* HSF gene and subsequent expression studies.

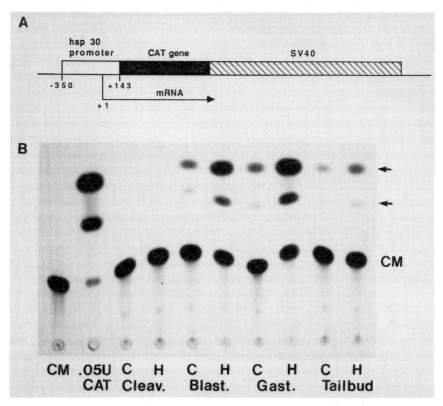

Fig. 10A,B. Expression of a microinjected *hsp70/cat* fusion gene in *Xenopus* embryos. **A** *Xenopus laevis hsp30/ cat* fusion gene (pXL30CB). This construct contains promoter sequences to 350 bp upstream of the transcription initiation site as well as 143 bp of *hsp30A* downstream DNA. **B** CAT activity in *Xenopus* embryos microinjected with the *hsp30/cat* fusion gene. Cleavage (*Cleav.*), blastula (*Blast.*), gastrula (*Gast.*), and tailbud embryos were maintained at either 22 °C (*C*) or 33 °C (*H*) for 1.5 h. Acetylated form of chloramphenicol (*CM*) are indicated by *arrows*. Positive control was 0.05 units Cat (Krone and Heikkila 1989)

5 Regulation of *hsp30* Gene Expression During Development

We have previously shown that heat shock-inducible *hsp30* gene expression is first detectable at the early tailbud stage of *Xenopus* embryogenesis (Krone and Heikkila 1989). A *Xenopus hsp30/cat* fusion gene was constructed (Fig. 10A) to determine if sequences that are required for the correct developmental regulation of endogenous *hsp30* gene expression are present in the promoter region of the *hsp30* alone genomic clone originally isolated by Bienz (1984a, b). The resultant construct contained approximately 350 bp of promoter sequence upstream of the transcription site as well as 143 bp of untranslated leader sequence. The construct was microinjected into fertilized eggs and expression during development was monitored by CAT assays and RNase protection analysis. While CAT activity was not detectable in control or heat shocked cleavage-stage embryos, heat shock induced a 50-100-fold increase in CAT activity levels in late blastula and gastrula stage embryos relative to controls (Fig. 10B). This result was surprising since the endogenous genes were not heat-activatable until the

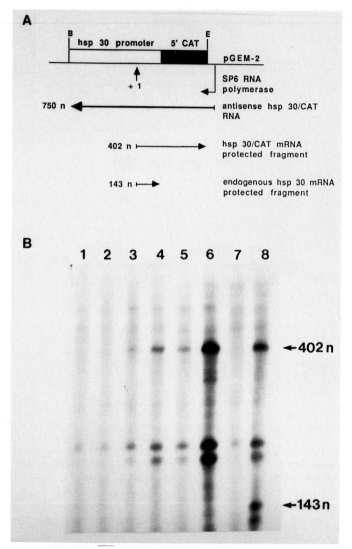

Fig. 11A,B. Transcription of the *hsp30/cat* fusion gene is correctly initiated in *Xenopus* embryos. **A** Antisense RNA probe utilized for RNase protection analysis of *hsp30/cat* and endogenous *hsp30* transcripts. The 750 bp Bam HI/Eco RI fragment from *hsp30/cat* spanning the transcriptional start site of the *hsp 30* promoter was cloned into pGEM-2 to give the plasmid pG230ČB. SP6 RNA polymerase in vitro transcription of this construct results in a 750 nucleotide antisense RNA probe which protects a 402 nucleotide correctly initiated *hsp30/cat* mRNA fragment as well as a 143 neuclotide portion of endogenous *hsp30* mRNA. **B** RNase protection analysis of *hsp30/cat* and endogenous *hsp30* mRNA in embryos microinjected with the *hsp30/cat* fusion gene. Ten μg of total RNA was isolated from either control (ss °C for 1.5 h) or heat-shocked (33 °C for 1.5 h) embryos, subjected to RNase protection analysis and separated on a 4% polyacrylamide/urea denaturing gel. *Arrows* indicate the positions of the 402 nucleotide *hsp30/cat* mRNA and 143 nucleotide endogenous *hsp30* mRNA protected fragments. *Lane 1* cleavage, control; *Lane 2* cleavage, heat shock; *Lane 3,* gastrula, control; *Lane 4* late blastula, heat shock; *Lane 5* gastrula, control; *Lane 6* gastrula, heat shock; *Lane 7* early tailbud, control; *Lane 8* early tailbud, heat shock (After Krone and Heikkila 1989)

tailbud stage. Furthermore, RNase protection analysis confirmed that the increases in CAT activity were the result of increased levels of correctly initiated *hsp-30/cat* transcription (Fig. 11). It was possible that a high copy number of h*sp30/cat* plasmid DNA might have overloaded the regulatory mechanism(s) controlling *hsp30* gene expression and caused premature transcription of the construct in pretailbud stage embryos. However, heat-inducible expression of *hsp30/cat* was detectable when as little as 10 pg (i.e., less than one copy per haploid genome equivalent at the gastrula stage) of plasmid had been injected. Studies with other genes such as human *globin* (Antoniou et al. 1988) and chicken *β-actin* (Lohse and Arnold 1988) have revealed transcriptional regulatory elements in the coding and 3' flanking region. In this context, a cis-acting regulatory element that is responsible for normal developmental expression may be located in the coding region or 3' end of the isolated *hsp30* genomic clone. Since such an element may interact with sequences present elsewhere on the gene and, moreover, correct spacing between such sequences may be crucial for proper developmental regulation, we examined the expression of the entire *hsp30* genomic clone in *Xenopus* embryos following microinjection into fertilized eggs. RNase protection analysis revealed that correct transcription of the entire construct was detectable in heat shocked gastrula stage embryos. Thus, neither *hsp30* coding sequences nor 5' and 3' flanking sequences present in the *hsp30* genomic clone isolated by Bienz (1984a, b) appear to be responsible for correct developmental regulation of heat-shock inducibility.

6 Isolation and Sequence Analysis of *hsp30* Genes from a *Xenopus laevis* Genomic Library

The results presented thus far indicate that any cis-acting DNA sequence elements which may be directly responsible for correct developmental regulation of *hsp30* gene expression are not present on the *hsp30A* genomic clone. Therefore, a *X. laevis* genomic library was screened using the *hsp30A* gene as a probe in order to isolate *hsp30* clones containing more extensive regions of flanking DNA (Krone et al. 1991). Five unique genomic clones were isolated, one of which, clone 29, was found to contain two complete *hsp30* genes, *hsp30C* and *hsp30D*, as well as a portion of a third gene, *hsp30E* (Fig. 12). Comparison of the DNA sequence of the *hsp30C* gene to the previously published sequence of *hsp30A* DNA (Bienz 1984b) revealed a high degree of similarity between the two genes. A total of only six mismatches are present in the promoter and coding DNA for which the sequences of both genes are known. A 2-bp insertion (5'-AC-3') in the promoter region as well as a 21-bp deletion (5'-AAGTGACACTGAGTGACACTG-3') in the coding region of *hsp30C* also distinguish it from the *hsp30A* gene. Thus the promoter and coding regions of the two genes are approximately 97% similar. In contrast, the *hsp30D* gene is only 75% similar to the aforementioned *hsp30* genes.

Transcription of the *hsp30C* gene, as determined by RNase protection analysis of tailbud RNA, is activated at the tailbud stage of development and is initiated at the same adenine residue at which *hsp30A* transcription begins (Bienz 1984b). This would be expected given that the promoters of both genes are virtually identical with respect to both identity and location of regulatory DNA sequence elements. Two TATA boxes,

Fig. 12. Restriction map of clone 29 genomic DNA insert. The size of the insert DNA is 14.3 kb. The *Eco* RI sites at either end of the genomic fragment (indicated by asterisk) are present within the polylinker regions of EMBL 4 DNA. *Arrows* indicate direction of transcription (After Krone et al., in prep.)

three HSE's (one single and two present as an overlapping doublet), and a downstream CCAAT box are all present in both genes at the same positions. The *hsp30C* gene also contains an HSE further upstream in the promoter region, but this element may also be present in the promoter region of the *hsp30A* gene, because this region of the *hsp30A* promoter was not contained in the clone isolated by Bienz (1984b). A polyadenylation signal (5'AATAAA-3') is located 121 bp downstream of the translational stop codon of the *hsp30C* gene. If termination and polyadenylation (200-300 adenine residues) occured 30 bp downstream from this signal, the resulting mRNA would have a length of approximately 1.1-1.2 kb. This agrees well with the size of *hsp30* mRNA determined by Northern blot analysis (Krone and Heikkila 1988). The 3' end of this message is AT-rich (76%; 119/156 bp) and also contains the sequences 5'UAUUUA-3', believed to be involved in regulation of mRNA stability (Shaw and Kamen 1986) and 5'UUUUUAU-3', which is involved in activation of polyadenylation during *Xenopus* oocyte maturation (Fox et al. 1989; McGrew et al. 1989). The region downstream of the putative mRNA coding sequence is also very AT-rich and contains two blocks of 6 and 25 direct repeated of the sequence 5'-AT-3'.

The *hsp30C* gene encodes a putative protein of 214 amino acids with a molecular mass of 24.2 kD. Interestingly, the protein encoded by the *hsp30A* gene is much smaller, being only 10 kD in size, but is encoded in the same reading frame as *hsp30C* and differs only by only 2 out of 90 amino acids. This truncated protein is formed due to the presence of 2 stop codons at positions 481 and 487 of *hsp30A* mRNA. These stop codons are present in a 21-bp section of the gene which is not present in the *hsp30C* gene and is generated by a triple repeat of the sequence 5'-AGTGACACTGA-3'. The putative 10-kD *hsp30A* protein also lacks the first 41 amino acids of *hsp30C* due to a transversion mutation which has converted the start codon used by the *hsp30C* gene into an ATT in the corresponding location of the *hsp30A* gene. It should be noted that the truncated *hsp30A* is not necessarily without a function. Perhaps it is analogous to one domain of the full sized protein which continues to function on its own.

Activation of the endogenous *hsp30* genes at the tailbud stage may occur either by means of a positive regulator or removal of an inhibitory system. The fact that the injected *hsp30* gene was expressed in both control and heat shocked pretailbud-stage embryos suggests that an inhibitory system may be responsible for regulation of the endogenous genes. Specific transcriptional repression of eukaryotic promoters is a topic which has become the focus of research in recent years (reviewed by Levine and Manley 1989). Although the results presented here suggest that a negative regulatory element seems unlikely within the 5' and 3' flanking regions or within the coding sequence of the *hsp30A* genomic clone that was used in the microinjection studies, they do not rule out the presence of such elements further upstream or downstream of the gene. Microinjection of promoter fusion constructs containing the entire *hsp30C* or *hsp30D* promoter region into *Xenopus* embryos and analysis of expression during development may help in unravelling this problem. At present, the chromosomal localization of the *hsp30* cluster is not known. Such a clustered structure could be in the developmental regulation of the *hsp30* genes given the similar linked arrangement of the *Drosophila* small heat shock genes, which exhibit a complex pattern of developmental regulation (Dura 1981; Sirotkin and Davidson 1982; Zimmerman et al. 1983; Mason et al. 1984). It is possible that a mechanism involving activation through a change in chromatin conformation (Goldman 1989) would be an efficient way by which all of the *Xenopus hsp30* genes could be coordinately regulated.

Acknowledgments. This research has been supported by a Natural Sciences and Engineering Research Council (NSERC) of Canada grant to J.J.H. P.H.K. and N.O. were the recipients of an NSERC and Ontario Graduate Scholarship, respectively.

References

Andrews GK, Harding MA, Calvet JP, Adamson ED (1987) The heat shock response in HeLa cells is accompanied by elevated expression of the *c-fos* proto-oncogene. Mol Cell Biol 7:3452–3458

Antoniou M, deBoer E, Habets G, Grosveld F (1988) The human ß-globin gene contains multiple regulatory regions:identification of one promoter and two downstream enhancers. EMBO J 7:377–384

Atkinson BG, Walden DB (eds) (1985) Changes in eukaryotic gene expression in response to environmental stress. Academic Press, Orlando, pp

Bienz M (1984a) Developmental control of the heat shock response in *Xenopus*. Proc Natl Acad Sci USA 81:3138–3142

Bienz M (1984b) *Xenopus* hsp70 genes are constitutively expressed in injected oocytes. EMBO J 3:2477–2483

Bienz M (1986) A CCAAT box confers cell -type-specific regulation on the *Xenopus* hsp70 gene in oocytes. Cell 46:1037–1042

Bienz M, Pelham HRB (1982) Expression of a *Drosophila* heat shock protein in *Xenopus* oocytes:conserved and divergent regulatory signals. EMBO J 1:1583–1588

Bond U, Schlesinger MJ (1985) Ubiquitin is a heat shock protein in chicken embryo fibroblasts. Mol Cell Biol 5:949–956

Brawerman G (1987) Determinants of messenger RNA stability. Cell 48:5–6

Browder, LW, Pollock M, Nickells RW, Heikkila JJ, Winning RS, (1989) Developmental regulation of the heat shock response. In:DiBeradino MA, Etkin LD (eds) Developmental biology. A comprehensive synthesis. Vol. 6. Adaptability in somatic cell specialization. Plenum, New York, pp 97–147

Ciechanover A, Finley D, Varshavski A (1984) The ubiquitin mediated proteolytic pathway and mechanisms of energy-dependent intracellular protein degradation. J Cell Biochem 24:27–53

Craig EA (1985) The heat shock response. CRC Crit Rev Biochem 18:239–280

Darasch SP, Mosser DD, Bols NC, Heikkila JJ (1988) Heat shock gene expression in *Xenopus laevis* A6 cells in response to heat shock and sodium arsenite treatments. Biochem Cell Biol 66:862–870

Didomenico BJ, Bugaisky GE, Lindquist S (1982a) Heat shock and recovery are mediated by different translational mechanisms. Proc Natl Acad Sci USA 79:6181–6185

Didomenico BJ, Bugaisky GE, Lindquist S (1982b) The heat shock response is self regulated at both the transcriptional and post transcriptional levels. Cell 31:593–603

Dura JM (1981) Stage dependent synthesis of heat shock induced proteins in early embryos of *Drosophila melanogaster*. Mol Gen Genet 184:381–385

Fox CA, Sheets MD, Wickens MP (1989) Poly (A) addition during maturation of frog oocytes:distinct nuclear and cytoplasmic activities and regulation by the sequence UUUUUAU. Genes Dev 3:2151–2162

Goldenberg CJ, Luo Y, Fenna M, Baler R, Weinmann R, Voellmy R (1988) Purified human factor activates heat shock promoter in a HeLa cell-free transcription system. J Biol Chem 263:19734–19739

Goldman MA (1988) The chromatin domain as a unit of gene regulation BioEssays 9:50–55

Heikkila JJ (1990) Expression of cloned genes and translation of messenger RNA in microinjected *Xenopus* oocytes. Int J Biochem 22:1223–1228

Heikkila JJ, Schultz GA (1984) Different environmental stresses can activate the expression of a heat shock gene in rabbit blastocyst. Gamete Res 10:45–56

Heikkila JJ, Kloc M, Bury J, Schultz GA, Browder LW (1985a) Acquisition of the heat shock response and thermotolerance during early development of *Xenopus laevis*. Dev Biol 107:483–489

Heikkila JJ, Miller JGO, Schultz GA, Kloc M, Browder LW (1985b) Heat shock gene expression during early animal development. In:Atkinson BG, Walden DB (eds) Changes in eukaryotic gene expression in response to environmental stress. Academic Press, Orlando, pp 135–158

Heikkila JJ, Browder LW, Gedamu L, Nickells RW, Schultz GA (1986) Heat-shock gene expression in animal embryonic systems. Can J Genet Cytol 28:1093–1105

Heikkila JJ, Darasch SP, Mosser DD, Bols NC (1987a) Heat and sodium arsenite act synergistically on the induction of heat shock gene expression in *Xenopus laevis* A6 cells. Biochem Cell Biol 65:310–316

Heikkila JJ, Ovsenek N, Krone P (1987b) Examination of heat shock protein mRNA accumulation in early *Xenopus laevis* embryos. Biochem Cell Biol 65:87–94

Horrell A, Shuttleworth J, Colman A (1987) Transcript levels and translational control of hsp70 synthesis in *Xenopus* oocytes. Genes Dev 1:433–444

Johnston RN, Kucey BL (1988) Competitive inhibition of hsp70 gene expression causes thermosensitivity. Science 242:1551–1554

Kimelman D, Kirschner M, Scherson T (1987) The events of the midblastula transition in *Xenopus* are regulated by changes in the cell cycle. Cell 48:399–407

Kingston RE, Schueltz TJ, Larin Z (1987) Heat-inducible human factor that binds to a human hsp70 promoter. Mol Cell Biol 7:1530–1534

Krieg PA, Melton DA (1985) Developmental regulation of a gastrula-specific gene injected into fertilized *Xenopus* eggs. EMBO J 4:3463–3471

Krieg PA, Melton DA (1987) An enhancer responsible for activating transcription at the midblastula transition of *Xenopus* development. Proc Natl Acad Sci USA 84:2331–2335

Krone P, Heikkila JJ (1988) Analysis of hsp 30, hsp 70 and ubiquitin gene expression in *Xenopus laevis* tadpoles. Development 103:59–67

Krone P, Heikkila JJ (1989) Expression of microinjected hsp70/CAT and hsp70/CAT chimeric genes in developing *Xenopus laevis* embryos. Development 106:271–281

Krone PH, Snow A, Ali A, Pasternak JJ, Heikkila JJ (1991) Comparison of regulatory and structural regions of the *Xenopus laevis* small heat shock protein gene family. Gene (in press)

Larson JS, Schuetz TJ, KIngston RE (1988) Activation in vitro of sequence-specific DNA binding by a human regulatory factor. Nature 335:372-375

Levine M, Manley JL (1989) Transcriptional repression of eukaryotic promoters. Cell 59:405–408

Lindquist S (1986) The heat-shock response. Annu Rev Biochem 55:1151–1191

Lohse P, Arnold HH (1988) The down-regulation of the chicken cytoplasmic ß-actin during myogenic differentiation does not require the promoter but involves the 3' end of the gene. Nucl Acids Res 16:2787–2803

Mason PJ, Hall LMC, Gausz J (1984) The expression of heat shock genes during normal development in *Drosophila melanogaster*. Mol Gen Genet 194:73–78

McGrew LL, Dworkin-Rastl E, Dworkin M, Richter JD (1989) Poly (A) elongation during *Xenopus* oocyte maturation is required for translational recruitment and is mediated by a short sequence element. Genes Dev 3:803–815

Mezger V, Bensaude O, Morange M (1989) Unusual levels of heat shock element-binding activity in embryonal carcinoma cells. Mol Cell Biol 9:3888–3896

Mosser DD, Theodorakis NG, Morimoto RI (1988) Coordinate changes in heat shock element binding activity and HSP70 gene transcription rates in human cells. Mol Cell Biol 8:4736–4744

Newport J, Kirschner M (1982a) A major developmental transition in early *Xenopus* embryos:I. Characterization and timing of cellular changes at the midblastula stage. Cell 30:675–686

Newport J, Kirschner M (1982b) A major developmental transition in early *Xenopus* embryos:II Control of the onset of transcription Cell 30:687–696

Nickells RW, Browder LW (1985) Region-specific heat-shock protein synthesis correlates with a biphasic acquisition of thermotolerance in *Xenopus laevis* embryos. Dev Biol 112:391–395

Nover L (1984) Heat shock response of eukaryotic cells. Springer-Berlin Heidelberg New York Tokyo

Nover L, Scharf K-D (1984) Synthesis, modification and structural binding of heat-shock proteins in tomao cell cultures. Eur J Biochem 139:303–313

Nover L, Scharf K-D, Neumann D (1989) Cytoplasmic heat shock granules are formed from precursor particles and are associated with a specific set of mRNAs. Mol Cell Biol 9:1298-1308

Ovsenek N, Heikkila JJ (1988) Heat shock-induced accumulation of ubiquitin mRNA in *Xenopus laevis* is developmentally regulated. Dev Biol 129:582–585

Ovsenek N, Heikkila JJ (1990) DNA sequence-specific binding activity of the heat shock transcription factor is heat-inducible before the midblastula transition of early *Xenopus* development. Development 110:427–433

Ovsenek N, Williams GT, Morimoto RI, Heikkila JJ (1990) Cis-acting sequences and trans-acting factors required for constitutive expression of a microinjected hsp70 gene after the midblastula transition of *Xenopus laevis* embryogenesis. Dev Genet 11:97–109

Ozkaynak E, Finley D. Solomon MJ, Varshavsky A (1987) The yeast ubiquitin genes:a family of natural gene fusions. EMBO J 6:1429–1439

Pelham HRB (1982) A regulatory upstream promoter element in *Drosophila* hsp70 heat-shock gene. Cell 30:517–528

Pelham HRB, Bienz M (1982) A synthetic heat-shock promoter element confers heat–inducibility on the herpes simplex virus thymidine kinase gene. EMBO J 1:1473–1477

Petersen NS, Mitchell HK (1982) Effects of heat shock on gene expression during development:induction and prevention of multihair phenocopy in *Drosophila*. In:Schlesinger MJ, Ashburner M, Tissieres A (eds) Heat shock:from bacteria to man. Cold Spring Harbour Laboratory, Cold Spring Harbour, NY, pp 345–352

Peterson R, Lindquist S (1989) The *Drosophila* hsp70 message is rapidly degraded at normal temperatures and stabilized by heat shock. Gene 72:161–168

Peterson R, Lindquist S (1989) Regulation of hsp synthesis by messenger RNA degradation. Cell Reg 1:135–149

Riabowol, KT, Mizzen LE., Welch WJ (1988) Heat shock is lethal to fibroblasts microinjected with antibodies against hsp70. Science 242:433–436

Roccheri MC, DiBernardo MG, Giudice G (1981) Synthesis of heat shock proteins in developing sea urchins. Dev Biol 83:173–177

Sadis S, Hickey E, Weber LA (1988) Effect of heat shock on RNA metabolism in HeLa cells. J Cell Physiol 135:377–386

Shaw G, Kamen R (1986) A conserved AU sequence from the 3' untranslated region of GM-CSF mRNA mediates selective mRNA degradation. Cell 46:659–667

Sirotkin K, Davidson N (1982) Developmentally regulated transcription from *Drosophila melanogaster* chromosomal site 67B. Dev Biol 89:196–210

Sorger PK, Lewis MJ, Pelham HRB (1987) Heat shock factor is regulated differently in yeast and HeLa cells. Nature 329:81–84

Steinbeisser H, Hofmann A, Stutz F, Trendelenberg MF (1988) Different regulatory elements are required for cell-type and stage specific expression of the *Xenopus laevis* skeletal muscle actin gene upon injection in *X. laevis* oocytes and embryos. Nucl Acids Res 16:3223–3238

Storti RV, Scott MP, Rich A, Pardue ML (1980) Translational control of protein synthesis in response to heat shock in *D. melanogaster* cells. Cell 22:825–834

Vriz S, Mechali M (1989) Analysis of 3'-untranslated regions of 7 *c-myc* genes reveals conserved elements prevalent in post-transcriptionally regulated genes. FEBS Lett 251:201–206

Wu C, Wilson S, Walker B, Dawid I, Paisley T, Zimarino V, Ueda H (1987) Purification and properties of *Drosophila* heat shock activator protein. Science 238:1247–1253

Zimarino V, Tsai C, Wu C (1990) Complex modes of heat shock factor activation. Mol Cell Biol 10:752–759

Zimmerman JL, Petri W, Meselson M (1983) Accumulation of a specific subset of D. melanogaster heat shock mRNAs in normal development without heat shock. Cell 32:1161–1170

9 Heat Shock Gene Expression During Mammalian Gametogenesis and Early Embryogenesis

Debra J. Wolgemuth and Carol M. Gruppi[1]

1 Introduction

1.1 Molecular Approaches to Studying Mammalian Germ-Cell Development

The progression of mammalian gametogenesis through mitotic and meiotic divisions, genetic recombination, and morphological differentiation involves varied and complex processes of cellular differentiation. Our laboratory has been interested in elucidating the genetic program that governs these events during mammalian germ-cell differentiation at the molecular level. This effort has been made increasingly feasible by advances in molecular biological approaches that facilitate the identification of specific genes which may play key roles in these processes. The advances have been at both the technical and strategic levels. For example, the use of techniques such as the polymerase chain reaction (PCR) has permitted the detection of specific gene expression in a very limited number of cells, such as in oocytes and early embryos (Rappolee et al. 1988). PCR can also be applied to the cloning of cDNAs from these stages (Welsh et al. 1990), allowing the identification and characterization of new genes which may be critical for gametogenesis and early embryogenesis.

The development of strategies to facilitate the identification of genes expressed at specific stages of germ cell differentiation and embryonic development has also been important. These strategies can be classified into three general approaches for the purposes of this discussion:

The first approach has entailed examining the expression of genes which are active in variety of tissues, including germ cell-containing tissues, but which exhibit unique patterns of expression at the molecular or cellular level that would suggest an important role in gametogenesis or early embryogenesis. Of particular interest in this category have been genes with potential function during cellular differentiation, such as proto-oncogenes, growth factors and their receptors, and cell-cycle genes. For example, the proto-oncogenes *c-abl* and *c-mos* have been suggested to have important roles in both male and female germ-cell differentiation because of their uniquely sized transcripts and the stage specificity of their expression in germ cells (e.g., Ponzetto and Wolgemuth 1985; Goldman et al. 1987; Propst et al. 1987; Mutter and Wolgemuth 1987; Mutter et al. 1988).

[1]Department of Genetics and Development and The Center for Reproductive Sciences, Columbia University College of Physicians and Surgeons, 630 West 168th Street, New York, NY 10032 USA

Results and Problems in Cell Differentiation 17
Heat Shock and Development
Hightower and Nover (Eds.)
© Springer-Verlag Berlin Heidelberg 1991

A second approach has involved the identification of genes expressed uniquely in germ cells. Such cell-specific genes can be isolated by subtractive hybridization screening of cDNA libraries and by enhancer-trap and gene-trap cloning in transgenic animal systems. The subtractive hybridization approach has been used by several laboratories studying early frog embryogenesis to identify genes expressed in stage- and spatially-regulated patterns (Rebagliati et al. 1985; Weeks et al. 1985). Testis-specific genes have also been isolated by differential screening and subtractive hybridization strategies, several of which have subsequently been shown to exhibit cellular specificity of expression within the germ line (e.g., Thomas et al. 1989).

A potentially more powerful method of identifying germ cell stage-specific genes involves the use of gene-trap or enhancer-trap approaches (Bellen et al.,1989; Gossler et al. 1989; Grossniklaus et al. 1989). In this approach, a readily expressed reporter gene, such as the bacterial *lacZ* gene, is inserted into constructs designed such that the expression of beta-galactosidase (ß-gal) in the resulting transgenic animals requires insertion of the construct within a gene (gene trap) or in the vicinity of the 5' regulatory region (enhancer trap) of a gene. The screening method involves selecting among the transgenic animals for those expressing the inserted ß-gal gene in a developmentally regulated pattern within the male (or female) germ line. Unlike subtractive hybridization, this technique may be used to identify genes expressed at low levels but which may be critical in regulatory functions. The enhancer-trap and gene-trap approaches have been widely used in the *Drosophila* system, utilizing P-element-mediated gene transformation, to identify genes expressed in the maternal germ line and during specific stages of embryogenesis (Bellen et al. 1989; Grossniklaus et al. 1989). Although there have been fewer applications of these approaches in the mammalian developmental system, most likely due to the highly specialized and labor intensive nature of generating transgenic mice, there is great potential for isolating genes important in development with this strategy (e.g. Kothary et al. 1988; Gossler et al. 1989).

Our laboratory also has used a third strategy to isolate genes with possible development-regulating function in the germ line. This approach utilizes genes with known development-regulating function in simpler systems to screen for related genes in the genomes of more complex organisms. For example, we used a gene containing a homeobox, a region of 61 amino acids, which is conserved among several *Drosophila* genes that are known to be important in regulating development and differentiation. Observations of the expression of homeobox-containing genes during oogenesis in lower vertebrates and in mammalian embryonal carcinoma cell lines of germ cell origin prompted us to consider if genes containing homeobox sequences were expressed during mammalian male germ-cell development (discussed in Wolgemuth et al. 1986). To test this hypothesis, a mouse testis cDNA library was screened with cloned homeobox-containing sequences from the *Antennapedia* gene of *Drosophila* under conditions of reduced stringency (Wolgemuth et al. 1986). We isolated a cDNA and subsequently the mouse homeobox-containing gene *Hox-1.4*, and observed interesting specificity of expression of this gene both in the adult male germ line and during embryogenesis. *Hox-1.4* has been shown to exhibit a highly restricted pattern of expression in the adult male germ line, being expressed as a distinctly sized transcript (different from the embryonic transcript) only after the germ cells have entered into meiosis (Wolgemuth et al. 1986,1987).

Each of the above approaches has been utilized in some manner in the study of the genes that form the focus of this Chapter, namely the heat shock genes. Members of the heat-shock gene families are expressed virtually ubiquitously in mammalian cells, but have been shown to exhibit molecular and cellular specificity of expression in the germ line. We have identified a unique member of the murine Hsp70 family whose expression is highest in spermatogenic cells in the pachytene stage of meiotic prophase (Zakeri et al. 1988). This gene would likely have been isolated using the differential screening or subtractive hybridization strategies outlined in the second approach, given its high levels and cellular specificity of expression. Finally, and as will be discussed in greater detail in this Chapter, the high level sequence similarity among the heat shock genes has facilitated the molecular cloning of multiple members of the gene families in a wide variety of species.

1.2 Brief Background and Classification of Hsp

Prokaryotic and eukaryotic organisms synthesize a distinct set of proteins in response to exogenous stress. These proteins are designated heat shock proteins (Hsp), although their synthesis can be induced by a variety of other conditions and agents such as anoxia, heavy metals, and viral infection (reviewed in Craig 1985; Lindquist 1986). The expression of *hsp* genes and their corresponding proteins is not limited to a stress response, since the genes for Hsp and Hsp-related family members are also induced during normal development and differentiation. Similarly to the situation for the bona fide heat shock response, developmental induction of the Hsp can be found in a wide variety of organisms, including flies, frogs, mice and humans (Craig 1985).

To facilitate reference to the Hsp, they have been divided into three families based on molecular weight (Pardue 1988): (1) the Hsp20 family, with hsp in the molecular weight range of 15-30 kD; (2) the Hsp70 family, which includes proteins that range in size between 65-75 kD; and (3) the Hsp90 family, with proteins in the range of 80-90 kD. The apparent molecular weights and the number of family members varies among species and even among family members of the same species. For example, the *Drosophila* genome contains seven *hsp* genes, *hsp83, hsp70, hsp68, hsp27, hsp26, hsp23,* and *hsp22.* Whereas the *hsp90* family has *hsp83* as its only member, the *hsp70* family consists of *hsp70* and *hsp68,* and the *hsp20* family contains *hsp27, hsp26, hsp23* and *hsp22.* In contrast, in mouse both the *hsp70* and *hsp90* families are multigene families. The murine *hsp70* family appears to contain at least five genes (Lowe and Moran 1986) while the murine *hsp90* family contains eight genes (Moore et al. 1987). However, only two of the genes in the *hsp90* family, *hsp86* and *hsp84,* encode functional proteins; the remaining genes are apparently pseudogenes (Moore et al. 1989). Two distinct *hsp90* genes have also been described in man: *hsp89α* and *hsp89β* (Rebbe et al. 1987; Hickey et al. 1989). Comparison of the nucleotide and amino acid sequences indicates that *hsp86* and *hsp84* are the mouse orthologues of the human *hsp89α* and *hsp89β* genes respectively (Hickey et al. 1989; Moore et al. 1989).

Although they will not be discussed in detail with respect to their expression in mammalian development in this review, it should be noted that there is another class of stress-induced proteins known as the glucose-regulated proteins (Grp). Grp are related to Hsp but are induced by glucose deprivation of cells, rather than by heat shock (rev. by Pelham 1986). As is the case for the Hsp, Grps are also expressed during normal growth and development in a wide variety of cell types (Pelham 1986). Grp are localized in the endoplasmic reticulum of nonstressed cells where they may also mediate protein-protein interaction, as has been proposed for the other Hsp (Pelham 1986). Grps are often expressed in secretory cells (Pelham 1986). For example, the mouse Sertoli cell line Tm4 expresses Grp78 at a very low basal level and at much higher levels in the presence of testosterone (Day and Lee 1989).

1.3 Significance of Conservation of Coding and Regulatory Regions in *hsp* Genes

Although the molecular weights of the Hsp vary within a family, they exhibit a high degree of structural conservation at both the nucleotide and amino acid level. Homology among the heat shock families is not limited to the coding region of the *hsp* genes, but also extends to the 5' regulatory region. Consensus sequences 5' to the promoter are required for proper stress and-or developmental induction of the *hsp* genes. This consensus sequence (5'CTnGAAnnTTCnAG3') is termed a heat shock element (HSE) and has been shown to be present in multiple copies in the 5' region of hsp genes of all species investigated (Pelham 1985; Lindquist 1986).

Deletion analysis of the 5' regulatory region of the *hsp26* gene of *Drosophila* has shown that different regulatory elements are required for its proper developmental and heat induction (Cohen and Meselson 1985). Sequences upstream of -341 are required for ovarian expression of *hsp26* but not for heat induced expression. Regions between -351 and -53 are required for heat induced expression but not for ovarian expression. *Hsp26* expression in *Drosophila* spermatocytes also requires 5' regulatory elements (Glaser et al. 1986). A *hsp26-lacZ* fusion construct that contained 2 kb of the 5' flanking region of the *hsp26* gene was introduced into the germ line of *Drosophila* by P-element mediated transformation. Expression of β-galactosidase (ß-gal) was observed in nurse cells and spermatocytes; this pattern is identical to endogenous *hsp26* expression. A second construct containing only 278 bp of 5' flanking region was able to direct expression of β-gal specifically to spermatocytes. These results imply that distinct 5' regulatory elements are involved in the expression of *hsp26* in the germ line of *Drosophila*.

Although not as well characterized, multiple 5' regulatory elements are also required for proper heat and developmental induction of heat shock genes in mammals. Analysis of transgenic mice carrying a mouse *hsp68-lacZ* construct revealed that there are distinct 5' regions responsible for proper heat and developmental induction of the mouse gene as well (Kothary et al. 1989). Approximately 800 bp of the 5' flanking region of *hsp68*, a region known to contain five HSE, was sufficient for proper heat induction of the *hsp68* gene but was not sufficient for developmental induction (Kothary et al. 1989).

1.4 Rationale for Examining Hsp Expression and Function in Germ Cells

Hsp have been shown to be expressed during development of the germ cell lineage in a wide variety of organisms. In yeast, Hsp26 is induced strongly during meiosis and ascospore formation (Kurtz et al. 1986). Hsp83, Hsp28 and Hsp26 are produced in the ovarian nurse cells of *Drosophila* and subsequently transported to the oocyte

(Zimmerman et al. 1983). As alluded to above, Hsp26 is expressed in the male germ cell lineage in *Drosophila* as well, specifically in spermatocytes (Glaser et al. 1986). As will be discussed in detail in this review, expression of members of both the *hsp70* and *hsp90* gene families has been shown in both the male and female mammalian germ cell lineages. For example, two members of the *hsp70* gene family have been shown to be expressed in a developmentally regulated pattern in germ cells of the mouse testis (Zakeri and Wolgemuth 1987; Zakeri et al. 1988). The expression of hsp genes in germ cells likely extends to other mammalian species and to the low molecular weight *hsp27* gene as well, since we have observed that members of the *hsp70* and *hsp20* families are expressed in the human testis (Zakeri and Wolgemuth 1987; C.M. Gruppi and D.J. Wolgemuth, unpublished observations).

The conservation of Hsp expression during oogenesis and spermatogenesis in evolutionary diverse organisms suggests a fundamental role for Hsp during germ cell development. The effect of hyperthermia on male germ cells has been recognized since the time of Hippocrates (cited in Rock and Robinson 1965). Histological studies in the early part of this century have demonstrated that it is the germinal compartment of the testis that is damaged by heat and that there may be a differential sensitivity to this damage within the germ cell populations (Moore and Chase 1923; Young 1927). For example, spermatogonia and spermatozoa appear to be the most resistant, whereas primary spermatocytes and early spermatids are the most sensitive (Young 1927).

The female germ line, like the mammalian male germ line, is also sensitive to hyperthermic stress. Ova exposed to hyperthermic stress during meiotic maturation often exhibit an atypical and degenerate morphology (Baumgartner and Chrisman 1981). Abnormalities include multinuclear eggs and an increase in size of the first polar body (Baumgartner and Chrisman 1981). In vitro, elevated temperature reduce the number of oocytes proceeding to metaphase II and decrease rates of fertilization (Lenz et al. 1983).

Mammalian embryos also incur developmental abnormalities when subjected to hyperthermia. High temperatures increase the incidence of abnormal cleavage in preimplantation embryos, causing retarded fetal development (Bellve 1972 1973; Ulberg and Sheean 1973) and birth defects (Webster and Edwards 1984; Germain et al. 1985).

2 Heat-Shock Gene Expression and Function During Mammalian Spermatogenesis

2.1 Key Features of Mammalian Spermatogenesis

Spermatogenesis, in particular in the mouse, provides a useful model system for identifying the genetic basis of cellular differentiation within a defined mammalian cell lineage. There are several features of spermatogenesis in the mouse that facilitate its use as a model system for the study of lineage-specific gene expression in mammals. The adult mouse testis contains virtually all the stages of germ cell differentiation, from mitotic stem cells to mature spermatozoa. However, since there is a gradual temporal progression of spermatogenic cell types during fetal and postnatal development (Nebel et al. 1961), cells at later stages of differentiation can be eliminated by utilizing animals at specific stages of fetal or postnatal development. Enriched populations of germ cell types can be prepared using simple and readily reproducible cell separation techniques (e.g. Bellve et al. 1977; Wolgemuth et al. 1985). These enriched cellular fractions can serve as a source of cells for molecular and biochemical analyses. There are also several sterile male mouse mutants in which germ cell development is arrested at different stages (Green 1981). Gonads can be obtained from animals homozygous or heterozygous for these mutations and used as a source of tissue containing (or lacking) specific somatic and-or germinal cells. Finally, the transgenic mouse system can be used to manipulate the expression of specific genes within the germ line, thus enabling an in vivo test system for detemining the molecular basis of the regulation of their expression and ultimately, of their function. As will be discussed in greater detail in the last section of this chapter, spermatogenic stage-specific promoters can be used to direct constructs in which *overexpression, ectopic expression*, and interference with expression could result.

2.2 Expression of the *hsp70* Gene Family

Two *hsp70* family members have been shown to exhibit unique patterns of expression during male germ cell differentiation in the mouse. An abundant 2.7 kb *hsp70* transcript is detected in the adult mouse testis in the absence of exogenous stress (Zakeri and Wolgemuth 1987). This *hsp70* transcript is unique to the testis and distinct from the *hsp70* transcripts of 3.5 kb and 2.4 kb seen in somatic tissues (Zakeri and Wolgemuth 1987). The testicular transcript, designated as *hsp70.1* is restricted to the germinal rather than the somatic compartment of the testis. The *hsp70.1* transcript first appears during the haploid stages of spermatogenesis and is stable throughout spermiogenesis (Zakeri and Wolgemuth 1987).

A second *hsp70* family member, designated *HSP 70.2* is also expressed at high levels in the mouse testis (Zakeri et al. 1988). The expression of *HSP 70.2* is also restricted to the germinal compartment of the testis, and like *HSP 70.1*, is ~2.7 kb in size (Zakeri et al. 1988). The developmental pattern of expression *HSP 70.2* is distinct from that of *HSP 70.1* in the germ cells of the testis, since *HSP 70.2* is expressed primarily in meiotic prophase cells (Zakeri et al. 1988). In rat a member of the *hsp70* gene family is also expressed at high levels in pachytene spermatocytes and is stable throughout spermiogenesis (Krawczyk et al. 1988). The levels of the *hsp70* mRNA appear not to be affected by hyperthermia (Krawczyk et al. 1988).

Studies of the expression of *hsp* at the protein level in the male germ line have focused on the *hsp70* family. Using antibodies against human Hsp70 and two-dimensional electrophoresis, Anderson et al. (1982) reported the detection of proteins in the 70-kD size range in both human and mouse testes, although the distinction among the heat inducible, constitutively expressed, and developmentally inducible forms was difficult to determine. Two dimensional electrophoretic analysis clearly revealed the presence of both a 72-kD and a 73-kD Hsp detected in the mouse testis as ^{35}S-methionine labeled proteins (Zakeri et al. 1990). A unique isoform of approximately 73-kD, designated 73T, is expressed in the germinal compartment of the testis. 73T is present in meiotic prophase and postmeiotic germ cells and is not induced by heat shock. The 72-kD Hsp is induced only slightly in the adult mouse testis. In contrast, induction of the 72-kD protein is substantial in the testis of germ cell deficient mice.

These data suggest that it is the somatic compartment, not the germinal compartment of the testis, that is responsible for the production of bona fide Hsp70 in response to heat shock. Other investigators have reported that the level of 72-kD protein is increased by heat shock in meiotic and post meiotic germ cells (Allen et al. 1988). It is possible that the exact conditions of heat shock varied slightly and resulted in differential response. Alternatively, the increased expression of the 72-kD protein might be due to somatic cell contamination of the enriched germ cell populations examined.

2.3 Expression of *hsp90* Genes

Members of the *hsp90* family also exhibit cellular specificity of expression in the murine testis. Two distinct *hsp90* transcripts, approximately 3.2 kb and 2.9 kb in size, are detected in the mouse testis (Gruppi et al. 1991). Based on the size of the putative corresponding proteins, the 3.2-kb transcript is likely to be the product of the *hsp86* gene, and the 2.9-kb transcript is likely the product of the *hsp84* gene.

Hsp86 is more abundant in the testis than is Hsp84. *Hsp86* is expressed primarily in the germinal compartment of the testis, particularly in germ cells in meiotic prophase, whereas *hsp84* is expressed primarily in the somatic compartment and at extremely low levels in the germinal compartment of the testis (Lee 1990; Gruppi et al. 1991). Both *hsp90* family transcripts are expressed in the testis of other mammals, suggesting an important role for Hsp90 in testicular function (Gruppi et al. 1991).

A monoclonal antibody raised against a human Hsp85 protein detected proteins in the 90-kD size class in a variety of human tissues, the highest levels being observed in the testis (Lai et al. 1984). An antibody generated against mouse Hsp90 was used to examine the tissue distribution of Hsp90 proteins in the mouse as well (Lai et al. 1984). Similarly to the pattern observed in human tissues, Hsp90 proteins were detected in a wide variety of murine tissues, with the highest levels found in the testis (Lai et al. 1984). The cellular localization within the human or mouse testis has not yet been determined.

3 Expression and Function of Hsp
During Mammalian Oogenesis and Early Embryogenesis

3.1 Key Features of Mammalian Oogenesis and Very Early Embryonic Divisions

Elucidating the genetic program that controls the differentiation of mammalian oocytes has been complicated by various factors, including the small size, limited numbers, and relative inaccessibility of the cells. The oocytes of virtually all mammals enter meiosis during embryonic development. This results in the lack of a renewing stem cell population and thus a continually diminishing pool of cells in the adult animal. The mid-gestation mouse embryo is not particularly amenable to experimental manipulation, making access to fetal ovaries which contain oocytes in meiotic prophase difficult. In addition, the absolute number of oocytes that can be obtained is limited. In the embryonic stages, there are hundreds of thousands of oocytes, but it is difficult to purify these cells away from the surrounding somatic cells. As development ensues, waves of atresia result in the loss of cells, to the point that the female mouse is born with at most ~12 000 oocytes (Peters 1969).

At about the time of pubertal development, a small number of cells enter into the growth phase, during which the cells increase in diameter and mass, thereby increasing the amount of RNA available for analysis. However, only a small number (< 50) of cells enter pre-meiotic growth or meiotic maturation at any one time. The cells in this stage are also difficult to isolate away from the surrounding follicular cells, a manipulation usually undertaken by hand. Use of superovulation can increase slightly the number of cells that enter the preovulatory phase. Ovulated eggs can be isolated in relatively pure (free of somatic granulosa cells) form, but again the ova are harvested individually. The same limitations apply to the early, preimplantation embryo. The embryos can be obtained, but the amount of RNA available for analysis frequently is in the nanogram to microgram range and represents the pooled samples of hundreds of female mice. Thus the mass of starting material is limited and the possibility for variation among developmental stages of different embryos from different females is high.

These limitations in obtaining female mammalian gametes are in marked contrast to the situation in working with male gametes. As mentioned previously, sufficient numbers of meiotic and postmeiotic cells at various stages of spermatogenesis can be obtained in enriched populations such that RNA and proteins from these cells can be analyzed at the molecular level. The most obvious limitation in the male system is the inability to manipulate various meiotic stages in vitro. Furthermore, the small size of the postmeiotic male germ cell (<10 μ in diameter) in contrast with that of the oocyte (>70 μ in diameter) makes microinjection and other forms of in vitro manipulation beyond the technical expertise of all but the most highly skilled.

3.2 Expression of Hsp in Oocytes and Early Embryos

Hsp86 and *hsp84* transcripts of approximately 3.2 kb and 2.9 kb, respectively, are detected in mouse ovary (Gruppi et al. 1991). The *hsp86* transcript is expressed at lower levels than the *hsp84* transcript. At the level of protein expression, unstressed preovulatory mouse oocytes and granulosa cells actively synthesize Hsp70 and Hsp89 (Curci et al. 1987). Upon heat shock the murine oocytes continue to synthesize the 70-kD and 89-kD proteins at the same level. Heat shocked granulosa cells synthesize de novo the 33-kD and 68-kD Hsp and exhibit increased expression of the 70-kD, 89-kD and 110-kD proteins.

In contrast to preovulatory oocytes, which are refractory to a heat shock response, growing oocytes synthesize de novo two Hsp68 isoforms (A. Curci, A. Bevilacqua, M.T. Fiorenza, and F. Mangia, per. comm.). The ability of mouse oocytes to mount a heat shock response appears to be highest during early follicular growth and declines and disappears completely prior to ovulation (*ibid*). These data suggest that mammalian oocytes have developmental "windows" for heat induction of the Hsp, similarly to previous observations on the early embryo and consistent with our observations on male mammalian germ cells (Zakeri et al. 1990).

Mammalian embryos, like those of flies, sea urchins, and frogs, express *hsp* during normal development. A member of the *hsp70* gene family is among the first products of the mouse zygotic genome and is expressed at high levels in the early mouse embryo

(Bensaude et al. 1983; Morange et al. 1984). Two electrophoretically distinct Hsp70 proteins are synthesized in the mouse preimplantation embryo and in undifferentiated embryonal carcinoma (EC) cells (Bensaude and Morange 1983; Barnier et al. 1987). Upon in vitro differentiation of EC cells, the levels of the two Hsp70 proteins decrease, suggesting that expression of the Hsp70 proteins correspond to developmental cues (Levine et al. 1984; Barnier et al. 1987). Developmental expression of Hsp70 family members was observed in placenta starting at day 8.5 and in yolk sac starting at day 11.5 (Kothary et al. 1987). No expression of this Hsp 70 family member was detected in embryonic tissues until day 15.5 (Kothary et al. 1987). Members of the *hsp90* gene are also expressed in the developing mouse conceptus. Transcripts from two distinct *hsp90* genes are seen at readily detectable and relatively constant levels in the embryonic and extra-embryonic compartments throughout midgestation (Lee 1990; Gruppi et al. 1991).

As in other developing systems, heat induction of *hsp* is stage-dependent in the mammalian embryo. Prior to the blastocyst stage in mouses, no induction of Hsp in response to heat shock is observed (Morange et al. 1984). After the blastocyst stage, a strong induction of the heat inducible *hsp70* family member can be detected (Morange et al. 1984). Similarly, heat shock treatment of 16-cell-stage rabbit embryos did not induce Hsp (Heikkila et al. 1985). In contrast, rabbit blastocysts subjected to heat shock treatment showed enhanced synthesis of a 70-kD Hsp and possibly enhanced synthesis of 95-kD and 28-kD proteins (Heikkila and Schultz 1984). The mechanism underlying the existence of this refractory period for heat induction of Hsp synthesis is unknown.

4 Summary and Speculation as to Function of Hsp in Mammalian Germ Cells and Embryos

4.1 Possible Functions of Hsp in General

Although there is much information on the structure and expression of Hsp, the exact function of these proteins remain elusive. It is generally assumed that Hsp expression "rescues" the cell from damage caused by exogenous stress. Speculation on the functions of Hsp during stress and developmental induction has been based primarily on the subcellular location of the Hsp and their association with other proteins.

One postulated function of Hsp in stressed and nonstressed cells may involve mediating protein-protein interactions. Hsp are induced by injection of denatured proteins into nonstressed *Xenopus* cells (Ananthan et al. 1986). Hsp90 is associated with the inactive form of several steroid receptors (Joab et al. 1984; Catelli et al. 1985; Sanchez et al. 1985). It is thought that dissociation of the steroid receptor from Hsp90 allows it to convert to its active form (discussed in Pratt 1987; Beato 1989). In contrast, an Hsp70 is present in both the inactive and active form of the progesterone receptor complex (Kost et al. 1989). Hsp90 has also been found to be associated with the *src* tyrosine kinase as it is transported from the nuclear to plasma membrane (Brugge 1986). It is postulated that binding of Hsp90 to the hydrophobic protein keeps the kinase soluble until it reaches the plasma membrane (Pelham 1985, 1986), essentially acting as a "molecular chaperone" (Pelham 1988). Hsp in general appear to facilitate movements of proteins within the cell and prevent protein complexes from forming insoluble aggregates.

Hsp may also function in the assembly and disassembly of macromolecular structures (Pelham 1986). An Hsp70 protein appears to function as an uncoating ATPase in the disassembly of clathrin trimers in coated vesicles (Chappell et al. 1986). In vitro studies have shown Hsp90 crosslinks actin fibrils in a calmodulin-dependent manner (Koyasu et al. 1986; Nishida et al. 1986). Hsp90 has been colocalized with microtubules in mammalian interphase cells and to mitotic spindles in dividing mammalian cells (Redmond et al. 1989). These observations further support the role of Hsp as mediators of protein-protein interaction in the cell.

4.2 Possible Functions of Heat Shock Genes in Male Germ Cell Differentiation

One can hypothesize several potential functions for Hsp during mammalian spermatogenesis since expression of members of the *hsp90* and *hsp70* correlates with discrete points in testicular maturation and with particular cell types. It has been speculated that Hsp are involved in the assembly and disassembly of macromolecular structures. These observations are significant with respect to spermatogenesis because of the major morphogenetic changes that occur during germ cell differentiation. Major protein structures like the synaptonemal complex are assembled and disassembled during meiotic prophase. During spermiogenesis structures including the acrosome and the flagellum are formed.

Particularly relevant to spermatogenesis and oogenesis is the association of Hsp90 and Hsp70 with steroid receptors. Hsp90 has been shown to be associated with various steroid receptors in the non-DNA binding state; disassociation of Hsp90 from the receptor converts it to its DNA binding state (Pratt 1987; Beato 1989). Since the testis and ovary are both a steroid hormone producing and responsive tissue, the Hsp90 family may function as a binding protein for various steroid receptors. Given the role of retinoic acid and its derivatives in the maintenance and progression of spermatogenesis (Wolbach and Howe 1925), it will be of interest to determine if Hsp90 proteins are also associated with the retinoic acid receptor (Giguere et al. 1987; Petrovich et al. 1987).

4.3 Molecular and Genetic Approaches for Identifying Function During Mammalian Gametogenesis

Genetic approaches have proved very powerful in elucidating the control of development in many systems. One approach would involve altering the expression of the gene in question during germ cell differentiation in a negative or positive manner in transgenic mice. Negative interference, accomplished by mutating the endogenous gene via homologous recombination or by overproducing an antisense orientation of the gene, would produce a loss-of-function mutation. Positive interference could be accomplished by ectopic or elevated levels of expression of a fully functional gene, under the direction of the endogenous or a heterologous promoter, yielding a gain-of-function phenotype.

The classical approach of generating mutations, which is now feasible in mice because of recent advances in the use of targeting recombination in embryonic stem cells (Robertson et al. 1986; Thomas and Capecchi 1987; Frohman and Martin 1989), involves disrupting the endogenous gene and examining the resulting phenotype in the heterozygous or homozygous state. Since most *hsp* genes are also expressed in various embryonic and adult tissues, a concern is that completely disrupting the function of these genes might result in embryonic ot neonatal lethality. Recently, Schwartzberg et al. (1989) reported the disruption of the ubiquitously expressed c-*abl* protooncogene and the generation of mice carrying the disrupted gene in the germ line. These animals did survive; although it will be most interesting to examine the resulting phenotype in these animals, especially with respect to reproductive functions. However, the fact that the heterozygous animals are fertile implies at very least that mice need only one functional c-*abl* gene in the germ line.

The usefulness of producing gain-of-function mutations is supported by our recent experiments in which we have manipulated the expression of *Hox-1.4* in transgenic mice and produced an interesting developmental abnormality (Wolgemuth et al. 1989). Elevated levels of *Hox-1.4* in the mesodermal germ layer of mid-gestation embryos correlates with the development of congenital megacolon in neonatal and adult animals, apparently due to abnormal differentiation of the enteric nervous system. Whether the abnormality arises in the migrating neural crest cells which differentiate into the enteric ganglia or to the micro-environment of the gut into which the cells migrate remains to be determined.

The experiment nonetheless underscores the usefulness of manipulating the expression of particular genes in a specific tissue or cell type. For example, the presence of high levels of uniquely sized *hsp70* transcripts in meiotic prophase spermatocytes and in haploid spermatids suggests that these genes may be important for these particular stages of spermatogenic differentiation or later gamete function. It therefore would be of interest to use spermatogenic stage-specific promoters to elevate levels of their protein products at the appropriate stages or to produce such proteins at a developmental stage in which they are not normally expressed. These experiments could be accomplished by using the protamine 1 promoter to drive expression in haploid cells (Peschon et al. 1987), whereas the phosphoglycerate kinase 2 promoter (Robinson et al, 1989) might be useful to achieve expression in meiotic prophase cells.

The fact that such experiments can be proposed is due to a synthesis of the results obtained by using the molecular approaches to identify genes involved in gametogenesis, outlined at the beginning of this Chapter. Protamine 1 was isolated using the second strategy, that of selecting for genes expressed only in the developing germ cells (Kleene et al. 1985). Various applications of the subtractive hybridization strategies are currently in use to identify genes expressed specifically in each spermatogenic cell type (e.g., Thomas et al. 1989). Such genes might then provide the additional promoters for manipulating Hsp expression throughout spermatogenesis.

Acknowledgments. This work was supported in part by grants from the NIH, HD05077, and from NASA, NAGW 1579 and NGT-50315.

References

Allen RL, O'Brien DA, Eddy EM (1988) A novel hsp70-like protein (P70) is present in mouse spermatogenic cells. Mol Cell Biol 8:828–832

Ananthan J, Goldberg AL, Voellmy R (1986) Abnormal proteins serve as eukaryotic stress signals and trigger the activation of heat shock genes. Science 232:522–524

Anderson NL, Giometti CS, Gemell MA, Nance SL, Anderson NG (1982) A two-dimensional electrophoretic analysis of the heat-shock-induced proteins of human cells. Clin Chem 28-4:1084–1092

Barnier JV, Bensaude O, Morange M, Babinet C (1987) Mouse 89 kD heat shock protein. Two polypeptides with distinct developmental induction. Exp Cell Res 170:186–194

Baumgartner AP, Chrisman CL (1981) Ovum morphology after hyperthermic stress during meiotic maturation and ovulation in the mouse. J Reprod Fer 61:91–96

Beato M (1989) Gene regulation by steroid hormones. Cell 56:335–344

Bellen HJ, O'Kane CJ, Wilson C, Grossniklaus U, Pearson RK, Gehring WJ (1989) P-element mediated enhancer detection: a versatile method to study development in *Drosophila*. Genes Dev 3:1288–1300

Bellve AR (1972) Viability and survival of mouse embryos following parental exposure to high temperature. J Reprod Fer 30:71–81

Bellve AR (1973) Development of mouse embryos with abnormalities induced by parental heat stress. J Reprod Fer 35:393–403

Bellve AR, Cavicchia JC, Millette CF, O'Brien DA, Bhatnagar YM, Dym M (1977) Spermatogenic cells of the prepuberal mouse. Isolation and morphological characterization. J Cell Biol 74:68–85

Bensaude O, Morange M (1983) Spontaneous high expression of heat-shock proteins in mouse embryonal carcinoma cells and ectoderm from day 8 mouse embryo. EMBO J 2:173–177

Bensaude O, Babinet C, Morange M, Jacob F (1983) Heat shock proteins, first major products of zygotic gene activity in mouse embryo. Nature 305:331–333

Brugge JS (1986) Interaction of the Rous sarcoma virus protein with the cellular proteins, pp 50 and pp 90. Curr Top Microbiol Immunol 123:1–22

Catelli MG, Binart N, Jung-Testas I, Renoir JM, Baulieu EE, Feramisco JR, Welch WJ (1985) The common 90-kD protein component of non-transformed '8S' steroid receptors is a heat-shock protein. EMBO J 4:3131–3135

Chappell TG, Welch WJ, Schlossman DM, Palter KB, Schlessinger MJ, Rothman JE (1986) Uncoating ATPase is a member of the 70 kilodalton family of stress proteins. Cell 45:3–13

Cohen RS, Meselson M (1985) Separate regulatory elements for the heat-inducible and ovarian expression of the *Drosophila hsp26* gene. Cell 43:737–746

Craig EA (1985) The heat shock response. Crit Rev Biochem 18:239–280

Curci A, Bevilacqua A, Mangia F (1987) Lack of heat-shock response in preovulatory mouce oocytes. Dev Biol 123:154–160

Day AR, Lee AS (1989) Transcriptional regulation of the gene encoding the 78-kD glucose-regulated protein GRP78 in mouse Sertoli cells: binding of specific factor(s) to the GRP78 promoter. DNA:301–310

Frohman MA, Martin GR (1989) Cut, paste, and save: new approaches to altering specific genes in mice. Cell 56:145–147

Germain MA, Webster WS, Edwards MJ (1985) Hyperthermia as a teratogen: parameters determining hyperthermia-induced head defects in the rat. Teratology 31:265–272

Giguere V, Ong ES, Segui P, Evans RM (1987) Identification of a receptor for the morphogen retinoic acid. Nature 330:624–629

Glaser RL, Wolfner MF, Lis T (1986) Spatial and temporal pattern of hsp26 during normal development. EMBO J 5:747–754

Goldman DS, Kisseling AA, Millette CF, Cooper GM (1987) Expression of c-*mos* RNA in germ cells of male and female mice. Proc Natl Acad Sci USA 84:4509–4513

Gossler A, Joyner AL, Rossant J, Skarnes WC (1989) Mouse embryonic stem cells and reporter constructs to detect developmentally regulated genes. Science 244:463–465

Green M (1981) Genetic variants and strains of the laboratory mouse. Gustav Fischer, New York, 476 pp

Grossniklaus U, Bellen HJ, Wilson C, Gehring W (1989) P-element mediated enhancer detection applied to the study of oogenesis in *Drosophila*. Development 107:189–200

Gruppi C, Zakeri ZF, Wolgemuth DJ (1991) Stage and lineage-regulated expression of two hsp90 transcripts during mouse germ cell differentiation and embryogenesis. Mol Reprod Dev (in press)

Heikkila JJ, Schultz GA (1984) Different environmental stresses can activate the expression of a heat shock gene in rabbit blastocysts. Gamete Res 10:45-56

Heikkila JJ, Miller JGO, Schultz GA, Kloc M, Browder LW (1985) Heat shock gene expression during early animal development. In: Atkinson BG, Walden DB (eds) Changes in eukaryotic gene expression in response to environmental stress. Academic Press, Orlando, pp 135–138

Hickey E, Brandon SE, Smale G, Lloyd D, Weber LA (1989) Sequence and regulation of a gene encoding a human 89-kilodalton heat shock protein. Mol Cell Biol 9:2615-2626

Joab I, Radanyi C, Renoir M, Buchou T, Catelli MG, Binart N, Mester J, Baulieu EE (1984) Common non-hormone binding component-transformed chick oviduct receptors of four steroid hormones. Nature 308:850–853

Kleene K, Distel RJ, Hecht NB (1985) Nucleotide sequence of a cDNA clone encoding mouse protamine 1. Biochemistry 24:719–722

Kost SL, Smith DL, Sullivan WP, Welch WJ, Toft DO (1989) Binding of heat shock proteins to the avian progesterone receptor. Mol Cell Biol 9:3829–3838

Kothary R, Clapoff S, Darling S, Perry MD, Moran LA, Rossant J (1987) Cell-lineage specific expression of the mouse hsp68 gene during embryogenesis. Dev Biol 121:342–348

Kothary R, Clapoff S, Brown A, Campbell R, Peterson A, Rossant J (1988) A transgene containing *lacZ* inserted into the *dystonia* locus is expressed in neural tube. Nature 335:435–437

Kothary R, Clapoff S, Darling S, Perry MD, Moran LA, Rossant J (1989) Inducible expression of an hsp68-*lacZ* hybrid gene in transgenic mice. Development 105:707–714

Koyasu S, Nishida E, Kadowaki T, Matsuzaki F, Iida K, Harada F, Kasuga H, Sakai H, Yahara I (1986) Two mammalian heat shock proteins, hsp90 and hsp100, are actin binding proteins. Proc Natl Acad Sci USA 83:8054–8058

Krawczyk Z, Mali P, Parvinen M (1988) Expression of a testis-specific hsp70 gene-related RNA in defined stages of rat seminiferous epithelium. J Cell Biol 107:1317–1323

Kurtz S, Rossi J, Petko L, Lindquist S (1986) An ancient developmental induction: heat-shock proteins induced in sporulation and oogenesis. Science 231:1154–1156

Lai BT, Chin NW, Stanek AE, Keh W, Lanks KW (1984) Quantitation and intracellular localization of the 85K heat shock protein by using monoclonal and polyclonal antibodies. Mol Cell Biol 4:2802-2810

Lee SJ (1990) Expression of hsp86 in male germ cells. Mol Cell Biol 10:3239–3242

Lenz RW, Ball GD, Leibfried ML, Ax RL, First NL (1983) In vitro maturation and fertilization of bovine oocytes are temperature-dependent processes. Biol Reprod 29:173–179

Levine RA, La Rosa GJ, Gudas LJ (1984) Isolation of cDNA clones for genes exhibiting reduced expression after differentiation of murine teratocarcinoma stem cells. Mol Cell Biol 4:2142–2150

Lindquist S (1986) The heat shock response. Annu Rev Biochem 55:1151–1191

Lowe DG, Moran LA (1986) Molecular cloning and analysis of DNA complementary to three mouse Mr=68000 heat shock protein mRNAs. J Biol Chem 261:2102-2120

Moore CR, Chase D (1923) Heat application and testicular differentiation. Anat Rec 26:344–345

Moore SK, Kozak C, Robinson EA, Ullrich SJ, Apella E (1987) Cloning and nucleotide sequence of the murine hsp84 cDNA and chromosomes assignments of related sequences. Gene 56:29–40

Moore SK, Kozak C, Robinson EA, Ullrich SJ, Apella E (1989) Murine 86- and 84-kDa heat shock proteins, cDNA sequences and chromosome assignments, and evolutionary origins. J Biol Chem 264:5343–5351

Morange M, Diu A, Bensaude O, Babinet C (1984) Altered expression of heat shock proteins in embryonal carcinoma cells and mouse early embryonic cells. Mol Cell Biol 4:730–735

Mutter GL, Wolgemuth DJ (1987) Distinct developmental patterns of c-*mos* protooncogene expression in female and male germ cells. Proc Natl Acad Sci USA 84:5301–5305

Mutter GL, Grills GS, Wolgemuth DJ (1988) Evidence for the involvement of the protooncogene c-*mos* in mammalian meiotic maturation and possibly very early embryogenesis. EMBO J 7:683–689

Nebel BR, Amarose AP, Hackett EM (1961) Calendar of gametogenic development in the prepuberal mouse. Science 134:832-833

Nishida E, Koyasu S, Sakai H, Yaharas I (1986) Calmodulin-regulated binding of the 90-kDa heat shock protein to actin filaments. J Biol Chem 261:16003–16036

Pardue ML (1988) The heat shock response in biology and human disease: a meeting review. Genes Dev 2:783–785

Pelham H (1985) Activation of heat shock genes in eukaryotes. Trends Genet 1:31-35

Pelham H (1988) Coming in from the cold. Nature 332:776-777

Pelham HRB (1986) Speculations on the functions of the major heat shock and glucose-regulated proteins. Cell 46:959–961

Peschon JJ, Behringer RR, Brinster RL, Palmiter RD (1987) Spermatid-specific expression of protamine 1 in transgenic mice. Proc Natl Acad Sci 84:5316–5319

Peters H (1969) The development of the mouse ovary from birth to maturity. Acta Endocrinol 62:98–116

Petrovich M, Brand NJ, Krout A, Chambon P (1987) A human retinoic acid receptor. Nature 330:444–450

Ponzetto C, Wolgemuth DJ (1985) Haploid expression of a unique c-abl transcript in the mouse male germ line. Mol Cell Biol 5:1791–1794

Pratt WB (1987) Transformation of glucocorticoid and progesterone receptors to the DNA-binding state. J Cell Biochem 35:51–68

Propst F, Rosenberg MP, Iyer A, Kaul K, Vande Woude GF (1987) C-mos proto-oncogene RNA transcripts in mouse tissues: structural features, developmental regulation, and localization in specific cell types. Mol Cell Biol 7:1629–1637

Rappollee DA, Brenner CA, Schultz R, Mark D, Werb Z (1988) Developmental expression of PDGF, TGF-alpha, and TGF-beta genes in preimplantation mouse embryos. Science 241:1823–1825

Rebagliati MR, Weeks DL, Harvey RP, Melton DA (1985) Identification and cloning of maternal mRNAs in Xenopus eggs. Cell 42:769–777

Rebbe NF, Ware J, Bertina RM, Modrich P, Stafford DW (1987) Nucleotide sequence of a cDNA for a member of the human 90-kDa heat-shock protein family. Gene 53:235–245

Redmond T, Sanchez ER, Bresnick EH, Schlesinger MJ, Toft DO, Pratt WB, Welsh MJ (1989) Immunofluorescence colocalization of the 90-kDa heat-shock protein and microtubules in interphase and miotic mammalian cells. Eur J Cell Biol 50:66–75

Robertson EJ, Bradley A, Kuehn M, Evans M (1986) Germ-line transmission of genes introduced into cultured pluripotential cells by retroviral vectors. Nature 323:445-447

Robinson MO, McCarrey JR, Simon MI (1989) Transcriptional regulatory regions of testis-specific PGK2 defined in transgenic mice. Proc Natl Acad Sci 86:8437–8441

Rock J, Robinson D (1965) Effect of induced intrascrotal hyperthermia on testicular function in man. Am J Obstet Gynecol 93:793–801

Sanchez ER, Toft DO, Schlessinger ML, Pratt WB (1985) Evidence that the 90-kDa phosphoprotein associated with untransformed cell glucocorticoid receptor is a murine heat shock protein. J Biol Chem 260:12398-12401

Schwartzberg P, Goff SP, Robertson EJ (1989) Germ-line transmission of a c-abl mutation produced by targeted gene disruption in ES cells. Science 246:799–803

Thomas KH, Wilkie TM, Tomashefsky P, Bellve AR, Simon MI (1989) Differential gene expression during mouse spermatogenesis. Biol Reprod 41:729–739

Thomas KR, Capecchi MR (1987) Site-directed mutagenesis by gene targeting in mouse embryo-derived stem cells. Cell 51:503–512

Ulberg LC, Sheean LA (1973) Early development of mammalian embryos in elevated ambient temperatures. J Reprod Fertil 19(Suppl):155-161

Webster WS, Edwards MJ (1984) Hyperthermia and the induction of neural tube defects in mice. Teratology 29:417–425

Weeks DL, Rebagliati MR, Harvey RP, Melton DA (1985) Localized maternal mRNA in Xenopus laevis eggs. Cold Spring Harbor Symp Quant Biol 50:21–29

Welsh J, Liu JP, Efstratiadis A (1990) Cloning of PCR-amplified total cDNA: construction of a mouse oocyte cDNA library. Genet Anal Techn Appl 7:5–17

Wolbach SB, Howe PR (1925) Tissue changes following deprivation of fat soluble vitamin A. J Exp Med 42:753-777

Wolgemuth DJ, Gizang-Ginsberg E, Engelmyer E, Gavin BJ, Ponzetto C (1985) Separation of mouse testis cells on a Celsep™ apparatus and their usefulness as a source of high molecular weight DNA or RNA. Gamete Res 12:1–10

Wolgemuth DJ, Engelmeyer E, Duggal RN, Gizang-Ginsberg EE, Mutter GL, Ponzetto C, Vivano C, Zakeri ZF (1986) Isolation of a mouse cDNA coding for a developmentally regulated, testis-specific transcript containing homeo box homology. EMBO J 5:1229–1235

Wolgemuth DJ, Viviano CM, Gizang-Ginsberg E, Frohman MA, Joyner AL, Martin GR (1987) Differential expression of the mouse homeobox-containing gene Hox-1.4 during male germ cell differentiation and embryonic development. Proc Natl Acad Sci USA 84:5813–5817

Wolgemuth DJ, Behringer RR, Mostoller MP, Brinster RL, Palmiter RD (1989) Transgenic mice over-expressing the mouse homeobox-containing gene *Hox-1.4* exhibit abnormal gut development. Nature 337:464–467

Young WC (1927) The influence of high temperature on the guinea-pig testis. J Exp Zool 49:459–499

Zakeri ZF, Wolgemuth DJ (1987) Developmental stage specific expression of the hsp70 gene family during differentiation of the mammalian male germ line. Mol Cell Biol 7:1791–1796

Zakeri ZF, Wolgemuth DJ, Hunt CR (1988) Identification and sequence analysis of a new member of the mouse hsp70 gene family and characterization of its unique cellular and developmental pattern of expression in the male germ line. Mol Cell Biol 8:2925–2932

Zakeri ZF, Welch WJ, Wolgemuth DJ (1990) Characterization and inducibility of hsp70 proteins in the male mouse germ line. J Cell Biol 111:1785–1792

Zimmerman JL, Petri W, Meselson M (1983) Accumulation of a specific subset of *D. melanogaster* heat shock mRNAs in normal development without heat shock. Cell 32:1161–1170

10 Heat Shock Protein Synthesis in Preimplantation Mouse Embryos and Embryonal Carcinoma Cells

Valérie Mezger[1], Vincent Legagneux[1], Charles Babinet[2], Michel Morange[1], and Oliver Bensaude[1]

1 Introduction

In numerous organisms, heat shock protein (Hsp) expression appears to be developmentally regulated. On the one hand, a high and spontaneous expression of Hsps at normal temperatures is observed at some stages of development. On the other hand, at some stages of development or differentiation, cells do not synthesize heat-inducible Hsp in response to a stress.

1.1 The Major Murine Heat-Shock Proteins

Mouse fibroblasts synthesize the following major Hsps: the synthesis of Hsp105 and Hsp70 is strongly stress-inducible and hardly detectable under normal culture conditions; the synthesis of Hsp86 is strongly stress-inducible, whereas synthesis of Hsc84, Hsc70, and Hsc60 is important under normal growth conditions and is only slightly stress-inducible. Hsp70 (also termed Hsp72 or Hsp68) and its cognate Hsc70 (also termed Hsc73 or Hsp70) are related to the bacterial protein DnaK. Hsc60 (also termed Hsp60, Hsp59, or Hsp58) is related to the bacterial protein GroEL. Hsc84 and Hsp86 belong to the 90-kD Hsp family and are related to the bacterial protein C62.5 (Ullrich et al. 1986, Barnier et al. 1987). The murine Hsp84 and Hsp86 are almost identical to the human Hsp89ß and Hsp89a, respectively (Hoffman and Hovemann 1988; Hickey et al. 1989; Rebbe et al. 1989). In murine established fibroblasts, Hsp84 behaves like a cognate (high constitutive synthesis in unstressed cells, weak stress induction) while Hsp86 displays strong stress-induced and weak constitutive synthesis (Barnier et al. 1987).

Here we focus on the expression of Hsps in the preimplantation mouse embryo and in mouse embryonal carcinoma (EC) cells which are often used as a model for the early embryonic cells. The preimplantation mouse embryonic cells and the EC cells are characterized by a high spontaneous synthesis of some Hsps and by a deficiency in heat-shock induced synthesis of the others at some very specific stages.

[1] Biologie Moléculaire du Stress, Département de Biologie, Ecole Normale Superieure, 46, rue d'Ulm 75230 Paris Cedex 05, France
[2] Unité de Génétique des Mammiféres, Institut Pasteur, 25 rue du Dr.Roux, 75724 PARIS Cedex 15, France.

Results and Problems in Cell Differentiation 17
Heat Shock and Development
Hightower and Nover (Eds.)
©Springer-Verlag Berlin Heidelberg 1991

1.2 Heat-Shock Protein Expression During Gametogenesis

Before questioning the status of Hsp expression during early development, it may be useful to review our knowledge of Hsp expression during gametogenesis. The details of heat-shock gene expression during mammalian gametogenesis are reviewed in another Chapter of this Volume by D. Wolgemuth, (Chapter 9). In brief, the spontaneous expression of some heat shock genes is notable during gametogenesis in various organisms. During rat and mouse spermatogenesis, the constitutive Hsc70 and at least two distinct but related testis-specific proteins accumulate (Anderson et al. 1982; Allen et al. 1988a, b). High levels of the Hsc84-Hsp86 have been noticed in human and mouse testis (Anderson et al. 1982). A particularly high level of *hsp86* mRNA expression is found in the testis (Lee 1990). The mRNA coding for proteins related to Hsp70 is also found in rat and human testes (Zakeri and Wolgemuth 1987; Allen et al. 1988a, b; Krawczyk et al. 1988; Zakeri et al. 1988a, b; Matsumoto and Fujimoto 1990).

During oogenesis, Hsc70 is expressed at high levels in the preovulatory oocyte (Curci et al. 1987). Its synthesis ceases shortly after germinal-vesicle breakdown (Barra and Bensaude, unpublished) and is undetectable in the ovulated oocyte at the time of fertilization. The synthesis of various other Hsps is important, and as a result, both Hsp86-84, Hsc70, and Hsc60 are major constituents of the ovulated oocyte together with the related glucose-regulated proteins Grp94 and Grp78 (Fig. 1). However, the shut-off of Hsp86-84 synthesis after meiosis is less dramatic than that of Hsc70. This is likely to be due to the instability of *hsc70* mRNAs which contrast with the stability of the *hsp86-84* mRNAs (Hickey et al. 1989; Legagneux et al. 1989). Lack of heat-shock induced Hsp expression is the other characteristic of gametogenesis. Thus, in mouse, Hsp70 is not inducible during late stages of spermatogenesis (Zakeri and Wolgemuth 1987) and oogenesis, i.e. in oocytes before ovulation (Curci et al. 1987).

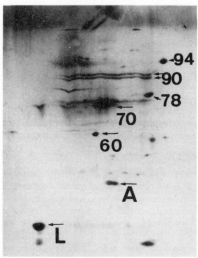

Fig. 1. The major proteins present in the unfertilized mature mouse oocyte. The proteins from 114 oocytes were separated by isoelectric focusing followed by SDS polyacrylamide gel electrophoresis. *94* – Grp94; *90* – Hsp86 and Hsc84 (which streak); *78* – Grp78; *70* – Hsc70; *60* – Hsp60; *A* Actin; *L* Lactate dehydrogenase. The acidic end of the isoelectric focusing gel is on the *right*

2 Heat Shock Protein Synthesis in Unstressed Early Embryonic Cells

2.1 Heat Shock Proteins, the First Major Products of Zygotic Transcription

In the mouse, the onset of embryonic transcription occurs one day after fertilization, at the two-cell stage in two phases with the synthesis of specific proteins. The synthesis of these proteins is inhibited by transcriptional inhibitors. The first phase is mainly characterized by the neosynthesis of a group of 70-kD proteins which occurs early after cleavage (ca. 32h post human-chorion gonadotropin hormone (hCG) injection) (Van Blerkom and Brockway 1975; Howe and Solter 1979; Flach et al. 1982; Bolton et al. 1984). The second phase takes place at the late two-cell stage (ca. 40 h post-hCG) and results in a complete change of proteins synthesized. Both phases are inhibited by a-amanitin (Flach et al. 1982), and in-vitro translation experiments demonstrate the appearance of the corresponding mRNA at the two-cell stage (Howlett and Bolton 1985). Proteins identical to both the heat-inducible and the cognate 70-kD heat-shock proteins correspond to the first phase 70-kD proteins (Bensaude et al. 1983). Synthesis of the Hsc70 and Hsp70 proteins is the most spectacular; little changes in Hsp86-84 synthesis are observed. However, it should be kept in mind that Hsc70 and Hsp86-84 synthesis were strong during oogenesis, that after meiosis Hsc70 synthesis has completely vanished, and that this is in constrast to what happens to Hsp86-84.

Both the paternal and the maternal genomes contribute to the transcription of the corresponding genes, since the same group of proteins are synthesized at the two-cell stage in both parthenogenetic and androgenetic embryos (Barra and Renard 1988). Experiments using nuclei transfer have demonstrated that aging of the egg cytoplasm directs the onset of heat-shock gene transcription (Barnes et al. 1987; Howlett et al. 1987). At the eight-cell stage, the mouse embryo does not synthesize the inducible Hsp70 even after heat-shock, but synthesizes very high levels of the cognate Hsc70. However, when an eight-cell stage nucleus is transferred into a one-cell embryo cytoplast (devoided of its pronuclei), the reconstructed one-cell embryos do not synthesize any Hsc70 in the hours which follow the manipulation. However, after allowing time for cell division, the reconstructed embryos synthesize both the inducible Hsp70 and the cognate Hsc70 at the right time relative to the development of the recipient cytoplast. Gene expression of the donor nucleus has been "reprogrammed" by the recipient cytoplast.

What turns on the transcription of the heat-shock genes in the mouse? The onset of embryonic transcription is controlled by genome replication in *Drosophila* and *Xenopus*. This does not seem to happen in the mouse both Hsp70s appear at the right time even if DNA replication or the first cell division are impeded with aphidicolin or X-ray irradiation (Howlett 1986; Grinfeld et al. 1987). However, such treatments block the appearance of the second wave of protein synthesis which requires transcription. Interestingly, reporter genes injected into a one-cell pronucleus at 20 h post-hCG are transcribed at the early two-cell stage (Martinez-Salas et al. 1989). Such delay is not observed when the reporter genes are injected into a two-cell nucleus. The same timings are obtained when DNA replication of the injected embryos is blocked by aphidicolin. Thus, the mouse embryo acquires a transcriptional competence for both the endogenous *hsp70* and *hsc70* genes and injected reporter genes at the early two-cell stage independent of DNA replication. A cascade of protein-kinases may be involved in acquiring this transcriptional competence, since some inhibitors of the cAMP dependent kinase block the appearance of the Hsps after the first cleavage (Poueymirou and Schultz 1989).

Does the transcription of *hsp70* and *hsc70* genes at the early two-cell stage involve specific transcription factors such as the heat shock factor (HSF)? Preliminary gel retardation assays from our laboratory suggest that heat-inducible HSF is present in the heat-shocked one-cell and two-cell embryos, but little or no active HSF is found in either unstressed one-cell or two-cell embryos (Mezger and Renard, to be published). Experiments with reporter genes injected into one-cell stage pronuclei suggest that minimal promoter elements are required for gene expression at the early two-cell stage. "Normal" regulation of transcription, involving specific transcription factors and enhancer sequences, would be established only later in development. Some negative regulatory factors are suggested to appear as components of the zygotic nuclear structure after the two-cell stage (Martinez-Salas et al. 1989). Transcription factors such as HSF would be required to overcome this negative regulation. However, the preeminence of 70-kD Hsp expression over that of house-keeping genes such as actin or the 90-kD Hsps remains unexplained.

2.2 High Spontaneous Expression of Hsps in the Preimplantation Mouse Embryo

Early embryogenesis displays spontaneous Hsp expression in various species. In the mouse, members of the 90-kD, 70-kD and 60-kD Hsp families are strongly synthesized and represent major proteins of the preimplantation embryo (Bensaude and Morange 1983; Morange et al. 1984; Barnier et al. 1987). At day 3 after fertilization, at the eight-cell stage, such proteins are the most actively synthesized proteins. In the 70-kD range, the cognate Hsc70 predominates; but, in contrast with the two-cell embryo, synthesis of the heat-inducible Hsp70 cannot be detected.

Both the heat-inducible Hsp86 and the cognate Hsc84 are constitutively expressed at very high levels by the mouse eight-cell embryo. Synthesis of Hsc70, and of both Hsc84 and Hsp86, remains elevated in the blastocysts; these proteins are found as major constituents of the embryonic ectoderm at day 8 after fertilization (Bensaude and Morange 1983).

2.3 Hsp Expression in Embryonal Carcinoma Cells

The study of the early mouse embryo can be approached by the use of multipotent cells originally found in tumors, the teratocarcinoma. These tumors are spontaneous tumors of the gonads or appear after ectopic implantation of early embryonic cells. Besides a large variety of differentiated tissues, teratocarcinomas contain malignant and multipotent stem cells called embryonal carcinoma cells which are presumably responsible for the malignancy of these tumors and their differentiation properties. Several embryonal carcinoma cell lines (EC) have been established from these tumors (Jakob and Nicolas 1987). Numerous properties of EC cells are similar to those of preimplantation mouse embryonic cells by morphological, biochemical and immuno-logical criteria.

The F9 EC and PCC4 Aza R1 (abbreviated PCC4) cell lines were derived from the same teratocarcinoma OTT6050, obtained by transplantation of a day 6.5 embryo and maintained by successive transplantations of embryoid bodies in the mouse strain 129. The PCC7-S-1009 line (abbreviated 1009) was derived from a spontaneous tumor in the testis of a 129 x C57BL-6J mouse. Some EC cells, such as F9 and 1009 are able to differentiate in vitro upon treatment with retinoic acid and dibutyryl cAMP (Pfeiffer et al. 1981; Strickland 1981). Permanent, differentiated cell lines have also been derived from EC cells (Jakob and Nicolas 1987).

Embryonic stem (ES) cell lines have been established directly from preimplanta-tion embryos and may be propagated undifferentiated on an irradiated fibroblasts feeder layer or in a conditioned medium (Evans and Kaufman 1981; Martin 1981).

3 Transcription of Heat Shock Genes in Unstressed EC Cells

3.1 Transcriptional and Posttranscriptional Regulation of Spontaneous Hsp Synthesis in EC Cells

Like the cells of the early embryo, all EC cell lines tested synthesize Hsp86, Hsc84, Hsc70 and Hsp59 at high rates that can be compared to the rates of synthesis found in heat shocked fibroblasts. Upon in vitro differentiation, synthesis of Hsps is reduced to values similar to those found in unstressed fibroblasts (Bensaude and Morange 1983; Morange et al. 1984; Barnier et al. 1987). The high levels of Hsp synthesis in EC cells correspond to high levels of the corresponding mRNAs; these mRNAs decrease severalfold upon differentiation. Both transcriptional and posttran-scriptional mechanisms appear to be involved. The transcription rate of *hsc70* genes is only two times higher in EC cells than in their differentiated derivatives (Giebel et al. 1988). In fibroblasts, mRNAs for Hsc70 display a relatively short half-life of approxi-mately 2 h but in EC cells this half-life is two to three fold longer. Therefore, mRNA stabilization contributes to the high expression of *hsc70* in F9 EC cells (Legagneux et al. 1989).

The mRNAs coding for Hsp86 were also found to be much more abundant at normal temperature in the undifferentiated F9 cells than in the differentiated F9 cells or in murine fibroblasts (Levine et al. 1984). However, these mRNAs are very stable both in differentiated cells and EC cells (Legagneux et al. 1989). In this case, a transcription rate six-fold higher in EC cells than in fibroblasts mainly accounts for the elevated amount of the corresponding protein.

3.2 HSE-Binding Activity in EC Cells

Because transcription is the regulated step for high constitutive *hsp86* expression, and since *hsp86* expression decreases during EC cell differentiation, the activity of the transcription factor(s) responsible for this expression is expected to be developmen-tally regulated. A candidate is the heat-shock transcription factor (HSF) which binds to heat-shock element (HSE) DNA sequences. This factor is generally thought to be activated by stress and therefore responsible for the stress activation of transcription of the heat-shock genes.

To address the question of HSF behavior in EC cell lines, an oligonucleotide containing two overlapping HSE was used in gel-shift assays. This oligonucleotide, HSE2, forms a strong binding site for HSF. The specificity of the interaction was investigated in competition experiments, in the presence of an excess of unlabeled HSE2 or an excess of a derived oligonucleotide carrying a point mutation in each HSEs (HSE2C). This mutation is known to lower the efficiency of a *Drosophila* heat shock promoter ten fold (Amin et al. 1988). The following observations were made:

1. Mouse fibroblasts, L or 3T6, cultivated at normal temperature, do not contain any significant HSE – binding activity. A 15-min, heat shock at 45 °C induces a strong activity for HSE binding.
2. In contrast to fibroblasts, all EC cell lines tested exhibit a relatively high HSE-binding activity even at normal temperature.
3. During in vitro differentiation of the 1009 EC cell line, the "constitutive" HSE-binding activity detected at normal temperature disappears. The pattern of HSE-binding of the differentiated 1009 derivatives is similar to that of fibroblasts.

Little is known about the contribution of the heat shock factor (HSF) to heat-shock gene expression in unstressed cells. The basal level expression of the yeast *hsp82* gene in the absence of stress requires the presence of the heat-shock element (McDaniel et al. 1989). A convincing example of HSF involvement in developmental HSP expression comes from the expression of one human *hsp70* gene *hsx70* under the control of hemin. In the human erythroleukemia cell line K562, the expression of the *hsx70* gene is activated during hemin-induced maturation of erythrocytes (Singh and Yu 1984). Theodorakis and coworkers showed that hemin induces the transcription of *hsx70* gene and that the HSE located between -107 and -94 relative to the initiation of transcription is required (Theodorakis et al. 1989). Moreover, hemin activates the HSE-binding properties of HSF as would a heat shock (see Chap. 11).

3.3 An EIa-Like Activity in EC Cells

The constitutive HSP expression in EC cells might be due to some factor distinct from HSF. Such a hypothesis has been proposed by Nevins. At first, it was found that adenovirus infection (like infection by many other viruses) stimulated the synthesis of some cellular proteins including a 70–kD Hsp (Nevins 1982). Indeed, the EIa adenovirus early gene was shown to activate the transcription of both a particular human *hsp70* gene and the human *hsp89a* gene (Nevins 1982; Kao and Nevins 1983; Simon et al. 1987; Hickey et al. 1989). Since mouse EC stem cells can sustain the growth of the d1312 adenovirus deletion mutant which lacks EIa, the existence of an EIa-like activity in such cells was postulated (Imperiale et al. 1984). When the EC cells are induced to differentiate by retinoic acid treatment, expression of the Hsps is lowered and their ability to support the growth of the d1312 mutant virus is lost. Furthermore, the EIa-responsive adenoviral promoter, EIIa, is very active in the undifferentiated EC cells (La Thangue and Rigby 1987).

In fibroblasts, the EIa gene product activates the transcription of the adenoviral promoters EIIa and EIV through the activation of two cellular factors E2F and E4F, respectively. EC stem cells appear to contain a cellular factor which binds on an E2F site of the EIIa promoter sequence. This factor disappears upon in vitro differentiation with retinoic acid (Reichel et al. 1987) A basal EIV transcription is also observed in murine EC cells. It requires the E4F binding sites, which are also involved in c-myc transactivation (Raychaudhuri et al. 1987). Since EC cells have high levels of c-myc protein, c-myc has been proposed as a candidate for the EIa-like activity (Onclercq et al. 1988) However, although both the E2F factor and the EC cell factor bind to the same region of the EIIa promoter there are clear differences between them (La Thangue et al. 1990). Two E2F molecules bind to the EIIa promoter in EIa trans-formed-cell extracts, whereas only one E2F site of the EIIa promoter is occupied in EC cell extracts (Jansen-Durr et al. 1989).

Extensive molecular dissection of a particular human *hsp70* gene which responds to serum stimulation and adenovirus infection has been achieved, demonstrating the role of a particular TATA box, a CCAAT and a purine rich element (Simon et al. 1988; Williams et al. 1989; Wu et al. 1986; see Chap 11). Neither E2F nor E4F binding sites have been characterized, and no HSEs are required in mediating the EIa response. However, EIa and probably the cellular EIa-like activity are expected to act on many different factors.

The EIa-like activity also seems to be present in preimplantation mouse embryos and oocytes which can sustain the transcription of either the EIIa gene after infection with the d1312 defective virus or the EIIa promoter coupled to various reporter genes (Suemori et al. 1988; Dooley et al. 1989). This EIa-like activity disappears at about the time of implantation.

3.4 High Levels of B2 Transcripts in Undifferentiated Mouse Embryonic Cells

Heat shock also strongly enhances the RNA polymerase III transcription of B2 sequences in various fibroblastic cell lines (Fornace and Mitchell 1986; Fornace et al. 1988). Since EIa also activates the RNA polymerase III transcription factor TFIIIC (Hoeffler et al. 1988), we discuss the developmental regulation of B2 transcripts.

In F9 EC cells, B2 sequences are transcribed at very high levels. This B2 gene transcription is down-regulated upon in vitro differentiation of the F9 cells into parietal endoderm cells (White et al. 1989). However, during F9 EC cell differentiation, the B2 transcript down-regulation appears to be due to a small decrease in RNA polymerase III activity but mainly to a reduction in transcription factor TFIIIB activity, and not TFIIIC.

B2 RNAs are also present in the mouse oocyte and increase in abundance upon fertilization (Vasseur et al. 1985; Taylor and Piko 1987). At day 4, they are abundant in the inner cell mass of the blastocyst.

As a conclusion, it is striking to note that during early development, after viral infection, or after a heat-shock, the transcription of a common set of genes is elevated though different mechanisms may be involved.

4 Defective Heat Shock Response in Early Embryonic Cells

4.1 Lack of Heat Shock Protein Inducibility
in the Early Preimplantation Mouse Embryo

It is a common feature of the higher eukaryotes, that very early stages of development are characterized by the lack of induced heat shock protein synthesis. However,
this lack of inducibility generally reflects the absence of embryonic transcription. As
soon as general transcription resumes, i.e. at the blastoderm stage for flies, at the
blastula stage for sea urchins and toads, the major heat shock genes become stress-
inducible (some heat shock genes remain noninducible). Acquisition of heat-shock
gene inducibility correlates with the development of an increased resistance of the
embryos to heat shock.

In the mouse heat shock genes are transcribed at the early two-cell stage. However,
heat shock does not affect the pattern of protein synthesis before the late morula stages,
and acquisition of a normal stress response correlates with an increased
thermoresistance and acquisition of thermotolerance (Muller et al. 1985). In particular,
the synthesis of heat-induced Hsp70, which is expressed at the two-cell stage, cannot
be obtained at the eight-cell stage (Wittig et al. 1983; Morange et al. 1984). Stress
inducibility of Hsp70 appears gradually after compaction of the eight-cell embryo
during the development of the morula into a blastocyst (Fig. 2).

All the blastomers of the eight-cell stage are equivalent. Since the blastocyst is
composed of two distinct cell types (the epithelial trophoblast cells on the outside and
the undifferentiated inner cell mass cells) the formation of the blastocyst corresponds
to the first differentiation process. One might have expected the heat-inducible phenotype to appear only in one cell type. Trophoblast cells were separated from inner cell
mass cells by microsurgery and both were found to have stress-induced Hsp70 synthesis (Fig. 3). A few transgenic mice have been obtained which carry the ß-galactosidase
gene under the control of a stress-inducible mouse *hsp70* promoter. Embryos from
these mice do not exhibit ß-galactosidase expression in response to a stress before the
blastocyst stage, where it appears simultaneously in both trophoblast and inner cell
mass cells (Kothary et al. 1989).

4.2 Inducible and Non Inducible Embryonal Cell Lines

EC cell lines can be divided into two classes according to their response to heat
shock. The first group is "inducible", it synthesizes the major thermoinducible heat
shock protein, Hsp70, in response to a stress (Bensaude and Morange 1983; Barnier et
al. 1987). The EC cell line F9 belongs to this group. Hsp86 synthesis is also stimulated
by stress in this cell line, but Hsp105 is not induced in these cells. Hsp105 cannot be
induced in the blastocyst either. After in vitro differentiation, the F9 cells synthesize
Hsp105 upon heat shock.

The second group displays a complete "non inducible" phenotype. The PCC4 and
1009 EC cells are not able to synthesize any thermoinducible Hsps in response to a
stress (Wittig et al. 1983; Morange et al. 1984). Such cell lines exhibit a behavior
similar to the eight-cell-stage-morula mouse embryos. However, upon differentiation

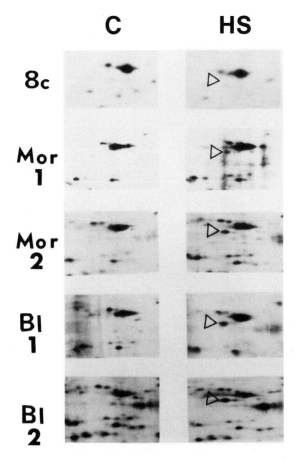

Fig. 2. Inducibility of hsp70 synthesis in response to heat-shock from the eight-cell stage embryo to the mature blastocyst. Various embryonic stages were analyzed : uncompacted eight-cell stage (*8c*); compacted eight-cell stage (*Mor 1*); 16- to 32-cell morulae (*Mor 2*); blastocyst at the beginning of blastocoele formation (*Bl 1*); and blastocyst with fully developed blastocoele (*Bl 2*). Embryos were labeled with (^{35}S)-methionine before (*C*) or after a 10-min heat shock at 44 °C (*HS*), and proteins were analyzed by isoelectric focusing followed by SDS polycrylamide gel electrophoresis. The spot with the highest intensity corresponds to the constitutive Hsc70. *Arrowheads* indicate the position of Hsp70

of 1009 cells, synthesis of Hsp70 as well as of Hsp86 and of Hsp105 becomes heat-inducible as in fibroblasts.

There does not seem to be a clear connection between the inducible phenotype and the differentiation potential of the embryonic cells in culture. Two embryonic stem (ES) cell lines were analyzed; cells from both the D3 and E14 lines were found to have an inducible phenotype like the F9 EC cells (V. Mezger, unpubl. data). In addition, a differentiated cell line (3-TDM 1) with a trophoblastic cell-type has been isolated from the pluripotent PCC3 EC cell. The parental EC cell line is heat-inducible, whereas the diferentiated derivative is non inducible (Mezger et al. 1989).

C   **HS**

ICM

Tro

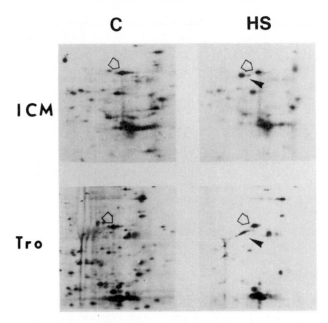

Fig. 3. Induction of Hsp70 synthesis in response to heat shock in trophectoderm (*Tro*) and in the inner cell mass (*ICM*). Trophectoderms were separated from inner cell masses by mechanical dissection of day-4 blastocysts. The cells were labeled with (^{35}S)-methionine and analyzed on 2-dimensional-gel, before (*C*) and after a 10-min heat shock at 44 °C (*HS*). The *open arrowheads* indicate the position of Hsc70; *closed arrowheads* point out the position of Hsp70

4.3 Noninducible EC Cells are Deficient in Transcriptional Transactivation of Heat Shock Genes by Stress

The mRNAs coding for Hsp70 strongly accumulate in stressed inducible cells, in fibroblasts or in the EC cell line F9. In noninducible cells, such as 1009 and PCC4, however, the Hsp70 mRNA level does not increase. The very low level detected in unstressed cells remains unchanged (Mezger et al. 1987). Analysis of transcription rates in run-on experiments demonstrates that *hsp70* and *hsp86* gene expression is strongly induced in F9 cells and in fibroblasts (Mezger et al. 1987, 1989; Legagneux et al. 1989). In contrast, the transcription rate of these genes is not increased by stress in the noninducible cell lines 1009 and PCC4. The absence of heat shock gene transcription stimulation is due to a deficiency in the *trans*-activation mechanism or transcription. This conclusion can be deduced from transfection experiments using a construct containing the CAT gene under the control of the regulatory region of the *Drosophila* thermoinducible *hsp70* gene. The CAT gene expression was strongly stimulated by stress in fibroblasts and in F9 EC cells, demonstrating that the construct we used contained all the sequences required for thermoinduction in mouse cells. Stimulation of CAT gene expression in the noninducible cells 1009 and PCC4 could not be achieved (Mezger et al. 1987). However, after in vitro differentiation of 1009 cells, CAT expression was increased upon a stress.

These results indicate that the noninducible cell lines are deficient in the *trans*-activation mechanism of heat shock genes by stress. This *trans*-activation mechanism is established during the in vitro differentiation of 1009 cells.

4.4 HSE-Binding Activity in Heat-Shocked EC Cells

Noninducible EC cells are expected to be either deficient in the HSF factor itself, or deficient in the mechanism of HSF activation by stress. The binding of heat shock factors to HSE DNA sequences was examined in gel-shift assays with various murine cell extracts (Mezger et al. 1989):

1. With an inducible EC cell line such as F9, the constitutive HSE-binding activity is strongly increased by heat shock.
2. In contrast, heat induction of HSE-binding activity is never observed in the noninducible EC cell lines PCC4, 1009, and the 3-TDM1 EC cell derivative. In contrast, a strong constitutive HSE-binding activity almost disappears after heat shock.
3. During *in vitro* differentiation of the 1009 EC cell line, the heat induction of HSE-binding activity is acquired. The pattern of HSE-binding activity of the differentiated 1009 derivatives is similar to that of fibroblasts.

Thus, transcriptional inducibility of the heat shock genes by stress correlates with an elevation of HSE-binding activity, whereas noninducibility correlates with the absence of increased HSE-binding activity upon heat shock. Though we cannot exclude that the noninducible cells are deficient in a form of HSF capable of being activated, it seems more likely that these cells are deficient in the mechanism of HSF activation by stress.

Few cell systems with deficient heat-shock responses have been described. One of them, murine erythroleukemia cells, also lack a heat-shock induced h*sp70* transcription. However, in this case stress activates the DNA binding properties of a heat shock factor (Hensold et al. 1990). The deficiency of the embryonic cells is clearly different from this.

5 Concluding Remarks

The functional significance of the developmental regulation of the 70-kD heat shock protein synthesis in the preimplantation mouse embryo remains unclear, but it appears promising as a tool to monitor the onset of embryonic transcription in the mouse and other mammals (Dr. Jean Paul Renard, personal communication), or to study the mechanism of a heat shock gene activation by stress. Since the heat activation of HSF cannot be obtained in some EC cells when undifferentiated, these cells may provide an interesting system to study the stress activation of the heat shock factor.

References

Allen RL, O'Brien DA, Eddy EM (1988a) A novel hsp70-like protein (P70) is present in mouse spermatogenic cells. Mol Cell Biol 8:828–832

Allen RL, O'Brien DA, Jones CC, Rockett DL, Eddy EM (1988b) Expression of heat shock proteins by isolated mouse spermatogenic cells. Mol Cell Biol 8:3260–3266

Amin J, Anathan J, Voellmy R (1988) Key features of heat shock regulatory elements. Mol Cell Biol 8:3761–3769

Anderson NL, Giometti CS, Gemmell MA, Nance SL, Anderson NG (1982) A two-dimensional electrophoretic analysis of the heat-shock-induced proteins of human cells. Clin Chem 28:1084–1092

Barnes FL, Robl JM, First NL (1987) Nuclear transplantation in mouse embryos: assessment of nuclear function. Biol. Reprod. 36:1267–1274

Barnier JV, Bensaude O, Morange M, Babinet C (1987) Mouse 89 kD heat shock protein. Two polypeptides with distinct developmental regulation. Exp Cell Res 170:186–194

Barra J, Renard JP (1988) Diploid mouse embryos constructed at the late 2-cell stage from haploid parthenotes and androgenotes can develop to term. Development 102:773–779

Bensaude O, Morange M (1983) Spontaneous high expression of heat–shock proteins in mouse embryonal carcinoma cells and ectoderm from day 8 mouse embryo. EMBO J 2:173–177

Bensaude O, Babinet C, Morange M, Jacob F (1983) Heat–shock proteins, first major products of zygotic gene activity in mouse embryo. Nature 305:331–333

Bolton VN, Oades PJ, Johnson MH (1984) The relationship between cleavage, DNA replication, and gene expression in the mouse 2-cell embryo. J Embryol. Exp. Morphol. 79:139–163

Curci A, Bevilacqua A, Mangia F (1987) Lack of heat-shock response in preovulatory mouse oocytes. Dev Biol 123:154–160

Dooley TP, Miranda M, Jones NC, DePamphilis ML (1989) Transactivation of the adenovirus EIIa promoter in the absence of adenovirus EIA protein is restricted to mouse oocytes and preimplantation embryos. Development 107:945–956

Evans MJ, Kaufman MH (1981) Establishment in culture of pluripotential cells from mouse embryos. Nature 292:154–156

Flach G, Johnson MH, Braude PR, Taylor RAS, Bolton VN (1982) The transition from maternal to embryonic control in the 2-cell mouse embryo. EMBO J 1:681–686

Fornace AJ, Mitchell JB (1986) Induction of B2 RNA polymerase III transcription by heat shock: enrichment for heat shock induced sequences in rodent cells by hybridization subtraction. Nucl Acids Res 14:5793–5811

Fornace AJ, Alamo I, Hollander MC, Lamoreaux E (1989) Induction of heat shock protein transcripts and B2 transcripts by various stresses in Chinese hamster cells. Exp Cell Res 182:61–74

Giebel LB, Dworniczak BP, Bautz EKF (1988) Developmental regulation of a constitutively expressed mouse mRNA encoding a 72-kDa heat shock-like protein. Dev Biol 125:200–207

Grinfeld S, Gilles J, Jacquet P, Baugnet-Mahieu L (1987) Late division kinetics in relation to modification of protein synthesis in mouse eggs blocked in the G2 phase after X-irradiation. Int J Radiat Biol 52:77–90

Hensold JO, Hunt.CR, Calderwood SK, Housman DE, Kingston RE (1990) DNA binding of heat shock factor to the heat shock element is insufficient for transcriptional activation in murine erythroleukemia cells. Mol Cell Biol 10:1600–1608

Hickey E, Brandon SE, Smale G, Lloyd D, Weber LA (1989) Sequence and regulation of a gene encoding a human 89-kilodalton heat shock protein. Mol Cell Biol 9:2615–2626

Hoeffler WK, Kovelman R, Roeder RG (1988) Activation of transcription factor IIIC by the adenovirus E1A protein. Cell 53:907–920

Hoffman T, Hovermann B (1988) Heat-shock proteins, Hsp84 and Hsp86 of mice and men: two related genes encode formerly identified tumour-specific transplantation antigens. Gene 74:491–501

Howe CC, Solter D (1979) Cytoplasmic and nuclear protein synthesis in preimplantation mouse embryos. J Embryol Exp Morphol 52:209–225

Howlett SK (1986) The effect of inhibiting DNA replication in the one-cell mouse embryo. Roux's Arch. Dev Biol 195:499–505

Howlett SK, Bolton VN (1985) Sequence and regulation of morphological and molecular events during the first cell cycle of mouse embryogenesis. J Embryol Exp Morphol 87:175–206

Howlett SK, Barton SC, Surani MA (1987) Nuclear cytoplasmic interactions following nuclear transplantation in mouse embryos. Development 101:915–923

Imperiale MJ, Kao H, Feldman LT, Nevins JR, Strickland S (1984) Common control of the heat shock gene and early adenovirus genes: evidence for a cellular E1A-like activity. Mol Cell Biol 4:867–874

Jakob H, Nicolas J (1987) Mouse teratocarcinoma cells. Methods Enzymol 151:66–81

Jansen-Durr P, Boeuf H, Kédinger C (1989) Cooperative binding of two E2F molecules to an E1A-responsive promoter is triggered by the adenovirus E1a, but not by a cellular E1a-like activity. EMBO J 8:3365–3370

Kao HT, Nevins JR (1983) Transcriptional activation and subsequent control of the human heat shock gene during adenovirus infection. Mol Cell Biol 3:2058–2065

Kothary R, Clapoff S, Darling S, Perry MD (1989) Inducible expression of an hsp68-lacZ hybrid gene in transgenic mice. Development 105:707–714

Krawczyk Z, Mali P, Parvinen M (1988) Expression of a testis-specific hsp70 gene-related RNA in defined stages of rat seminiferous epithelium. J Cell Biol 107:1317–1323

La Thangue NB, Rigby PWJ (1987) An adenovirus E1A–like transcription factor is regulated during the differentiation of murine embryonal carcinoma stem cells. Cell 49:507–513

La Thangue NB, Thimmapaya B, Rigby PWJ (1990) The embryonal carcinoma stem cell E1A-like activity involves a differentiation-regulated transcription factor. Nucl Acids Res 18:2929–2938

Lee SJ (1990) Expression of HSP86 in male germ cells. Mol Cell Biol 10:3239–3242

Legagneux V, Mezger V, Quélard C, Barnier JV, Bensaude O, Morange M (1989) High constitutive transcription of HSP86 gene in murine embryonal carcinoma cells. Differentiation 41:42–48

Levine RA, LaRosa GJ, Gudas LJ (1984) Isolation of cDNA clones for genes exhibiting reduced expression after differentiation of murine teratocarcinoma stem cells. Mol Cell Biol 4:2142–2150

Martin G (1981) Isolation of a pluripotent cell line from early mouse embryos cultured in medium conditioned by teratocarcinoma stem cells. Proc Natl Acad Sci USA 78:7634–7638

Martinez-Salas E, Linney E, Hassell J, DePamphilis ML (1989) The need for enhancers in gene expression appears during mouse development with the formation of the zygotic nucleus. Genes Dev 3:1493–1506

Matsumoto M, Fujimoto H (1990) Cloning of a hsp70-related gene expressed in mouse spermatids. Biochem Biophys Res Commun 166:43–49

McDaniel D, Caplan AJ, Lee MS, Adams CC, Fishel BR, Gross DS, Garrard WT (1989) Basal-level expression of the yeast HSP82 gene requires a heat shock regulatory element. Mol Cell Biol 9:4789–4798

Mezger V, Bensaude O, Morange M (1987) Deficient activation of heat shock gene transcription in embryonal carcinoma cells. Dev Biol 124:544–550

Mezger V, Bensaude O, Morange M (1989) Unusual levels of HSE-binding activity in embryonal carcinoma cells. Mol Cell Biol 9:3888–3896

Morange M, Diu A, Bensaude O, Babinet C (1984) Altered expression of heat shock proteins in embryonal carcinoma and mouse early embryonic cells. Mol Cell Biol 4:730–735

Muller WU, Li GC, Goldstein LS (1985) Heat does not induce synthesis of heat shock proteins or thermotolerance in the earliest stages of mouse embryo development. Int J Hyperthermia 1:97–102

Nevins JR (1982) Induction of the synthesis of a 70,000 dalton mammalian heat shock protein by the adenovirus E1a gene product. Cell 29:913–919

Onclercq R, Gilardi P, Lavenu A, Cremisi C (1988) c-myc products trans-activate the adenovirus E4 promoter in EC stem cells by using the same target sequence as E1A products. J Virol 62:4533–4537

Pfeiffer SE, Jakob H, Mikoshiba K, Dubois P, Guenet JL, Nicolas JF, Gaillard J, Chevance G, Jacob F (1981) Differentiation of a teratocarcinoma cell line: preferential development of cholinergic neurons. J Cell Biol 88:57–66

Poueymirou WT, Schultz RM (1989) Regulation of mouse preimplantation development: inhibition of synthesis of proteins in the two-cell embryo that require transcription by inhibitors of cAMP-dependent protein kinase. Dev Biol 133:588–599

Raychaudhuri P, Rooney R, Nevins JR (1987) Identification of an E1A-inducible cellular factor that interacts with regulatory sequences within the adenovirus E4 promoter. EMBO J 6:4073–4081

Rebbe NF, Hickman WS, Ley TJ, Stafford DW, Hickman S (1989) Nucleotide sequence and regulation of a human 90-kDa heat shock protein gene. J Biol Chem 264:15006–15011

Reichel R, Kovesdi I, Nevins JR (1987) Developmental control of a promoter-specific factor that is also regulated by the E1A gene product. Cell 48:501–506

Simon MC, Kitchner K, Kao HT, Hickey E, Weber L, Voellmy R, Heintz N, Nevins JR (1987) Selective induction of human heat shock transcription by the adenovirus E1A gene products, including the 12S E1A product. Mol Cell Biol 7:2884–2890

Simon MC, Fisch TM, Benecke BJ, Nevin JR, Heintz N (1988) Definition of multiple, functionally distinct TATA elements, one of which is a target in the hsp70 promoter for E1A regulation. Cell 52:723–729

Singh MK, Yu J (1984) Accumulation of a heat shock-like protein during differentiation of human erythroid cell line K562. Nature 309:631–633

Strickland S (1981) Mouse teratocarcinoma cell : prospects for the study of embryogenesis and neoplasia. Cell 24:277–278

Suemori H, Hashimoto S, Nakatsuji N (1988) Presence of the adenovirus E1A-like activity in preimplantation stage mouse embryos. Mol Cell Biol 8:3553–3555

Taylor KD, Piko L (1987) Patterns of mRNA prevalence and expression of B1 and B2 transcripts in early mouse embryos. Development 101:877–892

Theodorakis NG, Zand DJ, Kotzbauer PT, Williams GT, Morimoto RI (1989) Hemin-induced transcriptional activation of the HSP70 gene during erythroid maturation in k562 cells is due to a heat shock factor-mediated stress response. Mol Cell Biol 9:3166–3173

Ullrich SJ, Robinson EA, Law LW, Willingham M, Appella E (1986) A mouse tumor-specific transplantation antigen is a heat shock-related protein. Proc Natl Acad Sci USA 83:3121–3125

Van Blerkom J, Brockway GO (1975) Qualitative patterns of protein synthesis in the preimplantation mouse embryo. Dev Biol 44:148–157

Vasseur M, Condamine H, Duprey P (1985) RNAs containing B2 repeated sequences are transcribed in the early stages of mouse embryogenesis. EMBO J 7:1749–1753

White RJ, Stott D, Rigby PWJ (1989) Regulation of RNA polymerase III transcription in response to F9 embryonal carcinoma stem cell differentiation. Cell 59:1081–1092

Williams GT, McClanahan TK, Morimoto RI (1989) Ella transactivation of the human HSP70 promoter is mediated through the basal transcriptional complex. Mol Cell Biol 9:2574–2587

Wittig S, Hensse S, Keitel C, Elsner C, Wittig B (1983) Heat shock gene expression is regulated during teratocarcinoma cell differentiation and early embryonic development. Dev Biol 96:507–514

Wu BJ, Kingston RE, Morimoto RI (1986) Human hsp70 promoter contains at least two distinct regulatory domains. Proc Natl Acad Sci USA 83:629–633

Zakeri ZF Wolgemuth DJ (1987) Developmental-stage-specific expression of the hsp70 gene family during differentiation of the mammalian germ line. Mol Cell Biol 7:1791–1796

Zakeri ZF, Ponzetto C, Wolgemuth DJ (1988a) Translational regulation of the novel haploid-specific transcripts for the c-abl proto-oncogene and a member of the 70 kDa heat-shock protein gene family in the male germ line. Dev Biol 125:417-422

Zakeri ZF, Wolgemuth DJ, Hunt CR (1988b) Identification and sequence analysis of a new member of the mouse HSP70 gene family and characterization of its unique cellular and developmental pattern of expression in the male germ line. Mol Cell Biol 8:2925–2932

11 Transcriptional Regulation
of Human *Hsp70* Genes:
Relationship Between Cell Growth, Differentiation, Virus Infection, and the Stress Response

Benette Phillips and Richard I. Morimoto[1]

1 Introduction

Our understanding of the regulation of the 70-kD family of heat shock genes during human development is necessarily reliant on information obtained from animal studies and from human cell lines which can be induced to differentiate. It has been reported that expression of certain members of the mouse 70-kD heat shock family is regulated during embryogenesis (Bensaude et al. 1983; Bensaude and Morange 1983; Kothary et al. 1987; see Chap. 10) and during spermatogenesis (Krawczyk et al. 1987a; Zakeri and Wolgemuth 1987; Allen et al. 1988; Zakeri et al. 1988; see Chap. 9). Such studies suggest that expression of human heat-shock genes is developmentally regulated. One can also argue, based on our current understanding of the function of this family of genes, that the 70-kD heat shock proteins must play an important role during development and that their expression must, of necessity, be tightly regulated during this process. Heat shock proteins appear to participate in the assembly-disassembly of macromolecular complexes (Georgopoulos and Ang 1990), in transport of polypeptides into organelles such as mitochondria (Chirico et al. 1988; Deshaies et al. 1988; Kang et al. 1990), in trafficking of polypeptides through the endoplasmic reticulum (Dorner et al. 1987), in vesicular uncoating (Ungewickell 1985), in the protection of newly synthesized proteins (Beckmann et al. 1990), and in protein degradation (Chiang et al. 1989). All of these processes must be critical to the successful execution of developmental programs, during which cells alter their patterns of protein synthesis and secretion, respond to chemical and hormonal mediators, undergo morphological changes, migrate and interact with other cells, and undergo changes in their growth state. Considering also, as will be detailed in this chapter, the unique ability of the human *hsp70* genes to show altered expression in response to environmental stimuli, it seems quite likely that the expression of these genes is highly regulated during development.

The human *hsp70 gene* family codes for five, or possibly six, known proteins: (1) Hsp70, which is constitutively expressed but is also highly heat inducible; (2) P72 (clathrin uncoating ATPase, also commonly called Hsc70), which is constitutively expressed and only slightly heat inducible; (3) Grp78 (Bip), which resides in the lumen of the endoplasmic reticulum and is constitutively expressed but is readily induced by a variety of factors; (4) P75, a mitochondrial member of the family; and (5) one or

[1] Dept. of Biochemistry, Molecular Biology and Cell Biology, Northwestern University, 2153 Sheridan Road, Evanston, Illinois, 60208, USA

Results and Problems in Cell Differentiation 17
Heat Shock and Development
Hightower and Nover (Eds.)
©Springer-Verlag Berlin Heidelberg 1991

possibly two members which are strictly heat-inducible. The interest of our laboratory in the regulation of *hsp70* and *grp78* stems from the finding, both by us and by other investigators, that in cultured cells, the expression of these genes is remarkably sensitive to changes in the physiological state. These changes include imposed alterations in the cellular environment, e.g., raising the temperature or infecting with viruses, but also encompass fluctuations in cellular physiology which accompany transit through the cell cycle, growth, or differentiation. Although many of the conditions which alter expression of these genes were identified in the early phase of our investigations, the list of modulators is constantly growing. The intent of this Chapter is to update the list of factors to which endogenous *hsp70* and *grp78* are responsive, also including the few factors which have been shown to modulate expression of *p72*. We will then describe our attempts to understand the mechanisms by which several of these factors alter the rate of transcription of *hsp70* or *grp78*. Our laboratory has employed multiple avenues in our attempts to understand these mechanisms, and conclusions obtained using one approach have not always been confirmed by the results obtained using a different approach. The regulation of these genes appears to be very complex, and even when multiple approaches are used to investigate this regulation, an additional experiment will often yield unanticipated results.

2 Factors Which Alter the Expression of *Hsp70*, *Grp78*, and *P72*

2.1 Factors Which Alter Expression of *Hsp70*

Table 1 provides a compilation of factors which have been shown to alter the expression of the endogenous *hsp70*, *grp78*, and *p72* genes in human or simian cells. The level(s) of expression (transcription, message level, protein synthesis, protein level) at which changes have been observed is indicated in the text.

2.1.1 Determinants of Basal Expession

Before discussing the many factors which modulate *hsp70* expression in human cells, it should be noted that an examination of several different human cell lines, most of which are tumor derived, has revealed that basal levels of expression of *hsp70* vary greatly by cell type (B. Phillips, M. Myers, unpubl. data). The rate of *hsp70* transcription in the choriocarcinoma cell line JEG-3 is only 3% of that in HeLa cells and 1% of that in 293 cells (human embryonic kidney cells expressing the adenovirus E1A protein). Hsp70 mRNA and protein levels reflect this difference in transcription rates. Transcription rates and protein levels of *p72* and *grp78* are comparable in all three cell lines, however. Examination of total RNA from several tumor-derived and primary human cell lines revealed that these cell lines fell into one of two classes, exhibiting either no detectable *hsp70* RNA or showing levels comparable to that in HeLa cells. Neither the tissue of origin nor the transformation state appeared to be a determining

Table 1. Factors which modulate expression of *hsp70*, *grp78*, and *p72* in primate cells

Gene	Factor	References[a]
hsp70	Heat	1-3
	Heavy metals	1, 2, 4
	Amino acid analogs	1, 2
	Chloroethylnitrosoureas	5
	Infection by adenovirus	6-8
	Infection by HSV-1	
	Infection by cytomegalovirus	9
	Mock infection	
	Serum stimulation	10
	Position in cell cycle	11
	Interleukin-2	12
	Selected prostaglandins	13
	Hemin	14, 15
	N-methylformamide	16
	2-deoxyglucose	1
	A23187	1
grp78	glycosylation inhibitors	1
	2-deoxyglucose	1
	Heat	
	Heavy metals	1
	Amino acid analogs	1
	A23187	1
	Hemin	15
	Infection by paramyxovirus	16
p72	Infection by cytomegalovirus	9
	Serum withdrawal	18

[a] Key to references: (1) Wu et al. 1985; (2) Mosser et al. 1988; (3) Watowich and Morimoto 1988; (4) Wu et al. 1986a; (5) Schaefer et al. 1988; (6) Nevins 1982; (7) Kao and Nevins 1983; (8) Wu et al. 1986b; (9) Santomenna and Colberg-Poley 1990; (10) Wu and Morimoto 1985; (11) Milarski and Morimoto 1986; (12) Ferris et al. 1988; (13) Santoro et al. 1989; (14) Singh and Yu 1984; (15) Theodorakis et al. 1989; (16) Richards et al. 1988; (17) Peluso et al. 1978; (18) Chiang et al. 1989

factor in the basal level of expression. These cell lines are currently being analyzed further in our laboratory to compare transcription rates and protein levels.

Because 293 cells probably represent a case of uniquely high constitutive levels of expression, due to high levels of a viral transactivator to which the hsp70 gene responds, our preliminary results indicate that human cell lines fall into two classes with respect to *hsp70* expression. The question of what factor(s) influence the basal level of expression is obviously of great interest. A cellular factor with E1A-like activity has been postulated to exist in certain human cell lines, including HeLa (Imperiale et al. 1984); however, one of the cell lines with a higher basal level of transcription, IMR90, is derived from normal embryonic lung. We plan to analyze *hsp70* RNA levels in other primary human cell lines, and we have not discounted the possibility that in those cell lines with undetectable basal levels of expression, a transcriptional repressor is present.

2.1.2 Classical Stress-Response Inducers

The induction of human Hsp70 synthesis by heat shock, heavy metals, and amino acid analogs has been intensively studied (Wu et al. 1985, 1986b; Mosser et al. 1988; Watowich and Morimoto 1988). Induction takes place at the transcriptional level and does not require new protein synthesis except in the case of amino acid analog treatment, where newly synthesized misfolded proteins are the apparent signal for the transcriptional response (Kingston et al. 1987; Mosser et al. 1988). Transcriptional induction is transient, so that although the environmental stress is maintained, transcription peaks and then returns to control levels (or to a new basal level which is only slightly higher than control levels), following kinetics which vary with the nature of the inducer (Mosser et al. 1988). Superimposed on the transcriptional induction of *hsp70* during heat shock, the stability of the *hsp70* mRNA increases markedly upon heat shock. Heat Shock does not, however, appear to affect the translatability of the mRNA (Theodorakis and Morimoto 1987).

It is widely believed that the common signal generated by thermal stress, heavy metals, and amino acid analogs is protein damage. The stress response is not elicited by all cytotoxic agents, however, and may be dependent on a specific type or level of protein damage. This is indicated by the finding that of a group of antineoplastic drugs, all of which have cytocidal effects on human adenocarcinoma cells and many of which can alkylate proteins, only the chloroethylnitrosoureas induced *hsp70* transcription in these cells (Schaefer et al. 1988). For these drugs, which generate a reactive isocyanate moiety, induction was partially but not wholly dependent on protein synthesis, suggesting that carbamoylation of newly synthesized proteins may constitute a stress response signal (Kroes et al. 1991).

2.1.3 DNA Viruses

Our laboratory has carried out a systematic investigation of the induction of the *hsp70* family of genes during infection of primate cell by DNA viruses (B. Phillips et al. 1991). Although there are many reports on the induction of cellular genes encoding 70-kD stress proteins following viral infection, most of these studies have focused on a single virus and cell type. Our study has concentrated on the transcriptional induction of these genes in primate cells during infection by four DNA viruses which are members of four different viral classes: (1) adenovirus 5; (2) herpes simplex virus-1 (HSV-1); (3) simian virus 40 (SV40); and (4) vaccinia. (Primate cells are permissive or semi-permissive for the growth of all of these viruses). The response of these genes to viral infection has been analyzed at the protein synthesis level as well. We were particularly interested in assessing the specificity of induction of heat-shock gene expression by viral infection, both with respect to which of the viruses was capable of activating expression of the primate 70-kD heat shock genes and which of the 70-kD members was induced.

Confirming earlier reports from our laboratory and others (Nevins 1982; Kao and Nevins 1983; Wu et al. 1986b), the transcription rate of the *hsp70* gene in HeLa cells increases during the early phase of adenovirus infection. This induction is specific for *hsp70*; transcription of *p72* and *grp78* is not enhanced (Fig. 1). Two-dimensional gel analysis of ^{35}S-methionine labeled proteins shows the same selective enhancement of Hsp70 synthesis.

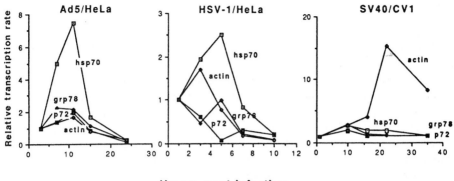

Hours post-infection

Fig. 1. Specificity of transcriptional induction of *hsp70* by DNA viruses. HeLa cells were infected with adenovirus 5 (*Ad5*) or herpes simplex virus-1 (*HSV-1*); monkey CV1 cells were infected with *SV40*. At selected times during infection, samples were taken for nuclear run-on analysis. Labeled transcripts were hybridized to immobilized plasmid DNA containing the coding regions for human *hsp70, p72, grp78,* or *B-actin* genes. Within each experiment, transcription rates for each gene are normalized to the value for that gene at 1 h post-infection

Previous studies using the adenovirus mutant d1312 (Nevins 1982; Kao and Nevins 1983) and studies from our laboratory using 12S and 13S mutant viruses (Wu et al. 1986a) identified the product of the E1A 13S message as the viral transactivator to which *hsp70* responds. The kinetics of *hsp70* induction parallel that of the adenovirus E1A responsive E3 gene; however, *hsp70* transcription begins to decline at approximately 12 h post-infection despite the fact that E1A levels increase throughout infection. This may be due to a sequestering of E1A once viral replication begins.

Despite the apparent ability of E1A to activate a wide range of promoters in cotransfection studies, *hsp70* remains one of the few reported endogenous genes which is E1A inducible. The induction occurs following infection of actively growing HeLa cells and thus is not related to a growth stimulatory effect of viral infection.

Infection of HeLa cells by HSV-1 also results in transcriptional induction of *hsp70* but not *grp78* or *p72* (Fig. 1). The onset of induction coincides with that of the viral thymidine kinase gene; furthermore, induction requires viral protein synthesis. This rules out a role for the virion-associated transactivator VP16 and suggests that activation is mediated by an HSV-1 immediate early protein. There are three HSV-1 immediate early proteins with known transactivating activity, and the identity of the protein(s) responsible for enhancement of *hsp70* transcription has not been determined.

The increase in transcription of *hsp70* following infection of HeLa cells with HSV-1 does not result in increases in Hsp70 mRNA levels or protein synthesis. Instead, one observes that in a background of declining mRNA levels and protein synthesis rates for most cellular genes, *hsp70* mRNA levels and synthesis rates remain constant, at least during the early phase of infection. HSV-1 induced degradation of host messages, with consequent effects on protein synthesis, is a well-established phenomenon, and the *hsp70* mRNA has been shown to be susceptible to such degradation (Kwong and Frenkel 1987). Presumably, the steady-state level of mRNA represents a balance between virus-induced transcription and degradative effects on host messages.

Another member of the human herpesvirus family, human cytomegalovirus (HCMV), has been reported to induce expression of the *hsp70* gene at both the mRNA level and at the level of protein accumulation in human foreskin fibroblasts (Santomenna and Colberg-Poley 1990). Increased levels of mRNA were observed during the early phase of infection (8-48 h). The mRNA levels of other 70K family members were not determined. Increases in accumulation of both P72 and Hsp70 were observed at both 48 and 72 h post-infection by Western blot analysis, and the levels of both proteins were highest at 72 h post-infection even though *hsp70* mRNA levels had returned to basal levels. It should be noted that the fibroblasts had been grown to confluency and then serum starved for 3 days prior to infection, and it is not stated whether these growth conditions were required for viral induction.

Although it has been reported that infection of monkey CV1 cells with SV40 resulted in an increase in the synthesis rate of Hsp70, as determined by both 1-D and 2-D gel electrophoresis (Khandjian and Turler 1983), nuclear run-on analysis performed in our laboratory has consistently failed to reveal transcriptional induction of *hsp70* (or of *grp78* or *p72*) during the first 48 h of SV40 infection (see Fig. 1). Transcription of the monkey *hsp70* gene is enhanced following infection of CV1 cells with either adenovirus or HSV-1, however. To rule out the possibility that there is post-transcriptional regulation of Hsp70 synthesis during SV40 infection, we performed 2-D gel analysis of protein synthesis levels and failed to confirm the earlier reported results. These same gels showed viral late protein synthesis commencing at 22 h post-infection, indicating that viral infection was proceeding normally.

Vaccinia virus also failed to enhance transcription of any of the HeLa 70 kD heat shock genes examined, despite cytopathic effects apparent soon after infection. Our results thus suggest that transcriptional induction of *hsp70* by viruses is not a general phenomenon and may be limited to those viruses encoding early proteins with promiscuous transactivating capabilities.

When performing nuclear run-on analysis of virus-infected HeLa cells, samples from mock-infected cultures were always included. Mock infections, designed to mimic the conditions of virus infection, were carried out using small volumes of serum-free medium with frequent rocking of the dishes during the 1-h infection period. Frequently, although not consistently, mock infection resulted in transcriptional induction of *hsp70*, although the magnitude of this induction was always less than that observed when cells were infected with virus. The basis for this induction is still not known, although using larger volumes to mock infect, using conditioned medium, or shortening the infection period to as little as 15 min, did not eliminate "mock induction." Such a phenomenon illustrates the extreme sensitivity of this gene to environmental pertubation and mandates the use of appropriate controls when examining *hsp70* regulation.

From the standpoint of developmental regulation of human *hsp70*, the most important result from our studies on the activation of this gene by virus infection is the responsiveness of *hsp70* to viral transactivators. Although cellular counterparts of these viral transactivators have not yet been identified, the responsiveness of *hsp70* to such factors suggests that this gene may be very sensitive to changes in levels of transcriptional modulators during development.

2.1.4 Cell Cycle Regulation, Growth Factors

Of more direct relevance to the postulated modulation of *hsp70* expression during human development is the cell-cycle specific regulation of its expression and its regulation by factors that have either growth-stimulatory or antiproliferative effects. The initial indication that *hsp70* is a growth regulated gene was the finding that serum stimulation of serum-starved HeLa or 293 cells resulted in tenfold increases in *hsp70* mRNA levels at 12 h post-serum stimulation (Wu and Morimoto 1985). Further analysis of this enhancement in 293 cells indicated that regulation was at the level of transcription. Subsequent experiments from our laboratory, in which mitotic detachment was used to obtain a synchronous population of HeLa cells, revealed a 10–15-fold increase in *hsp70* mRNA levels at the G_1-S border (Fig. 2); as a consequence, both the synthesis and absolute levels of Hsp70 protein are also temporally regulated during the cell cycle (Milarski and Morimoto 1986).

From the similar temporal patterns of expression, it seems quite likely that serum stimulation of *hsp70* expression reflects the entry of cells at the G_0 -G_1 border into the cell cycle, although there is no data directly addressing that point. The reported stimulation of *hsp70* mRNA levels and Hsp70 protein biosynthesis in quiescent IL-2 deprived human T-lymphocytes by IL-2 stimulation (Ferris et al. 1988), which also induces cell proliferation, may also reflect the entry of quiescent cells into the cell cycle. These results do indicate that growth regulation of *hsp70* can be observed in freshly isolated human cells.

There are also studies linking the nonproliferative or quiescent state with elevated expression of *hsp70*, however. Contradictory to the results on IL-2 stimulation, Kaczmarek et al. (1987) reported that any growth stimulus, defined as a stimulus which caused elevated c-fos expression, regardless of whether or not it led to proliferation, cause a sharp and rapid drop in the levels of *hsp70* mRNA in freshly isolated human peripheral blood mononuclear cells. There is also a recent report that in the human erythroleukemic cell line K562, treatment with prostaglandins which cause cessation of proliferation, but not with prostaglandins which lack antiproliferative activity, induces increased synthesis and accumulation of Hsp70 protein (Santoro et al. 1989). Whatever the resolution of these seemingly contradictory results, the sensitivity of *hsp70* to a perturbation in the growth state is quite apparent.

2.1.5 Agents Inducing Differentiation

Of the most direct relevance to developmental regulation of *hsp70* are two studies noting effects on *hsp70* expression of agents which induce differentiation of cultured human cells. Singh and Yu (1984) reported an accumulation of both *hsp70* mRNA and protein in K562 cells upon treatment with hemin, which causes these cells to differentiate and synthesize hemoglobin. These results have been confirmed and extended by our laboratory (Theodorakis et al. 1989), with nuclear run-on analysis indicating that activation occurs at the level of transcription. Our studies on the mechanism of hemin activation are discussed in detail below. In contrast, differentiation of HL-60 promyelocytic cells by N-methylformamide was reported to be accompanied by decreased synthesis of a 70-kD heat shock protein (Richards et al. 1988), although the 1-

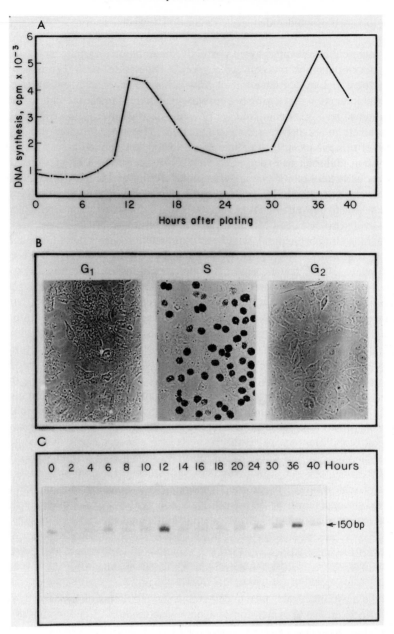

Fig. 2A-C. Cell synchrony and levels of *hsp70* mRNA. **A** DNA synthesis was measured by ^{3}H-thymidine incorporation at the indicated times after plating harvested mitotic HeLa cells. **B** Autoradiographic analysis of G_1, S, and G_2 populations of synchronized cells. **C** Relative levels of hsp70 mRNA throughout the cell cycle as determined by S1 nuclease analysis

D gel analysis presented did not allow a definitive identification of this protein as Hsp70.

2.1.6 Other Agents

During studies in our laboratory on the regulation of expression of *grp78* (detailed below), it was found that treatment of 293 cells with either the calcium ionophore A23187 or the glucose analog 2-deoxyglucose resulted in a decrease in *hsp70* mRNA levels. Nuclear run-on analysis revealed that this repression was not at the level of transcription, suggesting an effect on message stability (Watowich and Morimoto 1988).

2.2 Factors Which Alter the Expression of *Grp78*

Although the list of factors known to induce expression of endogenous *grp78* in primate cells is not as extensive as that just described for *hsp70*, it is becoming increasingly apparent that expression of this member of the 70-kD heat shock family is also very sensitive to environmental perturbation.

Consistent with Grp78 being a resident of the endoplasmic reticulum is the sensitivity of expression of its gene to factors which interfere with glycosylation of proteins in the endoplasmic reticulum, e.g., glucose analogs, tunicamycin, and castanospermine. Induction by these agents occurs at the level of transcription (Watowich and Morimoto 1988) with consequent increases in synthesis of Grp78 protein.

Other environmental perturbations, some of which also induce *hsp70* and whose effects are not specific to the endoplasmic reticulum, also enhance expression of *grp78*. These include heat shock, heavy metals, amino acids analogs, and the calcium ionophore A23187 (Watowich and Morimoto 1988). We have consistently observed transcriptional induction of *grp78* in K562 cells in response to a 42 °C heat shock, but this induction occurs with kinetics quite different from that of *hsp70*. Whereas induction of *hsp70* is quite rapid (within minutes), enhanced transcription of *grp78* is not observed until cells have been maintained at 42 °C for at least 2–3 h, and elevated *grp78* transcription is still apparent when the transcription rates of *hsp70* have returned to basal levels (L. Sistonen, unpubl. data).

Treatment of K562 cells with hemin, which, as discussed above, results in elevated transcription of *hsp70*, has a similar and quite potent effect on *grp78* transcription. In cells continuously exposed to hemin for 96 h, transcription of both genes increases and then decreases in parallel. Increased transcription is apparent as early as 3 h following hemin treatment (L. Sistonen, unpubl. data). The nature of the inducing signal and whether this is the same signal which triggers increased transcription of *hsp70* is not yet established.

Although none of the DNA viruses examined induced expression of *grp78*, infection of CV1 cells with the paramyxovirus SV5 results in elevated expression of this gene (Peluso et al. 1978). The mechanism of this induction, which occurs at the level of transcription, has been investigated in our laboratory (discussed in more detail in

Sect. 3) and appears to be due to a high flux of a specific viral protein through the endoplasmic reticulum. That such induction is highly specific is indicated by the fact that no transcriptional induction was observed during HSV-1 infection of primate cells, during which several species of viral glycoproteins are abundantly synthesized (B. Phillips et al. 1991).

It has previously been reported that RNA levels and rates of synthesis of Grp78 are increased following infection of CV1 cells with an SV40 vector expressing a mutant, malfolded form of influenza virus haemagglutinin which fails to be transported out of the endoplasmic reticulum. Such induction was not observed with a vector expressing the wild-type haemagglutinin, however (Kozutsumi et al. 1988), and there are no reported studies on the response of *grp78* to influenza virus infection.

2.3 Factors Which Alter the Expression of *P72*

Of the three *hsp70* family members whose regulation has been under study in our laboratory, *p72* appears to be the least responsive to imposed stresses or cyclic fluctuations in the cellular environment. Despite the presence in its promoter of a heat-shock consensus sequence which has been shown in other heat shock genes to mediate heat-induced transcriptional activation, the gene is only slightly, if at all, heat responsive. None of the viruses examined in our laboratory alters the expression of the endogenous gene in infected human or monkey cells; however, in a recent report on the induction of *hsp70* expression in human cytomegalovirus infected human diploid fibroblasts, increased accumulation of P72 protein was observed very late in infection (Santomenna and Colberg-Poley 1990). Further studies are required to confirm this result and to determine at what level regulation occurs.

In studies to determine whether P72 facilitates the translocation of proteins destined for lysosomal degradation, it was found that serum starvation of human lung fibroblasts resulted in increased levels of P72 in the cytoplasm. This observation was based on Western blot analysis of subcellular fractions, which showed very low levels of cytosolic P72 in fibroblasts maintained in serum and indicated that serum deprivation had only minimal effects on P72 levels in the mitochondrial plus lysosomal fraction (Chiang et al. 1989). Further studies to clarify these observations appear to be necessary in this case also.

3 Mechanisms of Activation

We describe below our ongoing efforts to understand the complexities of the regulation of the human *hsp70* and *grp78* genes. We begin with heat shock induction of *hsp70*, not only because an awareness of certain aspects of this regulation is necessary to appreciate our findings on hemin induction, but also because it is not unreasonable to suggest, as our findings on hemin induction indicate, that activation through a heat shock pathway could occur at certain stages of development. We conclude with a discussion of the mechanisms of viral activation of *hsp70* and *grp78*, which illustrate our attempts to understand the mechanisms by which *hsp70* responds to transactivating factors and to begin to dissect *grp78* regulatory pathways.

3.1 Heat Shock Induction

In this section we attempt to summarize and update studies on the mechanism of the transcriptional activation of the human *hsp70* gene by thermal stress. Where relevant, we incorporate information obtained from studies in yeast and *Drosophila*, two other systems where the mechanism is being intensively studied. To facilitate an understanding of the work described in this and the following sections, the sequence of the proximal 120 nucleotides of the human *hsp70* promoter and the elements in this region which are potential binding sites for known transcription factors are shown in Fig. 3.

The following are characteristics of the transcriptional activation of the human *hsp70* gene by heat which are relevant to an understanding of mechanism:

1.   Increased transcription is mediated by the interaction of a specific factor (heat shock transcription factor, HSF) with a specific promoter element, the heat shock element (HSE); (Kingston et al. 1987; Goldenberg et al. 1988; Mosser et al. 1988; Abravaya et al. 1991a,b).

2.   The HSE is presently viewed as an array of inverted repeats of NGAAN units (Xiao and Lis 1988; Amin et al. 1988); in the human *hsp70* gene, the array consists of five NGAAN sites (numbered in Fig. 3), three of which are perfect matches to the consensus. (The number and arrangement of these units in individual HSEs in different heat shock promoters can vary greatly, however).

3.   A deletion mutant of the human *hsp70* promoter which retains only the proximal two sites in the HSE is not heat shock responsive when transfected into HeLa cells (Williams and Morimoto 1990). These results are consistent with studies in *Drosophila* which indicate that functional heat shock elements comprise at least three contiguous NGAAN units (Amin et al. 1988). A linker scanner mutant in which the most proximal NGAAN site is lost shows reduced heat shock inducibility (Williams and Morimoto 1990). In vivo genomic footprinting reveals that during heat shock, all five sites in the HSE of the human *hsp70* promoter are occupied by protein (Abravaya et al. 1991a,b).

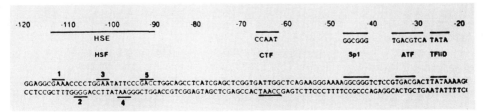

Fig. 3. Sequence of the proximal region of the human hsp70 promoter. Sites which are perfect or imperfect matches to consensus sites for known transcription factors are underlined. The consensus sequence and transcription factors which bind to these sites in vitro are indicated *above* each *underlined* site. NGAAN sequences, comprising an array of repeated inverted units characteristic of heat shock elements, are *underlined* and *numbered*

Taken together, these data suggest that all five of the NGAAN units in the HSE of the human *hsp70* promoter may be required for maximal heat shock responsiveness.

4.   Transfection studies indicate that in general, mutation of elements downstream of the HSE (collectively termed the basal promoter) reduces basal and heat induced expression to the same extent. Increasing the spacing between the HSE and the basal region of the promoter by half or full helical turns does not affect heat inducibility (Williams and Morimoto 1990). Genomic footprinting reveals that during heat shock, interactions of factors with the basal elements of the promoter are not perturbed (Abravaya et al. 1991).

5.   Besides HSF, a distinct HSE-binding activity (constitutive HSE-binding activity, CHBA), which in gel shift assays give rise to a complex with a faster mobility than that due to HSF, is present in extracts of non-heat-shocked HeLa cells. The DNA contacts of activated HSF and CHBA as determined by methylation interference are similar but not identical. During heat shcok, the levels of the constitutive HSE-binding activity decrease (Mosser et al. 1988). Genomic footprinting reveals that at 37 °C, the HSE of the human *hsp70* promoter is not occupied by protein (Abravaya et al. 1991). The observation that an HSE-specific binding activity is detected in non-heat-shocked cells in vitro but not in vivo may indicate that this factor is nonnuclear or is somehow prevented from binding the promoter.

6.   Prior to heat shock, HSF is present in an inactive (i.e., non-DNA binding) form (Kingston et al. 1987; Mosser et al. 1988). Thermal stress results in a very rapid activation of HSF, such that the factor can now bind to DNA and stimulate transcription from promoters containing heat-shock consensus elements. This activation does not require new protein synthesis, except when very mild heat-shock temperatures are employed (Zimarino et al. 1990).

   During a continuous 42 °C heat shock of HeLa cells, or during heat shock followed by recovery at 37 °C, the levels of HSF, as measured in whole-cell extracts by gel retardation analysis, and transcription of *hsp70* are very well correlated (Fig. 4) (Abravaya et al. 1991b).

   The requirement for protein synthesis during mild heat shock may indicate that damage to existing cellular proteins sufficient to trigger HSF activation does not occur under these conditions. Nascent polypeptides are known to be more susceptible to thermal denaturation, and damage to them may constitute the primary signal for HSF activation at lower heat shock temperatures.

7.   When HeLa cells are heat shocked at 42 °C, both HSF levels and *hsp70* transcription increase during the first 40 min of heat shock; then, although the cells are maintained at 42 °C, transcription gradually returns to control levels and activated HSF disappears (Mosser et al. 1988). This result suggests that the heat shock response is autoregulated in human cells, as has been postulated for *E. coli* (Tilly et al. 1983; Gross et al. 1990), yeast (Stone and Craig 1990), and *Drosophila* (DiDomenico et al. 1982).

8 .   In HeLa cell cytoplasmic extracts, HSF can be activated by heating, by lowering pH, or by the addition of calcium, urea, or nonionic detergents; all of these factors alter protein conformation. Agents that stabilize protein conformation, such as glycerol, inhibit in vitro activation (Larson et al. 1988; Mosser et al. 1990). Addition of dena-

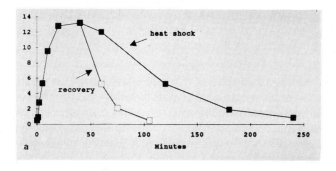

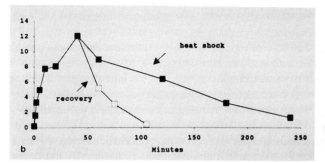

Fig. 4a,b. Correlation between **a** transcription rates of *hsp70* and **b** levels of activated heat shock transcription factor (HSF) during a continuous 42 °C heat shock or during heat shock and recovery. Suspension culture of HeLa cells were maintained at 42 °C for 4 h or were heat shocked at 42 °C for 45 min and allowed to recover at 37 °C. At the indicated times, samples were withdrawn for nuclear run-on analysis to measure the transcription rate of the *hsp70* gene and for gel shift analysis to determine HSF levels

tured protein to a HeLa cell extract does not result in HSF activation. Once in vitro activated, HSF remains in this state even upon termination or reversal of activation conditions (D. Mosser, unpubl. data).

The rate of in vitro activation is very dilution sensitive: only slight dilution of an extract results in pronounced effects on the rate of in vitro activation (K. Sarge, unpubl. data). This dilution sensitivity suggests a role for protein-protein associations in in vitro activation.

9. HSF from non-heat shocked HeLa cells, from heat shocked cells, and from cytoplasmic extracts which have been heated in vitro are chromatographically distinct when fractioned on an anion exchange column or on a gel filtration column. The non-DNA binding form (from control cells) is the least negatively charged and in vivo activated HSF is the most negatively charged. HSF from control cells fractionates on a gel filtration column with an apparent molecular weight of 80–100 kD, the in vitro activated form with an apparent size of 200–300 kD, and in vivo activated HSF with an apparent molecular weight of 500–600 kD. These results indicate that acquisition of DNA binding ability is associated with an increase in both negative charge and native size (K. Sarge, unpubl. data). Pertinent to these results are the data of Larson et al. (1988) suggesting that in vivo activated HSF is more highly phosphorylated than the in vitro activated factor.

10. Reports from two laboratories where cDNAs encoding human HSF have been isolated and sequenced indicate that two distinct cDNAs, which have similar DNA binding domains but which are otherwise quite distinct, have been obtained (Rabindran et al. 1991, Schuetz et al. 1991).

Based on the above considerations, it would be reasonable to propose that elevated temperatures induce oligomerization of HSF, either by direct conformational effects on HSF itself and-or by conformational effects on other cellular proteins which somehow trigger HSF oligomerization. During very mild heat shock, denaturation of nascent polypeptides may play a role in the signalling pathway. Oligomeric HSF, perhaps requiring additional modifications, can then bind to an element in which there is a tandem array of at least three NGAAN binding sites, thereby triggering transcriptional activation.

Such a scenario leaves many questions still to be resolved. One set of questions revolves around the pathway by which multimerization occurs and the composition, both with respect to number and identity, of the transcriptionally active multimer. Whether there is an autoregulatory loop which underlies the decrease in HSF levels during continuous heat shock is a related question. A second set of issues centers on the features of an HSE which contribute to its potency, both as regards the number of units in the array, the match of these units to the consensus, and the position of the array within the promoter, and whether the composition and position of the HSE are the sole determinants of the degree of transcriptional enhancement. Third, the mechanism by which HSF binding to the HSE results in transcriptional induction has not been elucidated. Such a mechanism must accomodate the positional flexibility revealed by our transfection studies as well as the failure of interactions of basal factors to be perturbed during heat shock. Such studies should be facilitated by the availability of cloned HSF. Finally, the function of the constitutive HSE-binding activity, its relationship, if any, to HSF and the possibility that there exists more than one functional HSF awaits clarification.

3.2 Hemin Induction

Treatment of the human erythroleukemic cell line K562 with hemin (a crystallized form of heme) induces these cells to synthesize both embryonic and fetal forms of hemoglobin. A report that Hsp70 protein synthesis was also enhanced by hemin treatment (Singh and Yu 1984), the only reported case so far induction of the human gene during a differentiation process, prompted our laboratory to examine the mechanism of *hsp70* activation. It was quickly established that activation occurred at the transcriptional level; the surprising finding from transfection studies was that only constructs retaining a heat shock element were hemin inducible. This result was strengthened when gel shift analyses revealed the presence of activated HSF in extracts of hemin treated K562 cells (Theodorakis et al. 1989).

We are currently conducting a more detailed characterization of HSF activation in hemin treated K562 cells. Preliminary results indicate that the correlation between HSF levels and *hsp70* transcription rates which is observed in HeLa cells maintained at 42 °C and which seems to apply to K562 cells heat shocked 42 °C as well (Fig. 5) is not maintained when K562 cells are treated with 30 μm hemin (L. Sistonen, unpubl. data).

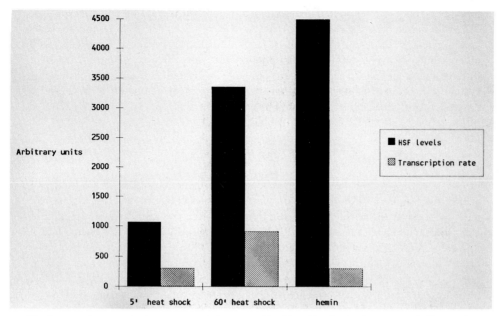

Fig. 5. Lack of correlation between transcription rates of *hsp70* and levels of activated heat shock transcription factor (HSF) in hemin treated K562 cells. K562 cells were heat shocked at 42 °C for 5 min or for 60 min or were treated with 30 μM hemin for 21 h. Samples were then taken for nuclear run-on analysis to measure the transcription rate of the h*sp70* gene and for gel shift analysis to determine HSF levels

Equivalent levels of transcriptional activation were observed in K562 cells which were hemin treated for 21 h or heat shocked at 42 °C for 5 min, yet extracts of hemin treated cells contained four times as much HSF (by gel shift assay) as did extracts of the heat shocked cells (Fig. 5). When hemin treated cells are heat shocked at 42 °C for 30 min, there is a 20-fold increase in the transcription rate of *hsp70* although HSF levels show only a twofold increase. This results raises an important question as to whether there exist multiple pathways for HSF activation, as has been suggested previously (Mosser et al. 1988), and-or whether there are multiple steps involved in a single pathway. Forms of HSF which may be indistinguishable in a gel retardation assay may not be equivalently active with regard to their ability to stimulate transcription. Comparisons of the biochemical properties, e.g., chromatographic characteristics, of HSF from hemin-treated and heat-shocked K562 cells may shed light on this question.

Other features of the pathway by which HSF is activated during hemin treatment have also been further characterized. Activated HSF is first detectable by 30 min after hemin addition. Activated HSF levels peak by 12 h of treatment and then high levels are maintained for at least 2 more days; transcription rates remain elevated during this period. Cycloheximide treatment abolishes both HSF activation and transcriptional induction (L. Sistonen, unpubl. data), a feature which distinguishes hemin-mediated activation of HSF from activation by heat shock and heavy metals.

It is important to clarify whether HSF activation is related to hemin induced differentiation and-or whether hemin treatment per se is stressful. While the rapid activation of HSF suggests a classical stress response, the pathway by which hemin treatment would elicit protein damage is not apparent. Furthermore, treatment of HeLa cells with hemin does not elicit *hsp70* (or *grp78*) activation, but we have not measured hemin uptake in these cells (B. Phillips, unpubl. data). Treatment of K562 cells with other agents known to induce differentiation may also help to determine whether differentiation and *hsp70* induction are obligatorily related. We are hopeful that this system will provide insights into both the mechanism of activation of HSF and a possible role of HSF activation during differentiation.

3.3 Viral Induction

As noted above, our laboratory has found viral induction of the 70-kD family of heat shock proteins to be highly specific, both with respect to virus and target gene. We have consistently observed that transcription of *hsp70* is induced early in infection of HeLa cells with either adenovirus or HSV-1 while transcription of *grp78* occurs late in infection of CV1 cells with the paramyxovirus SV5. Since viral effects on host gene expression are of interest not only from the standpoint of understanding gene regulation but also because of their influence on viral pathogenesis and the host immune response, our laboratory has been investigating the mechanisms underlying viral induction of *hsp70* and *grp78*.

3.3.1 Adenovirus

The most intensive examination has been conducted on adenovirus induction of *hsp70* transcription. Early studies revealed that induction of *hsp70* depends on an intact viral E1A gene (Nevins 1982) and specifically on the product of the E1A 13S mRNA (Wu et al. 1986a). That such induction can occur independently of viral infection was suggested by the high constitutive levels of *hsp70* expression in 293 cells (Nevins 1982) and more rigorously, by the observation that in IC4 cells, a HeLa cell line harboring a dexamethasone inducible E1A gene (Brunet and Berk 1988), induction of E1A is accompanied by an increase in *hsp70* mRNA levels (B. Phillips et al. 1991).

We then utilized transfection studies to delineate the sequences of the *hsp70* promoter required for E1A transactivation. A series of 5' deletion constructs revealed that sequences upstream of the CCAAT element were not required. The conclusion from transfections of promoter constructs in which linker scanner mutations disrupted various regions of the basal promoter, including the TATA element, revealed that basal and E1A induced expression were affected to the same degree (Williams et al. 1989). It should be noted that our results were confirmed by Taylor and Kingston (1990) but were not consistent with those of Simon et al. (1988), who suggested that the *hsp70* TATA element was the target for E1A. Our conclusion that E1A acts through a basal transcription complex without requirements for any specific promoter element or factor was identical to that drawn from a large number of similar studies on adenovirus promoters responsive to E1A (Murthy et al. 1985; Jakinot et al. 1987; Kornuc et al. 1990). Why then so few endogenous cellular promoters respond to E1A remains an important question.

Genomic footprinting of an E1A-responsive adenovirus promoter prior to and after E1A induction revealed identical patterns, suggesting that E1A mediated transcriptional induction did not perturb the binding of transcription factors to the promoter (Devaux et al. 1987). Interpretation of such an experiment, conducted during a virus infection, is complicated by the possibility that only a fraction of the total number of viral genomes are being actively transcribed, thus hindering detection of subtle differences in factor binding. This drawback can be avoided by conducting a similar analysis of the endogenous *hsp70* promoter. However, a comparison of footprinting patterns of the basal region of the *hsp70* promoter at various times after infection revealed that transcriptional induction was not accompanied by any detectable changes. It should also be noted that hypersensitivities to methylation, which could be suggestive of changes in DNA topology induced by interactions of E1A with other promoter bound proteins, were not observed. These footprinting results, considered together with our transfection data, supports current hypotheses for the mechanism of E1A action which propose as its target the general transcriptional machinery rather than specific promoter-bound factors (Flint and Shenk 1989).

Because of the clear involvement of E1A in the activation of *hsp70* transcription and our results from transfection studies indicating no requirement for the HSE in transcriptional induction, we were quite surprised to consistently find low levels of activated HSF in extracts of adenovirus-infected but not mock-infected HeLa cells. Activated HSF appears by 3 h post-infection and its levels increase up to at least 12 h post-infection (the latest time point examined), at which time *hsp70* transcription has peaked and is declining. Furthermore, extracts of 293 cells also contain substantial levels of activated HSF; however, preliminary results using IC4 cells induced by dexamethasone treatment to express E1A do not support a role for E1A in the activation of HSF (B. Phillips et al. 1991).

Although the contribution of HSF mediated induction to the transcriptional activation of *hsp70* during adenovirus infection is not known, these results exemplify the complexity inherent in the regulation of *hsp70* expression.

3.3.2 HSV-1

Only brief mention will be made of our studies of the mechanism underlying HSV-1 activation of the human *hsp70* gene. Although the kinetics of induction strongly suggest involvement of HSV-1 immediate early proteins, we have not yet tested viruses deleted in each of the immediate early genes (Albrecht et al. 1989) to establish which proteins are involved in *hsp70* induction. In contrast to our findings with adenovirus infection, there is no detectable HSF activation during HSV-1 infection. Genomic footprinting studies have revealed no perturbation in the binding of factors to basal elements of the promoter during the period when transcription of *hsp70* is activated, although later during infection, when there is a general shut-off of host transcription including that of the *hsp70* gene, a loss of factor binding to basal elements of the promoter is quite apparent (B. Phillips et al. 1991).

3.3.3 SV5

The common stimulus for induction of *grp78* expression appears to be the presence in the endoplasmic reticulum of malfolded proteins, or proteins which are incapable of undergoing proper glycosylation; thus, the induction of this gene during paramyxovirus infection is quite intriguing. Our laboratory has recently obtained results which link the synthesis of a specific viral glycoprotein, the hemagglutinin-neuraminidase (HN) protein, to enhanced transcription of *grp78* in infected CV1 cells. Grp78 has been previously shown to interact specifically and transiently with HN, one of two structural glycoproteins that are abundantly expressed during SV5 infection of CV1 cells (Ng et al. 1989). This association is detected concomitant with HN synthesis, following which *grp78* transcription is induced. When five SV5 proteins were individually expressed in CV1 cells using an SV40 vector, induction of *grp78* expression was observed only in response to expression of the viral HN protein (S. Watowich et al. 1991).

Since immunoprecipitation analysis indicates that very little unfolded or malfolded HN accumulates during SV5 infection, it appears that trafficking of large amounts of a viral polypeptide through the endoplasmic reticulum can stimulate *grp78* transcription, although it is clear that such induction is elicited only by specific viral glycoproteins. Having identified the specific viral component responsible for *grp78* induction, it is hoped that further studies of this induction will shed light on the mechanism of this activation.

4 Concluding Remarks

We have discussed in detail the regulation of expression of the human *hsp70* and *grp78* genes and have argued that because of their sensitivity to a wide variety of agents such as transcriptional activators, inducers of differentiation, and growth modulators, these genes are likely to be developmentally regulated. We also alluded to the recent accumulation of evidence suggesting vital functions for the 70-kD heat shock proteins in normal cellular activities. What we do not at present understand is the functional significance of changes in expression of these genes under conditions which are likely to be present during execution of a developmental program; e.g., we do not understand the significance of growth regulation of *hsp70* or of hemin induction of *hsp70* and *grp78* expression. One approach to addressing such a question is to assess the effects of a specific cellular phenomenon, e.g., hemin induced differentiation, of perturbing expression of either *hsp70* or *grp78*. Such studies are often difficult to interpret, however, when one is studying proteins with functions which impact on many cellular processes. Nonetheless, we believe that an appreciation of the regulation and role of the 70-kD heat shock genes in human development depends on trying to understand the functional significance of their regulation in vitro under conditions such as growth stimulation or induced differentiation.

References

Abravaya K, Phillips B, Morimoto RI (1991a) Heat shock-induced interactions of heat shock transcription factor and the human hsp70 promoter examined by in vivo footprinting. Mol Cell Biol 11:586–592

Abravaya K, Phillips B, Morimoto RI (1991b) Attenuation of the Heat Shock Response in HeLa Cells is Mediated by the Release of Bound HSF and is Modulated by Changes in Growth and Heat Shock Temperatures. Genes and Development (in press)

Albrecht MA, DeLuca NA, Byrn RA, Schaffer PA, Nammer SM (1989) The herpes simplex virus immediate-early protein, ICP4, is required to potentiate replication of human immunodeficiency virus in CD4+ lymphocytes. J Virol 63:1861–1868

Allen RL, O'Brien DA, Eddy EM (1988) A novel hsp70-like protein (P70) is present in mouse spermatogenic cells. Mol Cell Biol 8:828–832

Amin J, Ananthan J, Voellmy R (1988) Key features of heat shock regulatory elements. Mol Cell Biol 8:3761–3769

Beckmann RP, Mizzen LA, Welch WA (1990) Interaction of hsp70 with newly synthesized proteins:implications for protein folding and assembly. Science 248:850–854

Bensaude O, Babinet C, Morange M, Jacob F (1983) Heat shock proteins, first major products of zygotic gene activity in mouse embryos. Nature (Lond) 305:331–333

Bensaude O, Morange M (1983) Spontaneous high expression of heat shock protein in mouse embryonal carcinoma cells and ectoderm from day 8 mouse embryo. EMBO J 2:173–177

Brunet LJ, Berk AJ (1988) Concentration dependence of transcriptional transactivation in inducible E1a-containing human cells. Mol Cell Biol 8:4799–4807

Chiang HL, Terlecky SR, Plant CP, Dice JF (1989) A role for a 70-kilodalton heat shock protein in lysosomal degradation of intracellular proteins. Nature (Lond) 246:382–385

Chirico WJ, Waters MG, Blobel G (1988) 70K heat shock related proteins stimulate protein translocation into microsomes. Nature (Lond) 332:805–810

Deshaies RJ, Koch BD, Werner-Washburne M, Craig EA, Schekman R (1988) A subfamily of stress proteins facilitates translocation of secretory and mitochondrial precursor polypeptides. Nature (Lond) 332:800–805

Devaux B, Albrecht G, Kedinger C (1987) Identical genomic footprints of the adenovirus EIIa promoter are detected before and after E1A induction. Mol Cell Biol 7:4569–4563

DiDomenico BJ, Bugaisky GE, Lindquist S (1982) The heat shock response is self-regulated at both the transcriptional and posttranscriptional levels. Cell 31:593–603

Dorner AJ, Bole DG, Kaufman RJ (1987) The relationship of N-linked glycosylation and heavy chain-binding protein association with the secretion of glycoproteins. J Cell Biol 105:2665–2674

Ferris DK, Bellan AH, Morimoto RI, Welch W, Farrar WL (1988) Mitogen and lymphokine stimulation of heat shock proteins in T lymphocytes. Proc Natl Acad Sci USA 85:3850–3854

Flint J, Shenk T (1989) Adenovirus E1A protein:paradigm viral transactivator. Annu Rev Genet 23:141–161

Georgopoulos C, Ang D (1990) Properties of the *Escherichia coli* heat shock proteins and their role in bacteriophage lambda growth. In:Morimoto RI, Tissieres A, Georgopoulos C (eds) Stress proteins in biology and medicine. Cold Spring Harbor Laboratory Press, New York, pp 191

Goldenberg CJ, Luo Y, Fenna M, Baler R, Weinmann R, Voellmy R (1988) Purified human factor activates heat-shock promoter in a HeLa cell-free transcription system. J Biol Chem 263:19734–19739

Gross CA, Straus DB, Erickson JW (1990) The function and regulation of heat shock proteins in *Escherichia coli*. In:Morimoto RI, Tissieres A, Georogopoulos C (eds) Stress proteins in biology and medicine. Cold Spring Harbor Laboratory Press, New York, pp 167

Imperiale MJ, Kao HT, Feldman LT, Nevins JR, Strickland S (1984) Common control of the heat shock gene and early adenovirus genes:evidence for a cellular E1A-like activity. Mol Cell Biol 4:867–874

Jakinot P, DeVaux B, Kedinger C (1987) The abundance and in vitro DNA binding of three cellular proteins interacting with the adenovirus EIIa early promoter are not modified by the E1a gene products. Mol Cell Biol 7:3806–3817

Kaczmarek L, Calabretta B, Kao HT, Heintz N, Nevins J, Baserga R (1987) Control of hsp70 RNA levels in human lymphocytes. J Cell Biol 104:183–187

Kang PJ, Ostermann J, Shilling J, Neupert W, Craig EA, Pfanner N (1990) Requirement for hsp70 in the mitochondrial matrix for translocation and folding of precursor proteins. Nature (Lond) 348:137–143

Kao H, Nevins JR (1983) Transcriptional activation and subsequent control of the human heat shock gene during adenovirus infection. Mol Cell Biol 3:2058–2065

Khandjian EW, Turler H (1983) Simian virus 40 and polyoma virus induce synthesis of heat shock proteins in permissive cells. Mol Cell Biol 3:1–8

Kingston RE, Schuetz TJ, Larin Z (1987) Heat inducible human factor that binds to a human hsp70 promoter. Mol Cell Biol 7:1530-1534

Kornuc M, Kliewer S, Garcia J, Narrich D, Li C, Gaynor R (1990) Adenovirus early region 3 promoter regulation by E1A-E1B is independent of alterations in DNA binding and gene activation of CREB-ATF and AP1. J Virol 64:2004–2013

Kothary R, Perry MD, Moran LA, Rossant J (1987) Cell-lineage-specific expression of the mouse hsp68 gene during embryogenesis. Dev Biol 121:342–348

Kozutsumi Y, Segal M, Normington K, Gething M-J, Sambrook J (1988) The presence of malfolded proteins in the endoplasmic reticulum signals the induction of glucose regulated proteins. Nature (Lond) 332:462–464

Krawczyk Z, Wisniewski J, Biesiada E (1987a) An hsp70-related gene is constitutively highly expressed in testis of rat and mouse. Mol Biol Rep 12:27–34

Krawczyk Z, Szymik N, Wisniewski J (1987b) Expression of hsp70-related gene in developing and degenerating rat testis. Mol Biol Rep 12:35–41

Kroes RA, Abravaya K, Seidenfeld J, Morimoto RI (1991) Selective activation of human heat shock gene transcription by nitrosourea antitumor drugs mediated by isocyanate-induced damage and activation of heat shock transcription factor. Proc Natl Acad Sci USA 88:4825–4829

Kwong AD, Frenkel N (1987) Herpes simplex virus-infected cells contain a function(s) that destabilizes both host and viral mRNAs. Proc Natl Acad Sci USA 84:1926–1930

Larson JS, Schuetz TJ, Kingston RE (1988) Activation in vitro of sequence specific DNA binding by a human regulatory factor. Nature (Lond) 335:372–375

Milarski K, Morimoto RI (1986) Expression of human HSP70 during the synthetic phase of the cell cycle. Proc Natl Acad Sci USA 83:9517–9521

Mosser DD, Theodorakis NG, Morimoto RI (1988) Coordinate changes in heat shock element binding activity and hsp70 gene transcription rates in human cells. Mol Cell Biol 8:4736–4744

Mosser DD, Kotzbauer PT, Sarge KD, Morimoto RI (1990) In vitro activation of heat shock transcription factor DNA-binding by calcium and biochemical conditions that affect protein conformation. Proc Natl Acad Sci USA 87:3748–3752

Murthy SCS, Bhat GP, Thimmappaya B (1985) Adenovirus EIIa early promoter:transcriptional control elements and induction by the viral pre-early E1a gene, which appears to be sequence independent. Proc Natl Acad Sci USA 82:2230–2234

Nevins JR (1982) Induction of the synthesis of a 70 000 dalton mammalian heat shock protein by the adenovirus E1A gene product. Cell 29:913–919-

Ng DTW, Randall RE, Lamb RA (1989) Intracellular maturation and transport of the SV5 type II glycoprotein hemagglutinin-neuraminidase:specific and transient association with grp78-bip in the endoplasmic reticulum and extensive internalization from the cell surface. J Cell Biol 109:3273–3289

Peluso RW, Lamb RA, Choppin PW (1978) Infection with paramyxoviruses stimulates synthesis of cellular polypeptides that are also stimulated in cells transformed by Rous sarcoma virus or deprived of glucose. Proc Natl Acad Sci USA 75:6120–6124

Phillips B, Abravaya K, Morimoto RI (1991) Analysis of the Specificity and Mechanism of the Transcriptional Activation of the Human HSP70 Gene During Infection by DNA Viruses. J Virology

Rabindran SK, Giorgi G, Clos J, Wu C (1991) Molecular Cloning and Expression of a Human Heat Shock Factor, HSF1. Proc Natl Acad Sci USA 88:6906–6910

Richards FM, Watson A, Hickman JA (1988) Investigation on the effects of heat shock and agents which induce a heat shock response on the induction of differentiation of HL-60 cells. Cancer Res 48:6715–6720

Santomenna LD, Colberg-Poley AM (1990) Induction of cellular hsp70 expression by human cytomegalovirus. J Virol 64:2033–2040

Santoro MG, Garaci E, Amici C (1989) Prostaglandins with antiproliferative activity induce the synthesis of a heat shock protein in human cells. Proc Natl Acad Sci USA 86:8407–8411

Schaefer EL, Morimoto RI, Theodorakis NG, Seidenfeld J (1988) Induction of human stress response genes by chemicals that modify or damage DNA. Carcinogenesis 9:1733–1738

Schuetz TJ, Gallo GJ, Sheldon L, Tempst P, Kingston RE (1991) Evidence for Two Heat Shock Factor Genes in Humans. Proc Natl Acad Sci USA 88:6911–6915

Simon MC, Fisch TM, Benecke BJ, Nevins JR, Heintz N (1988) Definition of multiple, functionally distinct TATA elements, one of which is a target in the hsp70 promoter for E1a regulation. Cell 52:723–729

Singh MK, Yu J (1984) Accumulation of a heat shock-like protein during differentiation of human erythroid cell line K562. Nature (Lond) 309:631–633

Stone DE, Craig EA (1990) Self-regulation of 70-kilodalton heat shock proteins in Saccharomyces cerevisiae. Mol Cell Biol 10:1623–1632

Taylor ICA, Kingston RE (1990) E1a transactivation of human hsp70 gene promoter substitution mutants is independent of the composition of upstream and TATA elements. Mol Cell Biol 10:176–183

Theodorakis NG, Morimoto RI (1987) Posttranscriptional regulation of hsp70 expression in human cells:effects of heat shock, inhibition of protein synthesis, and adenovirus infection on translation and mRNA stability. Mol Cell Biol 7:4357–4368

Theodorakis NG, Zand DJ, Kotzbauer PT, Williams GT, Morimoto RI (1989) Hemin induced transcriptional activation of the HSP70 gene during erythroid maturation in K562 cells is due to a heat shock factor mediated stress response. Mol Cell Biol 9:3166-3173

Tilly K, McKittrick N, Zylicz M, Georgopoulos C (1983) The DnaK protein modulates the heat shock response of *Escherichia coli*. Cell 34:641–646

Ungewickell E (1985) The 70-kd mammalian heat shock proteins are structurally and functionally related to the uncoating protein that release clathrin triskelia from coated vesicles. EMBO J 4:3385–3391

Watowich SS, Morimoto RI (1988) Complex regulation of heat shock and glucose responsive genes in human cells. Mol Cell Biol 8:393–405

Watowich S, Morimoto RI, Lamb RA (1991) Flux of the Paramyxovirus Hemagglutinin-Neuraminidase Glycoprotein Through the Endoplasmic Reticulum Activates Transcription of the GRP78/BiP Gene. J Virology 65:3590–3597

Williams GT, McClanahan TK, Morimoto RI (1989) E1A-transactivation of the human HSP70 promoter is mediated through the basal transcription complex. Mol Cell Biol 9:2574–2587

Williams GT, Morimoto RI (1990) Maximal stress-induced transcription from the human hsp70 promoter requires interactions with the basal promoter elements independent of rotational alignment. Mol Cell Biol 10:3125–3136

Wu BJ, Morimoto RI (1985) Transcription of the human hsp70 gene is induced by serum stimulation. Proc Natl Acad Sci USA 82:6070–6074

Wu BJ, Hunt C, Morimoto R (1985) Structure and expression of the human gene encoding major heat shock protein hsp70. Mol Cell Biol 5:330–341

Wu BJ, Hurst HC, Jones NC, Morimoto RI (1986a) The E1A 13S product of adenovirus 5 activates transcription of the cellular human HSP70 gene. Mol Cell Biol 6:2994–2999

Wu BJ, Kingston RE, Morimoto RI (1986b) Human hsp70 promoter contains at least two distinct regulatory domains. Proc Natl Acad Sci USA 83:629–633

Xiao H, Lis JT (1988) Germline transformation used to define key features of heat-shock response elements. Science 239:1139–1142

Zakeri Z, Wolgemuth DJ (1987) Developmental-stage-specific expression of the HSP70 gene family during differentation of the mammalian male germ line. Mol Cell Biol 7:1791–1796

Zakeri ZF, Wolgemuth DJ, Hunt CR (1988) Identification and sequence analysis of a new member of the mouse hsp70 gene family and characterization of its unique cellular and developmental pattern of expression in the male germ line. Mol Cell Biol 8:2925–2932

Zimarino V, Tsai C, Wu C (1990) Complex modes of heat shock factor activation. Mol Cell Biol 10:752–759

12 Transforming Growth Factor-ß Regulates Basal Expression of the *hsp70* Gene Family in Cultured Chicken Embryo Cells

Ivone M. Takenaka, Seth Sadis, and Lawrence E. Hightower[1]

1 Introduction

During studies of stress protein induction by heavy metal ions (Whelan and Hightower 1985), it was noticed that the levels of newly synthesized heat shock proteins (Hsps) especially Hsp90 along with Hsp70 and-or its constitutive form, Hsc70, were reduced in chicken embryo cell (CEC) cultures incubated without serum relative to serum-containing control cultures. We suspected that growth factors in calf serum might be responsible for this effect. However, after a false start involving an early commercial preparation of platelet-derived growth factor (PDGF) that initially produced an effect on Hsp90 and Hsp70 accumulation that proved to be due to a contaminant, we set the observation aside. A visit to M. Sporn's laboratory in 1987 rekindled our interest in growth factors as potential regulators of heat shock gene expression. In particular, Sporn suggested to us that transforming growth factor-ß (TGF-ß), which only recently has been implicated in wound responses, might play such a role (Sporn et al. 1987). Shortly thereafter, I. Takenaka began her Ph.D. thesis research in the Hightower laboratory, and our renewed interest coupled with her prior experience with growth factors led us to carry out a systematic search among growth factors derived from platelets for any capable of stimulating stress protein accumulation in either CECs or an array of mammalian cell lines. TGF-ß and CECs proved to be the winning combination in our survey. The rates of accumulation of Hsp90 and three members of the Hsp70 family including Hsp70, Hsc70, and the 78 kDa glucose-regulated protein, Grp78, were all stimulated by TGF-ß in serum-free culture medium. A detailed analysis of the level of regulation of basal expression of the genes encoding these proteins was carried out, and a preliminary report on these data for the *hsp70* gene family is presented here.

Briefly, we show that bovine TGF-ß1 regulates *hsc70* gene expression posttranscriptionally and that Hsp70 family proteins are among a cohort of proteins induced by TGF-ß1 in serum-deprived CECs. It is proposed that the molecular chaperones of the Hsp70 family are induced by TGF-ß1 along with proteins that require

[1] Department of Molecular and Cell Biology, The University of Connecticut, Storrs, Connecticut 06269–3044, USA

Note: List of Abbreviations: *TGF-B*, Transforming growth factor-ß; *Hsp*, Heat shock protein; *hsp70* and *hsc70*, Genes that encode the 70 kD Hsp and its constitutive form, respectively; *Hsp70*, The product of the *hsp70* gene; *PDGF*, Platelet-derived growth factor; *Grp*, Glucose-regulated protein; *CEC*, Chicken embryo cells; *ECM*, Extracellular matrix; *EGF*, Epidermal growth factor; *DPP*, *Drosophila* decapentaplegic gene product; *SRE*, Serum-regulated element; *MEM*, Minimum essential medium; *MHC*, Major histocompatibility complex; *ER*, Endoplasmic reticulum.

Results and Problems in Cell Differentiation 17
Heat Shock and Development
Hightower and Nover (Eds.)
©Springer-Verlag Berlin Heidelberg 1991

chaperoning. A summary of current ideas on how Hsc70 might interact with other proteins is provided, based upon the Ph.D. thesis research of S. Sadis in our laboratory and upon recent work from other groups.

TGF-ß is notorious for causing different responses depending upon the cell type. Indeed, TGF-ß more frequently has growth inhibitory rather than proliferative effects in most cell types. TGF-ß is present in tissues containing rapidly dividing cells of mesodermal origin during embryonic development. Although TGF-ß may not be mitogenic itself in such tissues, it stimulates the production of other mitogenic growth factors such as PDGF and fibroblast growth factor (FGF). As a result, TGF-ß has been called a "master morphogen" and may function as a main switch to set in motion a cascade of events that first prepare and ultimately stimulate certain embryonic cell types into rapid proliferation and others into differentiation pathways. Part of this cascade is the stimulated production of extracellular matrix (ECM) proteins such as collagen. There is no evidence yet that Hsp70 family proteins interact directly with collagen or other matrix proteins; however, several links to matrix metabolism exist including the induction of a collagen-binding glycoprotein by heat stress in CEC (Nagata et al. 1986) and the increase of collagenase and stromelysin mRNA levels in heat shocked rabbit synovial fibroblasts (Vance et al. 1989). The translocation of several other secretory proteins across the endoplasmic reticulum (ER) is known to involve the activities of cytoplasmic Hsc70 and Grp78 located in the rough ER. It is possible that some of the secretory proteins induced by TGF-ß may require the chaperoning functions of these two proteins.

TGF-ß also stimulates chemotactic activity in some cell types that may be part of embryonic cell migration and tissue modeling. Such activities are likely to involve assembly and disassembly of cytoskeletal proteins, events that could conceivably increase cellular demand for cytoplasmic-nuclear members of the Hsp70 family. Proteins in the Hsp90 and Hsp70 families are associated with isolated cytoskeletal elements (Lim et al. 1984; Napolitano et al. 1985; Koyasu et al. 1986; Sanchez et al. 1988).

Although this introduction has been couched in terms of a possible developmental role for the regulation of basal expression of *hsp70* genes, this is based only on the fact that embryonic chicken cell cultures were the most responsive in our hands. This response need not be specific to developmental events, however. It is possible that TGF-ß may have similar regulatory effects on fibroblasts in somatic tissues during wound responses, for example, and in other situations in which alterations in cellular physiology require an increased capacity for the kinds of chaperoning functions provided by the Hsp70 family.

2 Biochemical and Biological Properties of TGF-ß

TGF-ß plays an important role in tissue formation and repair, as well as in cell proliferation and differentiation (Sporn et al. 1987; Roberts et al. 1990). TGF-ß is quite stable in acidic solutions, and is active as a mature 25-kDa homodimeric polypeptide. It is secreted by platelets and some tumor cells in a latent form, and subsequently activated by acid or alkaline treatments in vitro (Lawrence et al. 1985; Keski-Oja et al.

1989). The TGF-ß family is composed of several related molecules: TGF-ß1, a protein isolated from human and porcine platelets (Assoian et al. 1983); TGF-ß2, also called BSC-1, isolated from bovine bones and glioblastomas (Tucker et al. 1984; Wrann et al. 1987; Madisen et al. 1988); TGF-ß3, from humans and chickens (Derynck et al. 1988; ten Dijke et al. 1988; Kondaiah et al. 1990); TGF-ß4, from chicken chondrocytes (Jakowlew et al. 1988; ten Dijke et al. 1988, 1990; Lyons and Moses 1990); and TGF-ß5 from *Xenopus laevis* (Kondaiah et al. 1990). The TGF-ß superfamily also includes Mullerian inhibitory substance (Cate et al. 1986), inhibins, and activins (Mason et al. 1985) and a protein coded by the *Drosophila* decapentaplegic gene complex (DPP), involved in the orientation of dorsoventral development during embryogenesis (Padgett et al. 1987). The bone morphogenetic proteins (BMPs) that regulate cartilage and bone formation (Wozney et al. 1988), the *Xenopus vg-1* gene (Weeks and Melton 1987), and the product of the murine vgr-1 gene (Lyons et al. 1989a), which is related to DPP and *vg-1*, all have TGF–ß–like activity. These genes share 30–40% homology to TGF-ß1.

TGF-ß's are highly conserved. For example, there is suggestive evidence that for any particular TGF-ß compared between amphibian and mammalian species, sequence conservation will be greater than 95% (Kondaiah et al. 1990). The human, porcine, and bovine TGF-ß1's are highly conserved with a nearly identical amino-acid sequence. TGF-ß2 and TGF-ß3 are 80% and 72% homologous to TGF-ß1, respectively. TGF-ß5 shares about 70% amino acid sequence identity to TGF-ß1-2-3 and 4. In addition, TGF-ß superfamily members share seven highly conserved cysteine residues.

TGF-ß1 seems to play a role in hematopoiesis and inhibits endothelial cell proliferation. It plays an important part in bone remodeling and in the healing of fractured bones (Robey et al. 1987). TGF-ß1 effects on cellular differentiation are both stimulatory and inhibitory in various cell types and depend on the presence of other growth factors. TGF-ß alone is mainly an inhibitory factor and induces proliferation only in a few mesenchymal cell types such as fibroblasts and osteoblasts (Moses et al. 1990). It induces bronchial epithelial cells to undergo squamous differentiation, but blocks the development of NIH-3T3 fibroblasts into adipocytes. In mouse keratinocytes, TGF-ß acts as an inhibitor of cell proliferation. Within 1 h of TGF-ß treatment of these cells, c-myc expression is decreased but *β-actin* gene expression increases and *c-fos* expression is not affected (Coffey et al. 1988a,b). TGF-ß induces proliferation in suspension cultures but inhibits the same cell types when grown in monolayer cultures, indicating that the levels of ECM may play a role in cell growth control by this cytokine (Madri et al. 1988).

TGF-ß1 regulates the levels of ECM components such as collagen and fibronectin and the level of ECM proteases. Subcutaneous injection of TGF-ß1 into newborn mice produces a granulation response with increases in collagen type I and fibronectin levels at the site of injection (Roberts et al. 1983; Robey et al. 1987). In mesenchymal and epithelial cells in culture, TGF-ß increases the levels of collagen gene expression through a transcriptional activation of the collagen gene promoter (Rossi et al. 1988) and increases in mRNA stability. The maximum rate of mRNA accumulation occurs 16–24 h after the addition of TGF-ß to the culture (Penttinen et al. 1988). Cycloheximide blocks the accumulation of collagen and fibronectin mRNAs by TGF-ß in NIH-3T3 cells (Nilsen-Hamilton 1990). These results differ from those of Ignotz (Ignotz et al. 1987) who used different cell lines and found that the increase in fibronectin mRNA levels in rat NRK-49 fibroblasts and L6E9 myoblasts treated with TGF-ß was inhibited by actinomycin D but not by cycloheximide (Nilsen-Hamilton 1990).

The TGF-ß signaling pathway has not been elucidated. It does not seem to involve either phospholipid turnover or tyrosine kinases (Fanger et al. 1986). G-protein-dependent and -independent pathways have been reported (reviewed in Nilsen-Hamilton 1990). At the genomic end of the pathway, a common promoters element in genes known to be regulated by TGF-ß1 is the NF-1 binding site (Sporn et al. 1987), which is present in a large number of genes. A recent report shows that the AP-1 site (the Fos-Jun complex binding site) is present in the TGF-ß1 promoter where it is involved in the autocrine regulation of TGF-ß1 and in the induction of TGF-ß2 (Kim et al. 1990). Since TGF-ß1 regulates the levels of expression of several other cellular growth and differentiation-related genes such *c-myc, c-fos, c-jun, c-sis* (PDGF-α chain) and the epidermal growth factor (EGF) gene, it is possible that the effect of TGF-ß1 is mediated by a secondary messenger or that it triggers a unique signal pathway.

3 TGF-ß in Embryogenesis and Development

In situ hybridization studies showed that *TGF-ß1* mRNA is detected first in extraembryonic blood islands of the yolk sac 7 days after fertilization (Akhurst 1990). In 7–9-day embryos, *TGF-ß1* mRNA is localized in cardiac mesoderm and endothelial cells (Akhurst 1990). In 9–16-day mouse embryos, *TGF-ß1* mRNA is still mainly localized in the hematopoietic system (Wilcox and Derynck 1988) but has also been found in fetal bone, liver megakaryocytes, and in the overlying epithelia of these mesenchymal tissues (Lehnert and Akhurst 1988; Wilcox and Derynck 1988). High levels of transcripts remain in all heart valves until 8 days postpartum (Akhurst et al. 1990).

TGF-ß proteins are only detected after 9.5 days of murine development. Immuno-histochemical studies in 11–18-day mouse embryos showed TRF-ß 1 to be localized mainly in mesenchymal tissues (Heine et al. 1987). Other reports also revealed TGF-ß1 in developing bone and cartilage of the bovine fetus (Ellingsworth et al. 1986) and in human fetal tissues (Sandberg et al. 1988).

These data strongly suggest the TGF-ß1 is an important component in embryonic development. It probably exerts its effects in part by regulating the levels of ECM proteins, since ECM is extremely important in developing tissues, in bone development, and remodeling (see review, Nilsen-Hamilton 1990).

TGF-ß2 has properties similar to TGF-ß1. Both TGF-ß1 and 2 are present in the mineralized matrix of cartilage and bone (Carrington et al. 1988). TGF-ß2 induces mesoderm and ectoderm formation in *Xenopus* and its mRNA has been localized in mesenchymal tissues (Rosa et al. 1988). In addition to TGF-ß2, an mRNA encoding a novel TGF-ß, TGF-ß5, has been identified in *Xenopus laevis* (Kondaiah et al. 1990). It is developmentally regulated and highly expressed at early neurula (stage 14) and in tadpoles. *TGF-ß5* mRNA has also been found in many adult tissues such as kidney, lung, and heart.

Other proteins with TGF-ß-like activity also have been implicated in development. The more recently isolated *vg-1* gene of *Xenopus laevis* seems to play an important role in embryonic development and cell differentiation. *Vg-1* mRNA is present in oocyte, egg, blastula, and gastrula. It disappears in neurula and tadpole stages of the *Xenopus* embryo. These transcripts are maternally inherited and restricted to the vegetal endoderm and decline after gastrulation (Weeks and Melton 1987). The *vg-1* gene product also induces mesoderm formation in the vegetal pole of *Xenopus* oocytes. Transcripts of mouse *vgr-1* are present in a variety of embryonic, neonatal, and adult tissues (Lyons et al. 1989b). *Drosophila dpp* transcripts are present in embryo, larva, and adult, in several mesodermal and endodermal tissues as well as in dorsal ectodermal tissues (St. Johnson and Gelbart 1987). Mullerian inhibitory substance in mammals and the product(s) of the *dpp* complex in *Drosophila* have been implicated in embryonic remodeling of the developing male reproductive systems of mammals and fly embryos, respectively (Nilsen-Hamilton 1990). TGF-ß also plays a role in embryonic morphogenesis in *Drosophila* (Padgett et al. 1987).

In vitro evidence of a role for TGF-ß in development includes the acceleration in maturation of both follicle-enclosed oocytes and cumulus-oocyte complexes, as measured by an increase in the percentage exhibiting germinal vesicle breakdown upon TGF-ß treatment. TGF-ß at a concentration of 1 pM (25 pg/ml) increases oocyte maturation by 50% and at 100 pM causes a twofold increase in the maturation rate (Feng et al. 1988).

TGF-ß may play a major role in the initial formation of ECM and basement membranes in the early embryo during organogenesis. As tissues develop and depending upon the particular line of cellular differentiation, TGF-ß may have a growth-inhibiting effect that could be mediated by the accumulation of ECM. TGF-ß increases the levels of plasminogen activator inhibitor-1 (PAI-1) and tissue inhibitor of metalloprotease (TIMP), two inhibitors of ECM protein-degrading enzymes (Massague 1990). This is potentially a very important protective and morphogenic mechanism for the developing embryo (Nilsen-Hamilton 1990).

The complexity of responses that TGF-ß1 exerts on different cell types as an inhibitor, mitogenic factor, or differentiation factor suggests that TGF-ß may be a major cellular regulator of growth and differentiation during embryogenesis. (For reviews, see Sporn et al. 1987; Lyons and Moses 1990; Masssague 1990; Roberts et al. 1990).

4 Heat Shock Proteins Are Induced During Embryogenesis and in Highly Mitogenic Cells

A substantial body of information has accumulated indicating that certain members of Hsp families are developmentally regulated and are important components in normal embryonic development and organogenesis. Most of these studies have been carried out in *Drosophila melanogaster* and *Xenopus laevis*; however, more recently, searches for the role of inducible stress proteins and their cognates in mammalian development have begun. Elevation in the levels of Hsps have been associated with early stages of embryonic development and differentiation (Craig et al. 1983; Craig 1985; Banerji et al. 1987). Hsp90 and Hsc70 are relatively abundant proteins in rapidly dividing cells such as embryonic and tumor cells. High constitutive levels of h*sp90* and *hsp70* transcripts are found in unstressed mouse embryonal carcinoma cells and ecto-derm from day 8 mouse embryos (Bensaude and Morange 1983; Hahnel et al. 1986; Kothary et al. 1987; see also Mezger et al. Chapter 10).

During X*enopus laevis* embryonic development, Hsp90, Hsp70, and ubiquitin be-come inducible after the midblastula stage with the highest levels of induction occur-ring in neurulae, as measured by the accumulation of *hsp* mRNA. Hsp30, on the other hand, is not inducible until the early tailbud stage and may be under negative control (Heikkila et al. 1987; see Heikkila et al. Chapter 8, for additional references). Although unfertilized eggs and cleavage embryos contain low constitutive levels of *hsp70* mRNA, cleavage embryos are not able to induce Hsp's in response to heat or chemical stress. The accumulation of *hsp70* and *hsp90* mRNAs after midblastula in unstressed *Xenopus* embryos coincides temporally with the activation of generalized zygotic transcription (Newport and Kirshner 1982ab).

A similar type of response has been observed in early stages of *Drosophila* development (see Tanguay and Arrigo, Chapter 7 for additional references). In *Dros-ophila* embryos, Hsp's 22, 23, and 70 are neither expressed nor heat inducible until the 64–128 cell stage is reached (Zimmerman et al. 1983). A novel heat shock cognate gene, *hsc4*, has been recently characterized in *D. melanogaster*. This gene is more similar to the human *hsc70* gene than to the inducible *Drosophila hsp70*. Expression of *hsc4* is enhanced in cells actively undergoing endocytosis, rapid growth, and changes in shape (Perkins et al. 1990). The behavior of this gene offers the best analogy to date from a different developmental system for the responsiveness of the chicken *hsc70* gene described in Section 7 below.

In the freshwater snail *Lymnaea,* embryos become heat shock inducible during late blastula stage. At day 0 of development, there is a high constitutive level of a 70-kDa protein that appears to be Hsc70. Pre-blastula stage embryos also have a relatively high intrinsic thermotolerance (see Boon-Niermeijer, Chapter 1 for further references).

Germ-cell specific Hsp70-like proteins have been detected in immunoblots of mouse spermatogenic cells (Maekawa et al. 1989), and stable *hsp70* family transcripts specific to germ cells have been detected in mouse testis and haploid spermatogenic cells (Zakeri and Wolgemuth 1987; Zakeri et al. 1988). *Hsp70* and *hsp90* gene ex-pression during mammalian spermatogenesis, oogenesis, and in early embryos has been reviewed in this volume by Wolgemuth and Gruppi (Chapter 9). Among the Hsp90 family of proteins, germ cell-specific *hsp86* expression has also been observed in mouse testis (Lee 1990).

In contrast to mature cells, primitive avian erythroid cells do not respond to heat stress; however, high constitutive levels of Hsp70 are found in 3–4-day polychromatic primitive cells. As these cells mature, the rate of Hsp70 synthesis declines in parallel with overall protein synthesis (Banerji et al. 1987).

5 Regulators of Basal Expression
of Heat Shock Gene Families in Unstressed Cells

In the past, the heat shock genes were taken as good models to study the regulation of eukaryotic inducible genes. However, since the finding that members of Hsp families are present in high basal levels in "unstressed" cells and are induced during normal development, investigators have begun to focus on elucidating their normal cellular functions. One of the first observations of this kind linked the induction of small heat shock proteins in *Drosophila* early pupae to the molting hormone ecdysterone (see Tanguay and Arrigo, Chapter 7 for a thorough review of the developmental expression of the *Drosophila* genes encoding the small Hsp's). Steroid hormone receptor-binding sequences have recently been found in the *hsp23* and *hsp27* regulatory regions. This is consistent with previous observations that ecdysterone induces the corresponding proteins in tissue culture cells and isolated imaginal disks. Silver and coworkers have now observed effects of a steroid hormone on heat shock gene expression that may be related to development in a second system, the filamentous fungus *Achlya*. The steroid hormone antheridiol induces *Achlya hsp70* and *hsp90* mRNAs, and it is suggested that this induction may be related to changes in secreted proteins and increased hyphal branching (Silver et al. 1990).

For cells in culture, we somewhat arbitrarily define "unstressed" cells as those not subjected to chemical and physical stressors beyond the stress conditions that may already be present in culture. Cells in culture are likely to become at least partially adapted to certain stressors during cultivation; and the serum component of cell culture media contains potential stressors and response modifiers not present in blood plasma and uninjured tissue which may affect *hsp* and-or *hsc* gene expression.

Serum and several purified cytokines and lymphokines, as well as hormones are known to be physiological regulators of stress proteins in "unstressed" cultured cells. *Hsp70* gene expression is cell cycle-regulated in HeLa cells and *hsp70* mRNA levels peak at the G1-S boundary (Morimoto et al. 1986; Wu and Morimoto 1985). Increases in *hsp70* gene transcription, accumulation of cytoplasmic *hsp70* mRNA, and Hsp70 association with nuclei are some of the events that occur under serum stimulation in serum-deprived HeLa cells, especially during the S and G2 phases of the cell cycle (Milarski and Morimoto 1986). Hsp90 and Hsp70 levels are elevated by serum, and this effect is regulated by a promoter element designated the serum-regulated element (SRE) (Wu et al. 1987). Mitogenic factors such as phytohemagglutinin A and interleukin-2 also increase the levels of *hsp70* and *hsp90* gene expression in human T-cells (Ferris et al. 1988). Treatment of the human erythroleukemia cell line K562 with prostaglandins, which inhibit proliferation, induces Hsp70. Hemin also induces Hsp70 in K562 cells and appears to induce differentiation as well. These latter two effects are discussed in greater detail by Phillips and Morimoto (Chapter 11). In addition, they have provided an excellent summary table of factors that modulate the Hsp70 family in primate cells. They point out that p72, the human Hsc70, is the least sensitive member of the human Hsp70 family to cell cycle fluctuations and environmental stress. The effects known to date which modulate primate Hsc70 expression are limited to cytomegalovirus infection and serum withdrawal.

6 TGF-ß Rapidly Induces Hsc70 in Cultured Chicken Embryo Cells

In response to injury and stress in wound-healing responses, growth factors, especially platelet-derived factors, are released at high levels at the site of injury. This led us to speculate that some members of stress gene families might be regulated by cytokines which are also present in blood serum. Screening of several types of cells with a variety of growth factors revealed transforming growth factor-ß1 (TGF-ß1) to be the only factor of those tested capable of increasing significantly the levels of Hsp's and only in CEC, under our culture conditions. In the experiments to be described, CEC were allowed to reach confluency, i.e., proliferation was density-inhibited, and then serum was withdrawn during a subsequent incubation period to reduce the effects of serum factors on the embryonic fibroblasts.

Figure 1 shows the effect of TGF-ß1 on protein synthesis in CEC. For these experiments, primary cultures were prepared from 10-day-old chicken embryos (Sekellick and Marcus 1986). After 7 days of incubation, secondary cultures were prepared, grown to confluency, and treated as indicated in Fig. 1A. Compared to control cultures without serum (lane 1) TGF-ß1 (lane 2) and serum (lane 3) increased the rate of Hsp90 accumulation twofold and Hsp-Hsc70 protein accumulation threefold. The one-dimensional gels also show interesting changes in the 150-kDa region (arrowhead) which could be due to changes in secretion of procollagen $2\alpha(I)$ after treatment with either TGFD-ß1 or serum.

Since this work was done in unstressed CEC, we focused mainly on the effect of the TGF-ß1 on the constitutive form of the Hsp70 family, the cognate protein or Hsc70. The distinction between the inducible form Hsp70 and its cognate (Hsc70) is less clear in chicken cells than in cultured mammalian cells. Both forms are present at high basal levels in CEC, whereas in most cultured mammalian cells, the amount of Hsp70 is very low and Hsc70 is high in unstressed cells. Since these proteins have almost identical molecular masses by one-dimensional SDS-PAGE, they are marked as Hsp-Hsc70 in Fig. 1.

Based on Whelan and Hightower's previous observations, the accumulation of Hsp90 and Hsp-Hsc70 decreases when CEC are deprived of serum. As shown in Fig. 1A, serum deprivation was not required to demonstrate induction of Hsp's by either TGF-ß or fresh serum. However, preincubation without serum lowered basal levels of Hsps and increased the quantitative reproducibility of the stimulation by TGF-ß. Twelve hours of serum deprivation showed the lowest level of Hsp synthesis (data not shown). Therefore, subsequent experiments were carried out in CEC preincubated for 12 h without serum. Analysis of the time course of induction of Hsp's by TGF-ß1 showed that actin (Fig. 1B), whose synthesis closely followed total protein synthesis, increased more slowly than that of Hsp-Hsc70 (Fig. 1C), suggesting that Hsp-Hsc70 belong to a class of proteins the synthesis of which is stimulated rapidly by TGF-ß1. On the other hand, Grp78 belongs to a class of proteins for which increased synthesis is substantial but delayed (Fig. 1D). Two Hsp90 isoforms are members of the latter class as well (data not shown). None of these proteins undergoes significant changes in half-life under these conditions.

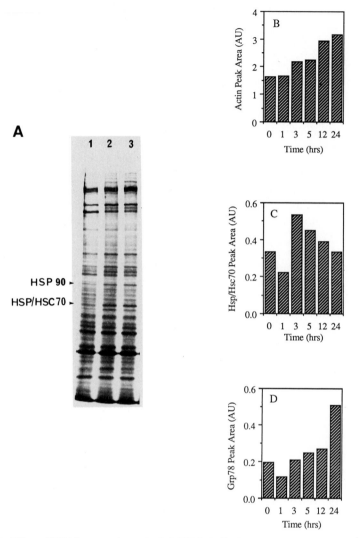

Fig. 1A-D. Effect of TGF-ß on protein synthesis in CEC. **A** Confluent chicken embryo cells (CEC) were washed with MEM to remove serum and further incubated with MEM (*lane 1*), or MEM supplemented with either bovine TGF-ß1 at 10 ng/ml (*lane 2*), or 6% calf serum (*lane 3*) for 5 h. Cells were labeled with [^{35}S]-methionine (20 μCi/ml) for 30 min and lysed with SDS-containing buffer (Hightower 1980). Equal volumes of lysates were separated on a 7.5:0.075% acrylamide:bis-acrylamide slab gel and a fluorograph of the gel is shown. Hsp-Hsc70:70-kDa heat shock protein and its cognate; HSP90: 90-kDa heat shock protein family. Small arrowhead indicates an approx. 150-kDa protein. **B, C** and **D** Time course of synthesis of selected proteins in TGF-ß-treated cells. CEC were pretreated for 12 h with MEM, then further incubated with TGF-ß1 (10 ng/ml) for 1,3,5,12, and 24 h. Cultures were labeled with [^{35}S]-methionine (20 μCi/ml) for 30 min at 37 °C and equal volumes of lysates were separated by SDS-PAGE. Fluorographs were obtained and Actin, Hsp70-Hsp70, and Hsp90 peak areas, normalized to the same absorbance units (AU) scale, were determined by densitometry. **B**, actin; **C**, Hsp-Hsp70, **D**, Grp78

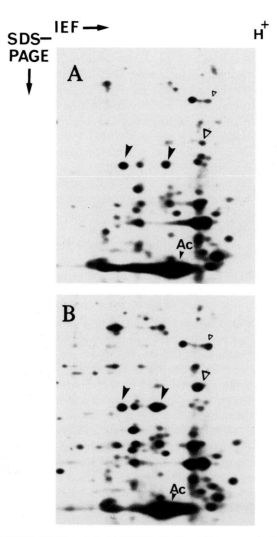

Fig. 2A,B. IEF-SDS-PAGE of CEC treated with TGF-ß1. Confluent 12-h serum-deprived CEC were further incubated with **A** MEM or **B** TGF-ß1 (10 ng/ml) for 5 h. Proteins were pulse labeled for 30 min with [^{35}S]-methionine (100 µCi/ml) at 37 °C and lysed in a modified O'Farrell buffer (O'Farrell 1975) containing 1.6%, pH 3.5–10.0 and 0.4%, pH 4-6 ampholines. Lysates were separated on isoelectric focusing (*IEF*)-acrylamide gels containing the same composition of *ampholines* as described above. 7.5% acrylamide : 0.075% bis-acrylamide was used in the second dimension SDS-PAGE. *Closed arrowheads* indicate the inducible Hsp70 (basic) and its cognate, Hsc70 (more acidic). *Large open arrowhead* points to Grp78 and small *open arrowhead* to Grp100. *Ac* indicates actin

TGF-ß1 regulates the production of ECM proteins such as collagen and fibronectin (Roberts et al. 1986 Penttinen et al. 1988;). Experiments done in mouse 3T3 preadipocytes showed a fivefold increase in the levels of fibronectin in the culture matrix 6 h after the addition of TGF-ß1. Fibronectin levels in the matrix attain their highest level at 9 h of treatment (Ignotz et al. 1987). A similar course of induction is observed for fibronectin levels in the medium. The basal level of collagenase-sensitive material is usually high in the matrix and in the medium; however, there is a significant increase in the levels of collagen in the matrix and in the medium when cultures are treated with TGF-ß1 for only 3 h (Ignotz et al. 1987). We propose that Hsc70 is very rapidly induced by TGF-ß1 in order to provide a chaperoning function for other induced proteins, possibly including ECM proteins.

The changes in protein accumulation in TGF-ß1 treated cells are more apparent with the increased resolution afforded by two-dimensional polyacrylamide gel electrophoresis, as shown in Fig. 2. Serum-deprived CEC were incubated with serum-free MEM or MEM containing TGF-ß1 at 10 ng/ml (TGF-ß) for 5 h. After radiolabeling with [^{35}S]-menthionine, cultures were solubilized in O'Farrell modified lysis buffer and equal amounts of RNase treated cell lysates were separated by isoelectric focusing-SDS-PAGE (IEF-SDS-PAGE). Fluorograms of proteins from serum-deprived CEC are shown in panel A, and those from TGF-ß1 treated CEC are shown in panel B. Addition of TGF-ß1 results in a general twofold stimulation of protein synthesis compared to serum-depleted cultures, but certain proteins, including members of the Hsp90 and Hsp70 families, are stimulated to higher rates than most other proteins. The family members located in the ER, Grp78 (large open arrowhead) and Grp100 (small open arrowhead), a member of the Hsp90 family, are both induced as is the cytoplasmic Hsc70 (the more acidic spot indicated by the closed arrowhead). A smaller increase in the levels of Hsp70 (the more basic spot indicated by the closed arrowhead) is sometimes observed as well.

Since the effect of TGF-ß1 on cell proliferation vary for different cell types, we examined whether TGF-ß1 is mitogenic in CEC. In the experiment shown in Fig. 3, confluent secondary cultures grown in NCI medium supplemented with 6% calf serum were preincubated in serum-free MEM for 12 h, at which time (marked by the arrow) some of the cultures were supplemented with either 6% calf serum (+S) or TGF-ß1 at 10 ng/ml (TGF-ß). The remainder of the serum-free cultures (-S) were allowed to continue incubation for an additional 12 h. Cultures were periodically labeled with [^{3}H] thymidine in serum-free MEM for 30 min at 37 °C, lysed in 0.3 N NaOH, and TCA-precipitated. Radioactivity incorporated into DNA was determined by liquid scintillation counting. It is clear that TGF-ß1 alone is not mitogenic in cultured CEC. Therefore, the changes in patterns of protein synthesis that we have observed under these conditions are not part of transit into S-phase, although they may be part of preparatory events during G1. They may also represent some of the proliferation-independent effects of TGF-ß.

Northern blot analysis of the effects of TGF-ß1 on *hsp70* and *hsc70* mRNA levels in CEC was performed, as shown in Fig. 4. Poly-A + RNA isolated from serum-deprived CEC treated for a further 5 h with MEM (lane 1) or MEM containing TGF-ß1 at 5 ng/ml (lane 2) or 10 ng/ml (lane 3) or 6% calf-serum (lane 4) or heat shocked at 44 °C for 1 h (lane 5), were separated in 1% formaldehyde agarose gels and blotted onto nylon membrane. Filters were hybridized with [^{32}P]-labeled cDNAs encoding

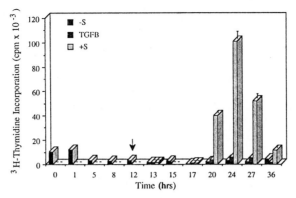

Fig. 3. [³H]-Thymidine incorporation as a measure of DNA synthesis. CEC were grown to confluency in Nutrient Colorado Inositol medium containing 6% calf serum and then incubated in MEM for 12 h were incubated further with MEM (*-S*) for 1, 5, 8, 12, 17, 20, 24, 27 and 36 h. In parallel, CEC pre-incubated in MEM for 12 h were incubated further with MEM containing TGF-ß1 (*TGF-ß*) at 10 ng/ml or 6% calf serum (*+S*) for 1, 5, 8, 12, 15, 24 h. Cells were labeled for 30 min with [³H]-thymidine (3 μCi/ml), lysed with 0.1 N NaOH, and TCA-precipitable material counted in a scintillation counter. Data are expressed as cpm per culture. *Arrow* indicates the time of TGF-ß1 or serum addition. *Error bars* are ±SEM

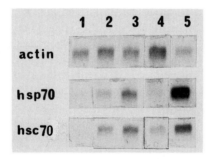

Fig. 4. *hsp70* and *hsc70* mRNA levels in TGF-ß-treated CEC. Cytoplasmic poly-A⁺RNA was isolated from 12-h MEM-pretreated CEC, further incubated with MEM (*lane 1*), TGF-ß1 at 5 ng/ml (*lane 2*), TGF-ß1 at 10 ng/ml (*lane 3*), or 6% calf serum (*lane 4*) for 5 h, or heat shocked for 1 h at 44 °C (*lane 5*). Cells were lysed in a buffer containing 1% v-v Triton N-101 and cytoplasmic poly A⁺RNA was prepared by oligo-dT chromatography. 1.5 μg of RNA was separated on a 1% formaldehyde agarose gel and blotted onto a nylon membrane. Membranes were hybridized to [³²P]-labeled cDNAs for 48 h at 45 °C in a high stringency formamide solution. Blots were washed with 2X SSC at 65 °C for 30 min. Autoradiograms were scanned and normalized to the *actin* signal. Actin: chicken *ß-actin* cDNA (2.3 kb) (Cleveland et al. 1980); hsp70: chicken inducible pC1.8 cDNA (2.6 kb) (Morimoto et al. 1986):, hsc70: rat cognate RC62 cDNA (2.6 kb) (O'Malley et al. 1985)

actin (Cleveland et al. 1980), chicken Hsp70 (Morimoto et al. 1986), and rat Hsc70 (O'Malley et al. 1985). Autoradiograms were obtained, *hsp70* and *hsc70* mRNA levels were quantified by densitometry, and these data were normalized to *actin* mRNA levels. These results showed a 5.2 and 4.5-fold increase in *hsp70* and *hsc70* nRNA levels, respectively. The larger increase in *hsp70* mRNA level compared to its small increase at the protein level (see Fig. 2) could be due to cross-hybridization between the *hsp70* probe and *hsc70* mRNA.

Addition of serum to serum-deprived CEC stimulates mRNA accumulation and overall protein synthesis. Several proteins including Hsp70 and Hsc70 accumulate very rapidly, reflecting the rapid increase in their mRNA levels. Since the levels of Hsp-Hsc70 protein and mRNA induced by serum are similar to those induced by TGF-ß1 alone, it is suggested that TGF-ß1 is the primary component in serum that causes these increases.

Further studies to elucidate the levels of *hsp-hsc70* gene expression regulated by TGF-ß1 were performed (Takenaka 1990). Run-on transcription assays were carried out and only very small increases of twofold or less in rates of transcription of *hsp70-hsc70* were found. The small increase in transcription of these genes cannot account for the observed increases in mRNA levels. Therefore, a posttranscriptional event is more likely to be the main target for regulation of *hsp-hsc70* gene expression by TGF-ß1. RNA stability assays in the presence of actinomycin-D showed no significant changes in the half-lives of *hsp-hsc70* cytoplasmic mRNA in the presence of TGF-ß1. However, actinomycin-D treatment blocked the elevation of *hsc70* mRNA levels. Since transcription of the *hsc70* gene cannot account for this effect, transcription of other genes might be necessary for the increase in levels of *hsc70* mRNAs in the presence of TGF-ß1.

TGF-ß1 had little effect on the overall translational activity in CEC as judged by a comparison of polysomal profiles; however, the level of *hsc70* mRNA in the polysomes specifically increased about threefold. From these and related studies, we concluded that in the presence of TGF-ß1 CEC undergo an overall increase in gene expression with some proteins being produced more rapidly than others. Hsc70 may be part of the early-recruited proteins due to its chaperoning role in the cell. We suspect that some of the proteins induced by TGF-ß require the kind of assistance in folding-assembly and unfolding-stabilization provided by the Hsp70 family of proteins as part of membrane translocation, for example. Since they are under the control of TGF-ß1 in embryonic chicken cells, we hypothesize that the up-regulation of certain proteins and their chaperones is an important part of the developmental effects of TGF-ß1.

7 The Hsc70 Molecular Chaperone Interacts with Diverse Polypeptide Sequences

The 70-kDa stress proteins have the novel property of associating with a host of diverse polypeptide sequences typically found in unfolded or extended polypeptides. However, the principles that govern these interactions are not yet known in detail. As one approach to this problem, we have recently analyzed the secondary structure of the mammalian Hsc70 protein using a combination of circular dichroism spectroscopy and secondary structure prediction. Our results, combined with the results of other investigators, have provided some interesting and unexpected insights into Hsc70 structure and function.

Polypeptide "substrates" for the 70-kDa stress proteins come in a variety of forms. For example, Hsc70 and Grp78 have specific binding sites for short, hydrophilic peptides (Flynn et al. 1989). In addition, Hsc70 interacts with newly synthesized polypeptide chains (Beckmann et al. 1990) and also with conformationally flexible regions of otherwise native proteins, such as clathrin (DeLuca-Flaherty et al. 1990). Hsc70 can distinguish between unfolded and folded versions of the same polypeptide, as Hsc70 binds to apo-but not to holocytochrome *c* (Sadis et al.; Sadis and Hightower, in prep.). The number of reported protein-protein interactions involving Hsc70 and Grp78 is large and likely to increase further.

The selective binding of the 70-kDa stress proteins to diverse, unfolded proteins suggests that they may function as molecular chaperones to increase the efficiency of protein folding and assembly (Ellis and Hemmingsen 1989; Rothman 1989). The results of recent studies support this view. For example, genetic evidence strongly suggests an essential role for Hsc70, Grp78, and mitochondrial Hsp70 in facilitating the translocation and subsequent folding of many secretory and mitochondrial proteins (Deshaies et al. 1988; Kang et al. 1990; Vogel et al. 1990). Biochemical studies have shown that Hsc70 can promote the dissociation of an aggregated dihydrofolate reductase fusion protein (Sheffield et al. 1990). The homologous DnaK protein of *Escherichia coli* acts similarly to dissociate aggregates of DnaA protein (Skowyra et al. 1990) and of RNA polymerase (Hwang et al. 1990). In addition, Grp78 promotes the solubilization of purified immunoglobulin heavy chains during their renaturation (Pelham 1990). These experiments, and those noted above showing the association of Hsc70 with nascent polypeptides (Beckmann et al. 1990) suggest that 70-kDa stress proteins may function during both early and late stages along the folding pathway for diverse polypeptides.

Recent studies have localized the polypeptide binding domain of the 70-kDa stress proteins to a region within the C-terminal 200 amino acids (out of a total of 640–650 amino acids). For example, a 44-kDa N-terminal fragment of mature Hsc70 generated by chymotrypsin digestion still displays ATP-binding and hydrolyzing properties but no longer binds to clathrin cages (Chappell et al. 1987). The results of mutagenesis studies of a human Hsp70 gene support the existence of an N-terminal ATP-binding domain and a C-terminal polypeptide-binding domain (Milarski and Morimoto 1989). Furthermore, these researchers find that a C-terminal 160 amino acid region still retains the ability to associate with denatured protein substrates present in the nucleolus of heat shocked cells. The results from X-ray crystallography studies performed by McKay and coworkers on the 44-kDa N-terminal fragment of Hsc70 have recently provided a detailed structure of the ATP-binding domain (Flaherty et al. 1990). Unfortunately, crystals of an intact 70-kDa stress protein have not yet been obtained. Thus, we have resorted to other methods to gain clues to the structure and properties of the C-terminal region in Hsc70 that contains the polypeptide binding domain.

Circular dichroism spectroscopy of full-length Hsc70 protein purified from bovine brain indicates the presence of a large fraction (approx. 40%) of α-helix in the native protein (Sadis et al. 1990). Analysis of the amino acid sequence and the secondary structure prediction indicates a high frequency of hydrophilic amino acids and an extensive α-helical prediction within the C-terminal polypeptide binding domain (Sadis et al. 1990). The prevalence of hydrophilic residues is best appreciated by analyzing the amino acid sequence with the Kyte-Doolittle hydropathicity program (Kyte and Doolittle 1982). To assess the potential significance of the α-helical prediction, we plotted the α-helical conformational parameters generated by the Garnier-

ALPHA—HELIX PREDICTION

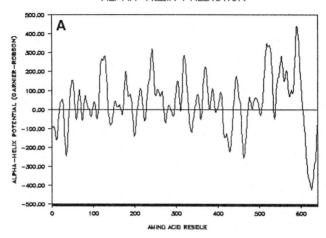

HYDROPATHICITY

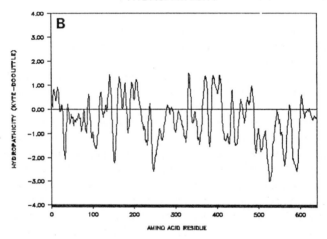

Fig. 5A,B. Helical conformational parameters and hydropathicity of the Hsc70 amino acid sequence. In **A**, the α-helical conformational parameters generated by the secondary structure prediction algorithm of Garnier et al. (1978) were plotted versus the rat Hsc70 amino acid sequence (O'Malley et al. 1985). Positive regions indicate amino acid sequences that are predicted to form α-helical secondary structures, negative regions have a poor tendency for α-helix formation. In **B**, the hydropathicity of the rat Hsc70 amino acid sequence was analyzed according to the methods of Kyte and Doolittle (1982) using a window size of seven residues. This data was then smoothed by taking a running average of every seven residues. Positive regions indicate areas that are rich in hydrophobic residues, negative regions indicate areas rich in hydrophilic residues. Note that the area generating the strongest hydrophilicity overlaps the area that generates the most extensive α-helical secondary structure prediction. This area comprises a major fraction of the amino acid sequence within Hsc70's polypeptide binding domain

Robson-Osguthorpe method (Garnier et al. 1978) of secondary structure prediction for the entire Hsc70 sequence. The results of these analyses are shown in Fig. 5. The amino acid sequence of the C-terminal region, especially between residues 480 and 600, contains by far the greatest concentration of hydrophilic residues within the entire protein (Fig. 5A). Previous studies have shown that α-helical regions of proteins tend to contain a high proportion of charged and polar amino acids (Chou and Fasman 1974; Garnier et al. 1978). Consistent with these observations, the α-helical conformational parameters within the C-terminal region were of the highest magnitude relative to the rest of the Hsc70 sequence (Fig. 5B). Based on these observations, we suspect that α-helices are structural components of Hsc70's polypeptide-binding domain. However, this idea must be directly tested by physical methods.

A small number of other proteins that bind diverse polypeptide have been identified and structural information is available for some of them. Interestingly, in every case, α-helices have been implicated as structural components within the polypeptide-binding domain. For example, the structure of the human HLA-A2 class I major histocompatability (MHC) antigen determined by X-ray crystallography shows that its binding pocket for diverse peptide antigens consists of a groove formed between two parallel α-helices positioned above an anti-parallel ß-sheet platform (Bjorkman et al. 1987ab). In the case of the 54-kDa protein of the signal recognition particle, predicted α-helices unusually rich in Met residues are postulated to bind to the diverse but generally hydrophobic signal peptide sequences (Bernstein et al. 1989). Structural analysis of calmodulin indicates a binding domain comprised of relatively short, clustered α-helices whose side-chains form a hydrophobic patch flanked by negatively charged amino acids (O'Neil and DeGrado 1990). These features assist calmodulin in binding to basic amphiphilic α-helices

Some other features of the MHC antigen deserve comment as they are similar to Hsc70's peptide-binding characteristics. The binding site HLA-A2 and predictions from sequence analysis of other MHC class I genes indicate that the peptide-binding pocket contains a number of polar and charged residues (reviewed by Bjorkman and Parham 1990). Protein modeling studies suggest that extended peptides approximately eight residues in lenght could easily fit into the binding pocket (Bjorkman et al. 1987b). However, the fit for peptides in an α-helical conformation would be quite tight, depending on the amino acid side chains. In addition, unfolded forms of proteins, such as lysozyme, ribonuclease, and myoglobin, can bind the class II MHC antigen, probably with much of their lenght protruding from one or both ends of the binding site (reviewed by Bjorkman and Parham 1990).

Similar to the MHC antigen but distinct from the polypeptide-binding domains of calmodulin or the putative-binding domain of the signal recognition particle, the predicted α-helices in Hsc70's binding domain contain predominantly charged and polar amino acid residues. Therefore, Hsc70 would be expected to have greatest affinity for peptide ligands having hydrophilic side chains. Since hydrogen bonding to the main amides and carbonyl oxygens of polypeptides with hydrophobic residues may also occur, Hsc70 may bind these ligands as well, but with lesser affinity. Hsc70's binding domain appears to select for peptide sequences present in unfolded proteins or in conformationally flexibles regions of native proteins. A channel- or groove-like domain formed by α-helices, perhaps generally similar to the binding domain of the MHC antigen, may restrict binding to relatively extended or disorded polypeptides.

8 Conclusion

In fetal development, the formation of new tissues by cell proliferation and differentiation consists of an extremely well-coordinated series of events. Germ cell maturation and embryonic development occur reproducibly at exact times with a set of specific genes being turned on, resulting in increases in corresponding mRNAs and proteins. What dictates these coordinated series of events is a major challenge for investigators of embryonic development. Evidence is accumulating in support of the idea that cytokines play important roles in regulating cell growth and differentiation during embryogenesis.

Among the known cytokines, TGF-ß is one of the most complex and is a major regulator of cell growth. TGF-ß also regulates the level of ECM by regulating the levels of proteases and protease inhibitors. High levels of proteases induced by mitogens cause the destruction of ECM which can be beneficial because it creates space for proliferating cells (Nilsen-Hamilton 1990). However, proteases can be harmful if overproduced or not closely regulated as occurs in certain inflammatory diseases. During development, the effect of TGF-ß in inducing protease inhibitors and ECM proteins may be protective. Interestingly, the levels of the metalloproteinases collagenase and stromelysin, proteases present in high levels in rheumatoid arthritis, increase in heat shocked rabbit synovial fibroblasts (Vance et al. 1989). In these experiments *hsp70* mRNA is rapidly induced in about 1 h and disappears by 3 h, whereas collagenase and stromelysin synthesis peak by 3 h. Thus, both heat shock and TGF-ß treatment result in rapid induction of Hsp70 family proteins followed by changes in components of the ECM. In the former case, this is part of the heat shock response, whereas in the latter case, the induction may be either part of normal developmental processes or a component of wound responses in somatic tissues.

The spectrum of effects of TGF-ß on cellular responses is determined in part by the presence of other growth factors, by the intracellular biochemical state, by the receptors present on the cell surface, and by the amounts of structural elements that affect regulatory molecules of cell proliferation, such as ECM and its proteases (Feng et al. 1988). According to these authors, TGF-ß1 regulates the expression of two important sets of genes: (1) growth factors and differentiation factors, and (2) ECM components including structural proteins, membrane-binding proteins, proteases, and protease inhibitors.

We propose that *hsp* genes may be part of a third set of genes regulated by TGF-ß since the evidence available today on these proteins has implicated Hsp's in protein assembly and transport; therefore, they may be necessary for cytoplasmic transport and membrane translocation or folding of ECM proteins into their mature conformations.

The increase in the levels of expression of *hsp* genes might be linked to stages of active cell division and differentiation, when there is an increase in proteins requiring molecular chaperones. Since the developmental link with Hsp's is primarily based on their tissue localization and temporal expression in embryonic development with the exception of ecdysterone induction in *Drosophila*, it is still unclear what mechanisms regulate the expression of *hsp* genes during embryogenesis. An overall increase in transcription has been observed in certain stages of embryo development (Bensaude and Morange 1983). Although the expression of different members of *hsp* gene families have been observed in certain stages of development (Zakeri et al. 1988; Maekawa et al. 1989; Lee 1990), the encoded proteins might have the same function of the Hsp's in other mature and differentiated cells or tissues. Alternatively, different Hsp70 isoforms may have evolved that are tailored to function in certain specialized cell types, perhaps recognizing different kinds of unfolded polypeptides. *Hsp70* expression has also been reported to be cell cycle regulated (Milarski and Morimoto 1986; Banerji et al. 1987). Again, the common theme here may be the need to accumulate Hsp's, especially Hsp-Hsc70 in preparation for the synthesis, intracellular trafficking and assembly of other cell cycle proteins.

Acknowledgments. We thank Michael Wright for assistance with graphics, Janet Frost for assistance with secondary structure prediction algorithms, and Carol White for helpful comments on the manuscript. This work was supported by grants from the National Institutes of Health (GM35334) and the National Science Foundation (DCB8916418). We benefited from the use of The Cell Culture Facility, University of Connecticut Biotechnology Center.

References

Akhurst RJ, Lehnert SA, Faissner A, Duffie E (1990) TGF beta in murine morphogenetic processes: the early embryo and cardiogenesis. Development 108:645–656

Assoian RK, Komoriya A, Meyers CA, Miller DM, Sporn MB (1983) Transforming growth factor S in human platelets. J Biol Chem 258:7155–7160

Banerji SS, Laing K, Morimoto RI (1987) Erythroid lineage-specific expression and inducibility of the major heat shock protein HSP70 during avian embryogenesis. Genes Dev 1:946–953

Beckmann RP, Mizzen LA, Welch WJ (1990) Interaction of hsp70 with newly synthesized proteins: implications for protein folding and assembly. Science 248:850–854

Bensaude O, Morange M (1983) Spontaneous high expression of heat shock protein in mouse embryonal carcinoma cells and ectoderm from day eight mouse embryo. EMBO J 2:173–177

Bernstein HD, Poritz MA, Strub K, Hoben PJ, Brenner S, Walter P (1989) Model for signal sequence recognition from amino-acid sequence of 54k subunit of signal recognition particle. Nature 340:482–486

Bjorkman PJ, Parham P (1990) Structure, function, and diversity of class I major histocompatibility complex molecules. An Rev Biochem 59:253–88

Bjorkman PJ, Saper MA, Samraoui B, Bennet WS, Strominger JL, Wiley DC (1987a) Structure of the human class I histocompatibility antigen, HLA-A2. Nature 329:506–512

Bjorkman PJ, Saper MA, Samraoui B, Bennet WS, Strominger JL, Wiley DC (1987b) The foreign antigen binding site and T cell recognition regions of class I histocompatibility antigens. Nature 329:512–519

Carrington JL, Roberts AB, Flanders KC, Roche NS, Reddi AH (1988) Accumulation, localization and compartmentation of transforming growth factor-ß during endochondral bone development. J Cell Biol 107:1969–1975

Cate RL, Mattaliano RJ, Hession C, Tizard R, Farber NM, Cheung A, Ninfa EG, Frey AZ, Gash DJ, Chow EP, Fisher RA, Bertonis JM, Torres G, Wallner BP, Ramachandran L, Ragin RC, Managanaro TF, MacLaughin DT, Honahoe PK, (1986) Isolation of bovine and human genes for Mullerian inhibiting substance and expression of the human gene in animal cells. Cell 45:685–698

Chappell TG, Konforti BB, Schmid SL, Rothman JE (1987) The ATPase core of a clathrin uncoating protein. J Biol Chem 262:746-751

Chou PY, Fasman GD (1974) Prediction of protein conformation. Biochemistry 13:222-245

Cleveland DW, Lopata MA, Mac Donald RJ, Cowan NJ, Rutter WJ, Kirshner MW (1980) Number and evolutionary conservation of α and ß-tubulin and cytoplasmic ß and γ-actin genes using specific cloned cDNA probes. Cell 20:95–105

Coffey RJ, Bascom CC, Sipes NJ, Graves-Deal R, Weissman BE, Moses HL (1988a) Selective inhibition of growth-related gene expression in murine keratinocytes by transforming growth factor beta. Mol Cell Biol 8:3088–3093

Coffey RJ, Sipes NJ, Bascom CC, Graves-Deal R, Pennington CY, Weissman BE, Moses HL (1988b) Growth modulation of mouse keratinocytes by transforming growth factors. Cancer Res 48:3180–3185

Craig EA (1985) The heat shock response. CRC Crit Rev Biochem 18:239–280

Craig EA, Ingolia TD, Manseau LJ, (1983) Expression of *Drosophila* heat shock cognate genes during heat shock and development. Dev Biol 99:418–429

DeLuca-Flaherty C, McKay DB, Parham P, Hill BL, (1990) Uncoating protein (hsc70) binds a conformationally labile domain of clathrin light chain LCa to stimulate ATP hydrolysis. Cell 62:875–887

Derynck R, Lindquist PB, Lee A, Wen D, Tamm I, Graycar J, Rhee L, Chen EY, (1988) A new type of transforming growth factor-ß. EMBO J 7:3737–3743

Deshaies R, Koch B, Werner-Washburne M, Craig EA, Scheckman R (1988) 70kD stress protein homologues facilitate translocation of secretory and mitochondrial precursor polypeptides. Nature 332:800–805

Ellingsworth LR, Brennan JE, Fok K, Rosen DM, Bentz H, Piez KA, Seyedin SM (1986) Antibodies to the N-terminal portion of cartilage-inducing factor A and transforming growth factors beta. Immuno-histochemical localization and association with differentiating cells. J Biol Chem 261:12362–12367

Ellis RJ, Hemmingsen SM (1989) Molecular chaperones: proteins essential for the biogenesis of some macromolecular structures. Trends Biochem Sci 14:339–343

Fanger BO, Wakefield LM, Sporn MB (1986) Structure and properties of the cellular receptor for transforming growth factor type-beta. Biochemistry 25:3083–3091

Feng P, Catt KJ, Knecht M (1988) Transforming growth factor-ß stimulates meiotic maturation of the rat oocyte. Endocrinology 122:181–186

Ferris DK, Harel-Bellan A, Morimoto RI, Welch WJ, Farrar WL (1988) Mitogen and lymphokine stimulation of heat shock proteins in T lymphocytes. Proc Natl Acad Sci USA 85:3850–3854

Flaherty KM, DeLuca-Flaherty C, McKay DB (1990) Three-dimensional structure of the ATPase fragment of a 70K heat-shock cognate protein. Nature 346:623–628

Flynn GC, Chappell TG, Rothman JE (1989) Peptide binding and release by proteins implicated as catalysts of protein assembly. Science 245:385–390

Garnier J, Osguthorpe DJ, Robson B (1978) Analysis of the accuracy and implications of simple methods for predicting the secondary structure of globular proteins. J Mol Biol 120:97–120

Hahnel AC, Gifford DJ, Heikkila JJ, Shultz GA (1986) Expression of the major heat shock protein (HSP70) family during early mouse embryo development. Teratog Carcinog Mutagen 6:493-510

Heikkila JJ, Ovsenek N, Krone P (1987) Examination of heat shock protein in mRNA accumulation in early *Xenopus laevis* embryos. Biochem Cell Biol 65:87-94

Heine UI, Munoz EF, Flanders KC, Ellingsworth LR, Lam H-YP, Thompson NL, Roberts AB, Sporn MB (1987) Role of transforming growth factor-ß in the development of the mouse embryo. J Cell Biol 105:2861–2876

Hightower LE (1980) Cultured animal cells exposed to amino acids analogues puromycin rapidly synthesize several polypeptides. J Cell Physiol 102:407–427

Hwang DS, Crooke E, Kornberg A (1990) Aggregated DnaA protein is dissociated and activated for DNA replication by phospholipase or DnaK protein. J Biol Chem 265:19244–19248

Ignotz RA, Endo T, Massague J (1987) Regulation of fibronectin and type I collagen mRNA levels by transforming growth factor-ß. J Biol Chem 262:6443–6446

Jakowlew SB, Dillard PJ, Sporn MB, Roberts AB (1988) Complementary deoxyribonucleic acid cloning of a messenger ribonucleic acid encoding transforming growth factor ß4 from chick embryo chondrocytes. Mol Endocrinol 2:1186–1195

Kang P-J, Ostermann J, Shilling J., Neupert W, Craig EA, Pfanner N (1990) Requirement for hsp70 in the mitochondrial matrix for translocation and folding of precursor proteins. Nature 348:137–143

Keski-Oja J, Laiho M, Lohi J (1989) Activation of latent cell derived transforming growth factor-ß by the plasminogen activator urokinase. J Cell Biol 107:50a

Kim S-J, Angel P, Lafyatis R, Hattori K, Kim KY, Sporn MB, Karin M, Roberts AB (1990) Autoinduction of transforming growth factor ß1 is mediated by the AP-1 complex. Mol Cell Biol 10:1492–1497

Kondaiah P, Sands MJ, Smith JM, Fields A, Roberts AB, Sporn M, Melton DA (1990) Identification of a novel transforming growth factor-ß (TGF-B5) mRNA in *Xenopus laevis*. J Biol Chem 265:1089–1093

Kothary R, Perry MD, Moran LA, Rossant J (1987) Cell lineage-specific expression of the mouse HSP68 gene during embryogenesis. Dev Biol 121:342–348

Koyasu S, Nishida E, Kadowaki T, Matsuzaki F, Iida K, Harada F, Kasuga MS H., Yahara I (1986) Two mammalian heat shock proteins, HSP90 and HSP100, are actin binding proteins. Proc Natl Acad Sci USA 83:8054–8058

Kyte J, Doolittle RF (1982) A simple method for displaying the hydropathic character of a protein. J Mol Biol 157:105–132

Lawrence DA, Pircher R, Julian P (1985) Conversion of a high molecular weight latent ß-TGF from chicken embryo fibroblasts into a low molecular weight active ß-TGF under acidic conditions. Biochem Biophys Res Commun 133:1026–1034

Lee SJ (1990) Expression of HSP86 in male germ cells. Mol Cell Biol 10:3239–3242

Lehnert SA, Akhurst RJ (1988) Embryonic expression pattern of TGF beta type-1 RNA suggest both paracrine and autocrine mechanisms of action. Development 104:263–273

Lim L, Hall C, Leung T, Whatley S (1984) The relationship of the rat brain 68 kDa microtubule-associated protein with synaptosomal plasma membranes and with the *Drosophila* 70 kDa heat-shock protein. Biochem J 224:677–680

Lyons K, Graycar JL, Lee A, Hashmi S, Lindquist PB, Chen EY, Hogan BLM, Derynck R (1989a) Vgr-1, a mammalian gene related to *Xenopus* Vg-1, is a member of the transforming growth factor ß gene superfamily. Proc Natl Acad USA 86:4554–4558

Lyons KM, Pelton RW, Hogan BLM (1989b) Patterns of expression of murine Vgr-1 and BMP-2a RNA suggest that transforming growth factor-ß-like genes coordinately regulated aspects of embryonic development. Genes Dev. 3:1657–1668

Lyons RM, Moses HL (1990) Transforming growth factors and the regulation of cell proliferation. Eur J Biochem 187:467-473

Madisen L, Webb NR, Rose TM, Marquardt H, Ikeda T, Twardzik D, Seyedin SM, Purchio AF (1988) Transforming growth factor ß2 cDNA cloning sequence analysis. DNA 7:1–8

Madri JA, Pratt PM, Tucker B (1988) Phenotypic modulation of endothelial cells by transforming growth factor-ß depends upon the composition and organization of the extracellular matrix. J Cell Biol 106:1375–1384

Maekawa M, O'Brien DA, Allen RL, Eddy EM (1989) Heat-shock cognate protein (hsc71) and related proteins in mouse spermatogenic cells. Biol Reprod 40:843–852

Mason AJ, Hayflick JS, Ling N, Esch F, Ueno N, Ying Y, Guillemin R, Naill H, Seeburg PH (1985) Complementary DNA sequences of ovarian follicle fluid inhibin show precursor structure and homology with transforming growth factor ß. Nature 318:659–663

Massagué J (1990) The transforming growth factor-ß family. Annu Rev Cell Biol b, 597–641

Milarski KL, Morimoto RI (1986) Expression of human HSP70 during the synthetic phase of the cell cycle. Proc Natl Acad Sci USA 83:9517–9521

Milarski KL, Morimoto RI (1989) Mutational analysis of the human hsp70 protein. J Cell Biol 109:1947–1962

Morimoto RI, Hunt C, Huang S, Burg KL, Banerji SS (1986) Organization, nucleotide sequence, and transcription of the chicken hsp70 gene. J Biol Chem 261:12692–12699

Moses HL, Yang EY, Pietenpol JA (1990) TGF-β stimulation and inhibition of cell proliferation: new mechanistic insights, Cell 63:245–247

Nagata K, Saga S, Yamada KM (1986) A major collagen-binding protein of chick embryo fibroblasts is a novel heat shock protein. J Cell Biol 103:223–229

Napolitano EW, Pachter JS, Chin SSM. Leim RKH (1985) ß-Internexin, a ubiquitous intermediate filament-associated protein. J Cell Biol 101:1323–1331

Newport J, Kirshner M (1982a) A major developmental transition in early *Xenopus* embryos: I. Characterization and timing of cellular changes at the midblastula stage. Cell 30:675–686

Newport J, Kirshner M (1982b) A major developmental transition in early *Xenopus* embryos: II. Control of the onset of transcription. Cell 30:687–696

Nilsen-Hamilton M (1990) Transforming growth factor-ß and its action on cellular growth and differentiation. In: Nilsen-Hamilton M (eds) Growth factors and development. Academic Press, Inc, San Diego, pp 96–136

O'Farrel PH (1975) High resolution two-dimensional electrophoresis of proteins. J Biol Chem 250:4007–4021

O'Malley K, Mauron A, Barchas JD, Kedes L (1985) Constitutively expressed rat mRNA encoding a 70-kilodalton heat-shock-like protein. Mol Cell Biol 5:3476–3483

O'Neil KT, DeGrado WF (1990) How calmodulin binds its targets: sequence independent recognition of amphiphilic α-helices. Trends Biochem Sci 15:59–64

Padgett RW, St Johston RD, Gelabert RW (1987) A transcript from *Drosophila* pattern gene predicts a protein homologous to the transforming growth factor-ß family. Nature 325:81–84

Pelham RHB (1990) Functions of the hsp70 protein family: an overview. In: Morimoto RI, Tissieres ACG (eds) Stress proteins in biology and medicine. Cold Spring Harbor Lab, Cold Spring Harbor, p 287–300

Penttinen RP, Kobayashi S, Bornstein P (1988) Transforming growth factor ß increases mRNA for matrix proteins both in the presence and in the absence of changes in mRNA stability. Proc Natl Acad Sci USA 85:1105–1108

Perkins LA, Doctor JS, Zhang K, Stinson L, Perrimon N, Craig EA (1990) Molecular and developmental characterization of the heat shock cognate 4 gene of *Drosophila* melanogaster. Mol Cell Biol 10:3232–3238

Roberts AB, Anzano MA, Meyers CA, Wideman J, Blacher R, Pan YE, Stein S, Lehrman R, Smith JM, Lamb LC, Sporn MB (1983) Purification and properties of a type ß transforming growth factor from bovine kidney. Biochemistry 22:5692–5698

Roberts AB, Sporn MB, Assoian RK, Smith JM, Roche NS, Wakefield LM, Heine UI, Liotta LA, Falanga V, Kehrl JH, Fauci AS (1986) Transforming growth factor type-beta rapid induction of fibrosis and angiogenesis in vivo and stimulation of collagen formation in vitro. Proc Natl Acad Sci USA 83:4167–41171

Robert AB, Flanders KC, Heine UI, Jakowlew S, Kondaiah P, Kim SJ, Sporn MB (1990) Transforming growth factor-ß: multifunctional regulator or differentiation and development. Phil Trans R Soc Lond B 327:145–154

Robey PG, Young MF, Fladers KC, Roche NS, Kondaiah P, Reddi AH, Termine JD, Sporn MB, Roberts AB (1987) Osteoblasts synthesize and respond to TGF-beta in vitro. J Cell Biol 105:457–463

Rosa F, Roberts AB, Danielpour D, Dart LL, Sporn MB, Dawid IB (1988) Mesoderm induction in amphibians: the role of TGF-ß2-like factors. Science 239:783–785

Rossi P, Karsenty G, Roberts AB, Roche NS, Sporn MB, David I (1988) A nuclear factor 1 binding site mediates the transcriptional activation of a type I collagen promoter by transforming growth factors-ß Cell 52:405–414

Rothman JE (1989) Polypeptide chain binding proteins: catalyst of protein folding and related processes in cells. Cell 59:591–601

Sadis S, Raghavendra K, Schuster TM, Hightower LE (1990) Biochemical and biophysical comparison of bacterial DnaK and mammalian Hsc73, two members of an ancient stress protein family. In: Villafranca JJ (eds) Current research in protein chemistry. Academic Press, Lond New York, p 339

Sanchez ER, Redmond T, Scherrer LC, Bresnick EH, Welsh MJ, Pratt WB (1988) Evidence that the 90kD heat shock protein is associated with tubulin containing complexes in L cell cytosol and in intact PtK cells. Mol Endocrinol 2:756–760

Sandberg M, Vuorio T, Hirvonen H, Alitalo K, Vuorio E (1988) Enhanced expression of TGF-ß and c-fos mRNAs in the growth plates of developing human long bones. Development 102:461–470

Sekellick MJ, Marcus PI (1986) Induction of high titer chicken interferon. Methods Enzymol 119:115–125

Sheffield WP, Shore GC, Randall SK (1990) Mitochondrial precursor protein: effects of 70-kilodalton heat shock protein on polypeptide folding, aggregation, and import competence. J Biol Chem 265:11069–11076

Silver JC, Brunt SA, Armavil V (1990) Steroid hormone regulated expression of certain *hsp70* and *hsp85* genes in Achlya. In: Heat Shock Int Workshop, Ravello, Italy, IIGB Press, Naples, p 185

Skowyra D, Georgopolous C, Zylicz M (1990) The *E. coli* dnaK gene product, the hsp70 homolog, can reactivate heat-inactivated RNA polymerase in an ATP hydrolysis-dependent manner. Cell 62:939–944

Sporn MB, Roberts AB, Wakefield LM, de Crombrugghe B (1987) Some recent advances in the chemistry and biology of transforming growth factor-beta. J Cell Biol 105:1039–1049

St Johnston RD, Gelbart WM (1987) Decapentaplegic transcripts are localized along the dorsal-ventral axis of the *Drosophila* embryo. EMBO J 6:2785–2791

Takenaka IM (1990) Regulation of basal expression of heat shock genes by transforming growth factor-ß Ph D Thesis Univ Connecticut , Storrs, Conn

ten Dijke PT, Hansen P, Iwata KK, Pieler C, Foulkes JG (1988) Identification of another member of the transforming growth factor type ß gene family. Proc Natl Acad Sci USA 85:4715–4719

ten Dijke P, Iwata KK, Thorikay M, Schwedes J, Stewart A, Pieler C (1990) Molecular characterization of transforming growth factor ß3. Ann NY Acad Sci (in press)

Tucker RF, Shipley GD, Moses HL, Holley RW, (1984) Growth inhibitor from BSC-1 cells is closely related to the platelet type-ß transforming growth factor. Science 226–705–707

Vance BA, Kowalski CG, Brinckerhoff CE (1989) Heat shock of rabbit synovial fibroblasts increases expression of mRNAs for two metalloproteinases, collagenase and stromelysin. J Cell Biol 108:2037–2043

Vogel JP, Mistra LM, Rose MD (1990) Loss of BiP-GRP78 function blocks translocation of secretory proteins in yeast. J Cell Biol 110:1885–1895

Weeks DL, Melton DA (1987) A maternal mRNA localized to the vegetal hemisphere in *Xenopus* eggs codes for a growth factor related to TGFß. Cell 51:861–867

Whelan SA, Hightower LE (1985) Differential induction of glucose-regulated and heat shock proteins: effects of pH and sulfhydryl-reducing agents on chicken embryo cells. J Cell Physiol 125:251–258

Wilcox JN, Derynck R (1988) Developmental expression of transforming growth factors alpha and beta in the mouse fetus. Mol Cell Biol 8:3415–3422

Wozney JM, Rosen V, Celeste AJ, Mitsock LM, Whitters MJ, Kriz RW, Hewick RM, Wang EA (1988) Novel regulators of bone formation: molecular clones and activities. Science 242:1528–1534

Wrann M, Bodmer S, de Martin R, Siepl C, Hofer-Warbinek R, Frei K, Hofer E, Fontana A (1987) T cell suppressor factor from human glioblastoma cells is a 12.5 kD protein closely related to transforming growth factor ß EMBO J 6:1633–1666

Wu BJ, Morimoto RI (1985) Transcription of the human hsp70 gene is induced by serum stimulation. Proc Natl Acad Sci USA 82:6070–6074

Wu B, Williams GT, Morimoto RI (1987) Detection of three protein binding sites in the serum regulated promoter of human gene encoding 70kD heat shock protein. Proc Natl Acad Sci USA 84:2203–2207

Zakeri Z, Wolgemuth DJ (1987) Developmental-stage-specific expression of the HSP70 gene family during differentiation of the mammalian male germ line, Mol Cell Biol 7:1791–1796

Zakeri Z, Wolgemuth DJ, Hunt CR (1988) Identification and sequence analysis of a new member of the mouse HSP70 gene family and characterization of its unique cellular and developmental pattern of expression in the male germ line. Mol Cell Biol 8:2925–2932

Zimmerman JL, Petri W, Meselson M (1983) Accumulation of a specific subset of *D. melanogaster* heat shock mRNAs in normal development without heat shock. Cell 32:1161–1170

13 Cell Growth, Cytoskeleton, and Heat Shock Proteins

I. Yahara,[1] S. Koyasu,[1] K. Iida,[1] H. Iida,[2] F. Matsuzaki,[3] S. Matsumoto,[1] and Y. Miyata[1]

1 Cyclic AMP and Expression of Heat Shock Proteins in the Budding Yeast

A distinctive property of cultured mammalian cells existing in the resting state (G_0) may be an unusually long delay between the shift of the culture from nonpermissive to permissive conditions for growth and the initiation of DNA synthesis. For instance, lymphocytes that had divided recently responded quickly to a stimulus, but lost this ability and became cells that responded slowly when incubated in the absence of the stimulus (Kumagai et al. 1981). We supposed that this alteration would be a result of the entrance to G_0. This feature associated with various G_0 cells was also seen with yeast cells arrested by certain *cdc* and other temperature-sensitive mutations under nonpermissive temperatures (Iida and Yahara 1984a).

Growth arrests of *Saccharomyces cerevisiae* cells in early G_1 phase achieved by various means were classified into two types, according to the mode of growth recovery after release of the restraints against growth (Iida and Yahara 1984a). The first type, including arrests caused by *cdc25, cdc33, cdc35*, and *ils1* mutations at the nonpermissive temperature and also by sulfur starvation, showed a subsequent delay in the onset of budding and DNA synthesis when shifted back to permissive conditions. The second type, including those caused by *cdc28* and *cdc4* mutations and by α-factor, did not affect the mode of growth recovery after the shift to permissive conditions, irrespective of the time that cell proliferation had been restricted. Judging from the criterion described above, growth arrests of the first type appear to allow yeast cells to enter a resting state equivalent to the G_0 state of higher eucaryotes.

Using this system in the yeast, we have made an attempt to identify proteins specifically or preferentially synthesized during the transition from G_1 to G_0 or in G_0 (Iida and Yahara 1984b). We have found that at least nine proteins were synthesized specifically in G_0 yeast cells ; six of them were identified with those known as heat shock proteins (Hsps) and their isotypes. One of these Hsps was identified as an isoform of a glycolytic enzyme, enolase 1 (Iida and Yahara 1985). Preexposure of yeast cells to mild heat shock which allows them to synthesize and accumulate Hsps render the cells heat shock resistant (Iida and Yahar 1984c). The synthesis of a certain class of Hsps in the G_0 state seems to be causally related to the fact that G_0 yeast cells

---

[1] The Tokyo Metropolitan Institute of Medical Science, Bunkyo-ku, Tokyo 113, Japan
[2] National Institute for Basic Biology, Okazaki 444, Japan
[3] Institute of Neuroscience, National Center of Neurology and Psychiatry, Kodaira-shi 187, Japan.

Results and Problems in Cell Differentiation 17
Heat Shock and Development
Hightower and Nover (Eds.)
©Springer-Verlag Berlin Heidelberg 1991

Table 1. Survival fractions of yeast cells exposed to heat shock

Strain	Genotypes	Survival fractions[a]
A364A	*HSR1*	0.03
A364A preheated	*HSR1*	28
A364A resting	*HSR1*	63
A364A sulfur-starved	*HSR1*	22
H204	*hsr1*	24
H204 x X2180B	*HSR1/hsr1*	0.05

[a] Cells were exposed to 52 °C for 5 min, after which the colony-forming ability was determined. Survival fractions are expressed as % of the colony forming ability before the heating. Data from Iida and Yahara (1984c).

are highly tolerant against high temperatures, which are otherwise lethal to growing yeast cells (Table 1) (Iida and Yahara 1984c). To confirm this point, we have isolated several heat shock-resistant mutants from ethyl methane sulfonate-treated yeast cells. Among them, we have characterized the most highly heat shock-resistant mutant. Since phenotypes revealed by this mutation are attributed to a single nuclear mutation, the mutated gene in the mutant was designated *hsr1* (a recessive mutation). One of us has further analyzed this mutation and found that the *HSR1* locus is identical to the *CYR1* (or *CDC35*) which encodes adenylate cyclase (Iida 1988). The *hsr1* mutations were caused by insertion of a retrotransposon Ty into either the 5'-coding or noncoding region close to the putative initiation codon of the *CYR1* gene. The *hsr1* mutants showed approximately one-third of the adenylate cyclase activity of *HSR1* strains.

Different from the growth of mammalian cells, that of yeast cells is critically regulated by the level of a cAMP (Matsumoto et al. 1982). Lowering the level of cAMP causes a decrease in the growth rate and sometimes also cell growth arrest. As we have shown previously (Iida and Yahara 1984c), exponentially growing cells of *hsr1* have an approximately two times longer G_1 period than *HSR1* cells. Since there is no evidence that a cAMP-responsive element (CRE) is negatively involved in the expression of *hsp* genes, the low level of cAMP in growing *hsr1* cells affects indirectly the expression of Hsps. Only few Hsps are expressed in growing *hsr1* cells, although they are included in the set of Hsps relatively highly expressed in G_0 yeast cells (Iida and Yahara 1984b,c). This difference may be due to different degrees to which the levels of cAMP are lowered in the two systems. Alternatively, transition into the G_0 state, which was in fact caused by lowering the level of cAMP, increased the expression of some Hsps through an unknown mechanism.

2 Heat Shock-Induced Reorganization of Cytoskeletal Structures

In addition to the induction of Hsps, an exposure of cultured mammalian cells to heat shock causes changes in physiological states of these cells, including reorganization of cytoskeletal structures (Iida et al. 1986). Cultured cells of a mouse fibroblast cell line, C3H-2K have in the interphase well-developed stress fibers consisting of numerous actin filaments. Upon exposure of C3H-2K cells to 42 or 43 °C for 30–90 min, the stress fibers disintegrated; instead, actin paracrystal-like structures, called actin rods, formed in nuclei (Iida et al. 1986). The actin rods were similar in morphology to those induced by treatment with 5–10% (v/v) DMSO (Fukui 1978) or with cytochalasin D (Yahara et al. 1982). The induction of intranuclear actin rods by the heat treatment was fully reversible. Intranuclear actin rods by heat shock also in other cell lines including HeLa, NRK, 3Y1, L, CHO, BALB/c 3T3 and Swiss 3T3, although the efficiencies of the induction differed among these cell lines (Iida et al. 1986).

Electron-microscopic observations clearly showed that actin rods were decorated by heavy meromyosin, forming typical arrowhead structures, which suggested that these structures consist of actin filaments. We have found that phalloidin, a fungal metabolite, does not bind actin rods (Fig. 1), raising the possibility that an actin-binding protein preoccupies the binding site of phalloidin on actin molecules in the filaments. We have tested this possibility using specific antibodies directed against various actin-binding proteins, including myosin, α -actinin, tropomyosin, filamin, and vinculin. The results indicated, however, that these actin-binding proteins are not included in intranuclear actin rods. Finally, we found that cofilin, an actin-binding protein, is a component of the rod (Nishida et al. 1987).

Cofilin is a 21-kD actin-binding protein distributed in a variety of mammalian cells and tissues (Nishida et al. 1984). It binds to actin filaments in a 1:1 molar ratio of cofilin to actin monomer in the filament, shortens the average length of the filament, and increases the steady-state concentration of monomeric actin to a limited extent

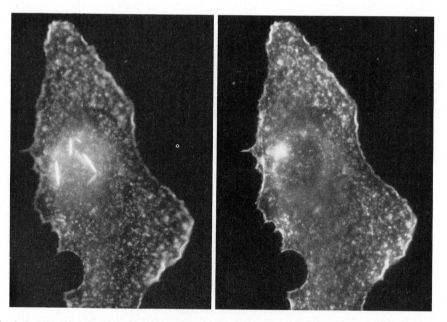

Fig. 1. Inability of phalloidin to bind to intranuclear actin rods. Mouse C3H-2K cells exposed to 43 °C for 60 min were processed for immunofluorescence visualization of cytoskeleton, and stained with anti-actin antibody by the indirect immunoflourescence method (*left*) and rhodamine-labeled phalloidin (*right*). The rods were stained with actin antibody but not with phalloidin

(Nishida et al. 1984). Cofilin reversibly controls actin polymerization and depolymerization in a pH-sensitive manner (Yonezawa et al. 1985). As expected, cofilin and phalloidin compete with each other for binding to actin filaments (Nishida et al. 1987).

Cofilin is almost uniformly distributed in the whole cell space; but it is not bound to stress fibers, probably because tropomyosin inhibits binding of cofilin to actin molecules contained in stress fibers. When cultured cells were heated as described above, cofilin was induced to translocate into nuclei and organized to the rods together with actin (Nishida et al. 1987). These results suggest that cofilin may be involved in reorganization of cellular actin structures in response to various stimuli.

To better understand the structure and function of cofilin, we have isolated cDNA clones encoding cofilin from a cDNA library constructed from porcine brain mRNA (Matsuzaki et al. 1988). The deduced amino acid sequence of porcine cofilin is 166 residues long and contains a sequence of Lys-Lys-Arg-Lys-Lys which is very similar to the nuclear location signal of SV40 large T antigen. Since actin does not contain a nuclear location signal-like sequence, a complex of cofilin and actin might be induced to translocate to the nucleus using the putative signal by heat shock. Murine cofilin also consists of 166 amino acids and only 2 amino acid residues differ between murine cofilin and the porcine counterpart (Moriyama et al. 1990). The well-conserved amino acid sequence of cofilin implicates functional importance of this protein. Comparison of the cofilin sequence with that of other actin-binding proteins has further revealed several sequence similarities in the carboxy-terminal one-third of cofilin sequence. This suggests that cofilin interacts with actin through its carboxy-terminal domain.

Treatment of erythroleukemia cells with DMSO induces erythropoiesis, which involves alterations in gene expression (Friend et al. 1971). Heat shock also induces the expression of several sets of specific proteins called Hsps and suppresses the synthesis of other proteins to some extent (Ashburner and Bonner 1979). As described above, both DMSO and heat shock induce translocation of actin and cofilin from the cytoplasm to the nucleus. The similarities between DMSO treatment and heat shock prompt us to hypothesize an involvement of intranuclear actin or/and cofilin in gene expression, although there is no direct evidence for this hypothesis.

3 Hsp90 is an Actin-Binding Protein

We have isolated a heat shock-resistant variant of a Chinese hamster ovary (CHO) cell line that expresses severalfold increased amounts of Hsp90 and has a relatively flat morphology as compared to the parental strain (Yahara et al. 1986). In addition, the variant always forms larger colonies than those of the parental CHO strain, although these two strains proliferate at the same growth rate. This fact is attributed to a relatively high motility of the variant as compared to the parental strain.

We have investigated the intracellular distribution of Hsp90 in cultured KB cells using the indirect immunofluorescence method (Koyasu et al. 1986). Membrane ruffles were efficiently induced on KB cells in response to insulin. We have found that a bulk of Hsp90 was present in the cytoplasm around the nucleus and, in addition, a part of Hsp90 was localized in ruffling membranes, suggesting an abundance of Hsp90 in these regions. It is well known that there are networks of actin filaments in ruffling membranes. Taken these results together with those obtained with the above heat shock-resistant variant of CHO cells, we assumed that Hsp90 might interact with actin filaments. This turned out to be the case (Koyasu et al. 1986).

The results clearly showed that Hps90 was coprecipitated with polymerized actin (Fig. 2), indicating that Hsp90 binds to actin filaments under physiological ionic conditions. In addition, we have shown by viscometric analysis that Hsp90 cross-links actin filaments (Koyasu et al. 1986). The binding is saturable in a molar ratio of about one Hsp90 dimer to ten actin in the polymerized form (Nishida et al. 1986). A dissociation constant (K_D) of the binding was calculated to be 2.3×10^{-6}M. This value is relatively high, suggesting that the interaction is not strong. However, it is quite likely that the interaction occurs in cells because of abundance of the two proteins.

Tropomyosin inhibits the binding of Hsp90 to actin filaments (Koyasu et al. 1986; Nishida et al. 1986). This explains the fact that Hsp90 is not distributed along stress fibers which consists of actin filaments bound with tropomyosin (Lazarides 1975). Calmodulin inhibits the binding of Hsp90 to actin filaments in a Ca^{2+} ion dependent manner. The equilibrium gel filtration method using a Sephadex G-100 column equilibrated with 40 µg-ml calmodulin revealed that calmodulin binds Hsp90 only in the presence of Ca^{2+}. This result that Ca^{2+}-calmodulin inhibits the binding by interacting with Hsp90.

Hsp90 has been shown to associate transiently with avian sarcoma viral transforming proteins such as pp60[v-src] (Oppermann et al. 1981; Brugge et al. 1981). The pp60[v-src] is released from Hsp90 during the acquisition of the tyrosine kinase activity that includes acylation by myristic acid, phosphorylation and association with the inner side of the plasma membrane. It has been also revealed that Hsp90 is associated with various steroid hormone receptors and dioxin receptors in the cytoplasm (Dougherty et al. 1984; Renoir et al. 1984; Perdew 1988). The binding of ligands to these receptors induces dissociation from Hsp90 and translocation of the receptors into the nucleus . It is possible that in these cases Hsp90 functions as a carrier protein, interacting with cytoskeleton and helping in the location, transport and stabilization of functionally key proteins. In fact, we have recently proved that the cytoplasmic glucocorticoid receptor binds actin filaments through the Hsp90 moiety of the receptor (Miyata and Yahara 1991).

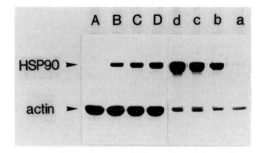

Fig. 2. Binding of Hsp90 to polymerized actin. Purified actin from rabbit skeletal muscles was incubated in the standard buffer solution with or without purified Hsp90, and centrifuged. The pellet (*A-D*) and supernatant (*a-d*) fractions were analyzed by SDS-polyacrylamide gel electrophoresis. Hsp90 alone did not sediment under the conditions used (See Nishida et al. 1986)

We have found that Hsp100 (= Grp94), an endoplasmic reticulum (ER) protein belonging to the Hsp90 family, also binds actin filaments in a Ca^{2+} - calmodulin-sensitive manner (Koyasu et al. 1986, 1989). As suggested from the amino acid sequence, Grp94 may be an ER membrane-spanning protein, a part of which is exposed to the cytoplasmic space (Mazzarella and Green 1987). If this prediction is correct, the cytoplasmic domain of Grp94 may interact with actin filaments, and may function in the dynamic organization of ER. However, the extractability of the protein from intact cells suggests that Grp94 does not have characteristics of a membrane-protein (Munro and Pelham 1987; Koyasu et al. 1990). Furthermore, while Grp94 has a KDEL sequence in its carboxy-terminal end which has been identified as the ER-retention signal (Munro and Pelham 1987), all the other ER proteins possessing the KDEL sequence are proteins locating in the lumen of ER. For these reasons, it is not clear at present whether the actin-binding property associated with Grp94 is biologically significant.

References

Ashburner M, Bonner JJ (1979) The induction of gene activity in *Drosophila* by heat shock. Cell 17:241–254

Brugge JS, Erikson E, Erikson RL (1981) The specific interaction of the Rous sarcoma virus transforming protein, pp60[src], with two cellular proteins. Cell 25:363–372

Dougherty JJ, Puri RK, Toft DO (1984) Polypeptide components of two 8S forms of chicken oviduct progesterone receptor. J Biol Chem 259:8004–8009

Friend C, Scher W, Holland JG, Sato T (1971) Hemoglobin synthesis in murine virus-induced leukemia cells in vitro: Stimulation of erythroid differentiation by dimethylsulfoxide. Proc Natl Acad Sci USA 68:378–382

Fukui Y (1978) Intranuclear actin bundles induced by dimethyl sulfoxide in interphase nucleus of *Dictyostelium*. J Cell Biol 76:146–157

Iida H (1988) Multistress resistance in *Saccharomyces cerevisiae* is generated by insertion of retrotransposon Ty into the 5'-coding region of the adenylate cyclase gene. Mol Cell Biol 8:5555–5560

Iida H, Yahara I (1984a) Specific early-G_1 blocks accompanied with stringent response in *Saccharomyces cerevisiae* lead to growth arrest in resting state similar to the G_0 of higher eucaryotes. J Cell Biol 98:1185–1193

Iida H, Yahara I (1984b) Durable synthesis of high molecular weight heat shock proteins in G_0 cells of the yeast and other eucaryotes. J Cell Biol 99:199–207

Iida H, Yahara I (1984c) A heat shock-resistant mutant of *Saccharomyces cerevisiae* showing constitutive synthesis of two heat shock proteins and altered growth. J Cell Biol 99:1441–1450

Iida H, Yahara I (1985) Yeast heat-shock protein of Mr 48,000 is an isoprotein of enolase. Nature 315:688–690

Iida K, Iida H, Yahara I (1986) Heat shock-induction of the intranuclear actin rods in cultured mammalian cells. Exp Cell Res 165:207–215

Koyasu S, Nishida E, Kadowaki T, Matsuzaki F, Iida K, Harada F, Kasuga M, Sakai H, Yahara I (1986) Two mammalian heat shock proteins, HSP90 and HSP100, are actin-binding proteins. Proc Natl Acad Sci USA 83:8054–8058

Koyasu S, Nishida E, Miyata Y, Sakai H, Yahara I (1989) HSP100, a 100-kDa heat shock protein, is a Ca^{2+}-calmodulin-regulated actin-binding protein. J Biol Chem 264:15083–15087

Kumagai J, Akiyama H, Iwashita S, Iida H, Yahara I (1981) In vitro regeneration of resting lymphocytes from stimulated lymphocytes and its inhibition by insulin. J Immunol 126:1249–1254

Lazarides E (1975) Tropomyosin antibody: the specific localization of tropomyosin in nonmuscle cells. J Cell Biol 65:549–561

Matsumoto K, Uno I, Ohshima Y, Ishikawa T (1982) Isolation and characterization of yeast mutants deficient in adenylate cyclase and cAMP-dependent protein kinase. Proc Natl Acad Sci USA 79:2355–2359

Matsuzaki F, Matsumoto S, Yahara I, Yonezawa N, Nishida E, Sakai H (1988) Cloning and characterization of porcine brain cofilin cDNA: cofilin contains the nuclear transport signal sequence. J Biol Chem 263:11564–11568

Mazzarella RA, Green M (1987) ERp99, an abundant, conserved glycoprotein of the endoplasmic reticulum, is homologous to the 90-kDa heat shock protein (hsp90) and the 94-kDa glucose regulated protein (GRP94). J Biol Chem 262:8875–8883

Miyata Y, Yahara I (1991) Cytoplasmic 8S glucocorticoid receptor binds to actin filaments through the 90-kDa heat shock protein moiety. J Biol Chem 266:8779–8783

Moriyama K, Matsumoto S, Nishida E, Sakai H, Yahara I (1990) Nucleotide sequence of mouse cofilin cDNA (Record). Nucl Acids Res 18:3053

Munro S, Pelham HRB (1987) A C-terminal signal prevents secretion of luminal ER proteins. Cell 48:899–907

Nishida E, Maekawa S, Sakai H (1984) Cofilin, a protein in porcine brain that binds to actin filaments and inhibits their interactions with myosin and tropomyosin. Biochemistry 23:5307–5313

Nishida E, Koyasu S, Sakai H, Yahara I (1986) Calmodulin-regulated binding of the 90 kDa heat shock protein to actin filaments. J Biol Chem 261:16033–16036

Nishida E, Iida K, Yonezawa N, Koyasu S, Yahara I Sakai H (1987) Cofilin is a component of intranuclear and cytoplasmic actin rods induced in cultured cells. Proc Natl Acad Sci USA 84:5262–5266

Oppermann H, Levinson W, Bishop JM (1981) A cellular protein that associates with the transforming protein of Rous sarcoma virus is also a heat shock protein. Proc Natl Acad Sci USA 78:1067–1071

Perdew GH (1988) Association of the *Ah* receptor with the 90-kDa heat shock protein. J Biol Chem 263:13802–13806

Renoir JM, Buchou T, Mester J, Radanyi C, Baulieu EE (1984) Oligomeric structure of molybdate-stabilized, nontransformed 8S progesterone receptor from chicken oviduct cytosol. Biochemistry 23:6016–6023

Yahara I, Harada F, Sekita S, Yoshihira K, Natori S (1982) Correlation between effects of 24 different cytochalasins on cellular structures and cellular events and those on actin in vitro. J Cell Biol 92:69–78

Yahara I, Iida H, Koyasu S (1986) A heat shock-resistant variant of Chinese hamster cell line constitutively expressing heat shock protein of Mr 90,000 at high level. Cell Struct Funct 11:65–73

Yonezawa N, Nishida E, Sakai H, (1985) pH control of actin polymerization by cofilin. J Biol Chem 260:14410–14412

14 Expression of Heat Shock Genes (*hsp70*) in the Mammalian Nervous System

1 Introduction

Studies on heat shock proteins have dealt with a wide range of organisms including *E. coli, Saccharomyces, Drosophila* and vertebrate cells grown in culture (Schlesinger et al. 1982; Pardue et al. 1989), however, comparatively little work has been carried out on intact thermoregulating animals. Obviously, it is of interest to ascertain whether the heat shock response is physiologically relevant. For example, are heat shock genes turned on in the mammalian nervous system following feverlike temperatures, ischemia, or tissue wounding and if so, which cell types show induction? As will be shown in this article, initial experiments in this area demonstrated the prominent induction of a 70-kD heat shock protein (Hsp70) when labeled brain proteins isolated from hyperthermic animals were analyzed. Recently, in situ hybridization and immunocytochemistry have been utilized to map out the pattern of expression of both constitutively expressed and stress-inducible members of the *hsp70* multigene family. Different types of neural trauma have been found to induce characteristic cellular responses in the mammalian brain with regard to the type of brain cell that responds by inducing Hsp70 and the timing of the induction response. The pattern of induction of Hsp70 has been found to be a useful early marker of cellular injury in the nervous system and may identify previously unrecognized areas of vulnerability.

Cells respond to a sudden elevation in temperature by a transient inhibition of ongoing RNA and protein synthesis and the induction of a set of genes encoding heat shock proteins which may play important roles in cellular repair and-or protective mechanisms (for reviews see Craig 1985; Lindquist, 1986; Schlesinger 1986; Subjeck and Shyy 1986; Lindquist and Craig 1988; Pelham 1989; Welch 1990). This review will focus on neural expression of the *hsp70s,* a multigene family which is composed of constitutively expressed and stress-inducible members (Welch and Feramisco 1982; Moran et al. 1983; Wu et al. 1985; Watowich and Morimoto 1988; Welch et al. 1989; Welch 1990). To date very few studies have been carried out in the nervous system on expression of the *hsp90s* and the low-molecular weight *hsp20s*.

[1] Department of Zoology, University of Toronto, Scarborough Campus, West Hill, Ontario, Canada M1C 1A4

2 Early Studies on Brain Heat-Shock Proteins

Induction of Hsp70 in the mammalian brain was initially detected by analyzing labeled brain proteins which had been resolved by one-or two-dimensional gel electrophoresis following either in vivo injection of labeled amino acids or cell-free translation of isolated brain polysomes. Studies in our laboratory demonstrated that feverlike temperatures, induced by the psychotropic drug LSD, rapidly induce synthesis of Hsp70 in the adult and fetal rabbit brain (Freedman et al. 1981; Brown 1983, 1985a,b; Cosgrove and Brown 1983) concomitant with a transient disaggregation of brain polysomes (Brown et al. 1982). Induction of Hsp90 has also been noted in the adult brain (Freedman et al. 1981). Whole body hyperthermia (Currie and White 1981) and amphetamine-induced hyperthermia (Nowak 1988) have also been shown to induce Hsp70 in the rodent brain. Other studies indicate that ischemia induces Hsp70 in the gerbil brain (Nowak 1985) and in the rat brain (Dienel et al. 1986; Kiessling et al. 1986; Jacewicz et al, 1986; Dwyer et al. 1989).

3 Induction of Heat Shock Proteins in the Visual System

Induction of Hsp70 and Hsp90 has been demonstrated in the rabbit retina both in vivo and in vitro following heat shock (Clark and Brown 1982, 1986a). The induced retinal Hsp70 copurifies with twice-cycled microtubules and also with purified intermediate filaments, is precipitated by antibodies prepared against purified Tau proteins and binds to calmodulin (Clark and Brown 1986b). In agreement with the last point, DNA sequencing studies have recently revealed that members of the Hsp70 family contain a highly conserved calmodulin-binding domain (Stevenson and Calderwood 1990).

Altered expression of the induced retinal Hsp70 has been noted in the presence of agents which affect microtubule stability (Clark and Brown 1987). Taxol, an antimitotic agent which stabilizes microtubules, was found to reduce the level of Hsp70 which is synthesized in response to elevated temperature. Colchicine, a potent microtubule destabilizing agent, did not induce synthesis of Hsp70 in the absence of elevated temperature; however, under heat shock conditions, synthesis was elevated in the presence of the drug. A microtubule-associated protein isolated from brain has been reported to be a constitutively expressed member of the Hsp70 family (Lim et al. 1984; Whately et al. 1986).

Hsp70 protein is induced in rabbit retinal ganglion neurons and transported down the optic nerve (Clark and Brown 1985). However, this is by slow axonal transport (Clark and Brown 1985), specifically with slow component b, which includes such transported molecules as actin microfilaments, spectrin, clathrin, and calmodulin (Tytell and Barbe 1987). A constitutively expressed Hsp70 which is a clathrin uncoating ATPase has also been reported to be axonally transported as part of slow component b (de Waegh and Brady 1989).

Immunocytochemistry and in situ hybridization have been utilized to map out the pattern of induction of Hsp70 and its mRNA in the retina after hyperthermia (Tytell et al. 1989a). A strong induction in photoreceptor cells was noted. It has been reported that protection against light-induced degeneration of retinal photoreceptors is conferred by prior whole-body hyperthermia (Barbe et al. 1988; Tytell et al. 1989b).

4 Analysis of *hsp70* mRNAs
in the Mammalian Nervous System

The rabbit genome contains a family of *hsp70* genes which include both constitutively expressed and heat shock-inducible members (Brown et al. 1985). As shown in Fig. 1a, Northern blot analysis of RNA isolated from rabbit brain 1 h after elevation of body temperature by 2–3 °C revealed the massive induction of a 2.7-kb mRNA species, while in control animals the presence of a constitutively expressed 2.5-kb mRNA was apparent (Sprang and Brown 1987). *Hsp70* nucleic acid probes have recently been reported which can distinguish transcripts in brain which are derived from either inducible or constitutively expressed members of the *hsp70* gene family (Miller et al. 1989; Brown and Rush 1990; Nowak et al. 1990). Use of an inducible *hsp70* transcript-specific riboprobe revealed that induction of the 2.7-kb brain transcript in hyperthermic rabbits is transient and parallels the rise and fall in body temperature (Brown and Rush 1990). Northern blotting has also been utilized to demonstrate that stereotaxic injection of kainic acid into the striatum of rats induces *hsp70* mRNA within 2 h (Uney et al. 1988).

5 Regional Differences in Expression of *hsp70* genes
in Brain Detected by In Situ Hybridization

An interesting question which required investigation was whether there were regional and cell type differences in the expression of *hsp70* genes in the nervous system. If the ability of brain cells to survive heat shock and other stresses is related to their capacity to induce Hsps, an analysis of the pattern of expression of heat shock genes in the brain may further the understanding of the selective vulnerability of certain brain cells to various forms of trauma and neurogenetic diseases. In situ hybridization can be used to identify which cells in the brain demonstrate constitutive or heat-shock inducible *hsp70* gene expression. In this experimental procedure, a labeled nucleic acid probe is hybridized to a brain tissue section under conditions in which the probe can bind to target RNA with high specificity (Uhl 1986). By subsequent autoradiography one can identify those cells in a complex tissue which are expressing the gene of interest.

In situ hybridization with *hsp70* riboprobe revealed striking regional differences in the expression of constitutive and inducible heat shock genes in the rabbit brain 1 h after drug-induced hyperthermia (Sprang and Brown 1987). Constitutive expression of an *hsp70* gene was observed in several neuronal enriched areas such as hippocampal regions CA1 to CA4 (Fig. 1b,e), and the Purkinje and granule cell layers of the

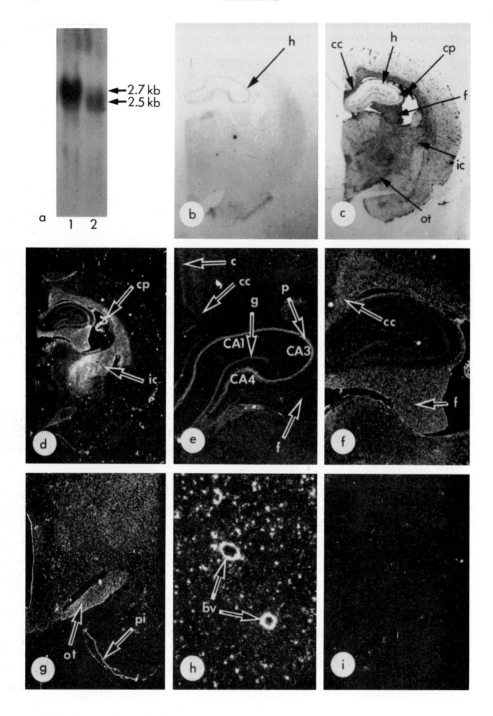

cerebellum (Fig. 2a-c). One hour after hyperthermia, inducible expression of an *hsp70* gene was noted in certain neuronal enriched regions, but not in others. For example, induction was noted in the granule cell layer of the cerebellum (Fig. 2d-f) but not in hippocampal neurons (Fig. 1c,d,f). A dramatic induction of heat shock mRNA was noted 1 h after hyperthermia in fiber tracts throughout the rabbit forebrain (Fig. 1c,d,f) and in the cerebellum (Fig. 2f), a pattern consistent with a strong glial response to heat shock. Strong induction was also detected in the choroid plexus which lines the ventricles of the brain (Fig. 1d), in the microvasculature (Fig. 1h) and in the pia mater, a cellular layer which surrounds the cerebellum and cerebral cortex (Figs. 1g, 2e).

Induction of *hsp70* mRNA in glial, granule, and pial cells of the rat cerebellum has also been noted using in situ hybridization following hyperthermia (Morrison-Bogorad et al. 1989; Blake et al. 1990). Higher levels of hsp70 protein have been reported in nonneuronal cells compared to neuronal cells of isolated ganglia and associated connective cells of *Aplysia* following heat shock (Greenberg and Lasek 1985).

In the isolated squid giant axon, Hsp70 is synthesized in adaxonal glial cells and rapidly exported into the axon (Tytell et al. 1986). Release of Hsp70 from cultured rat embryo cells has also been reported (Hightower and Guidon, 1989). Induction of *hsp70* mRNA in glial cells in fiber tracts of the mammalian brain (Sprang and Brown 1987) and export of the resultant protein into adjacent axons could provide a "fast response" mechanism to ensure the delivery of the heat shock protein to regions of the neuron which are distant from the cell body.

To extend the investigation of heat shock gene expression in the brain to the cellular level, in situ hybridization utilizing plastic-embedded tissue sections has been carried out (Masing and Brown 1989). Decreased section thickness compared to frozen sections, enhanced tissue integrity, and examination at increased magnification have facilitated analysis of cell types which are engaged in the expression of *hsp70* genes in the rabbit cerebellum after hyperthermia. A prominent induction of *hsp70* mRNA was observed in oligodendroglia in fiber tracts of the deep white matter of the cerebellum 1 h after hyperthermia. Intravenous injection of LSD has been used as a convenient means of producing a rapid and predictable increase in body temperature in rabbits (Brown et al. 1982; Brown 1985a,b). A recent in situ hybridization study indicates that induction of *hsp70* mRNA in the rabbit cerebellum is due to hyperthermic effects of this psychotropic drug (Manzerra and Brown 1990). Induction was not present when LSD-induced hyperthermia was blocked.

←

Fig. 1a-i. Regional differences in the induction of an *hsp70* gene in the rabbit brain after hyperthermia. **a** Northern blot analysis. *Lane 1* brain RNA from hyperthermic animal (body temperature elevated 2–3 °C above normal for 1 h); *lane 2* brain RNA from control animal; the blot containing 15 µg of total RNA per lane was probed with ^{32}P-labeled *hsp70* antisense riboprobe (Sprang and Brown 1987). **b-h** In situ hybridization of coronal sections of the rabbit forebrain with ^{35}S-labeled *hsp70* antisense riboprobe (Sprang and Brown 1987). **b,c** X-ray film autoradiograms of brain from control and hyperthermic rabbits respectively, mag. 3x. **d-i** Dark field microscopy of autoradiograms prepared with liquid emulsion. **d** Coronal sections of brain from hyperthermic animal, 3x. **e,f** Hippocampus from control and hyperthermic animals, respectively, mag. 12x. **g** Optic tract from hyperthermic animal, mag. 12x. **h** Cerebral microvasculature: blood vessels in cross-section (indicated by *arrows*) from heat shocked animal, mag. 72x. **i** Control for nonspecific binding: hippocampus from heat shock animal probed with ^{35}S-labeled *hsp70* sense riboprobe, mag. 12x. *bv* blood vessel; *c* neocortex; *cc* corpus callosum; *cp* choroid plexus; *f* fimbria; *g* granule cell layer; *h* hippocampus; *ic* internal capsule; *ot* optic tract; *p* pyramidal cell layer; *pi* pia mater (Sprang and Brown 1987)

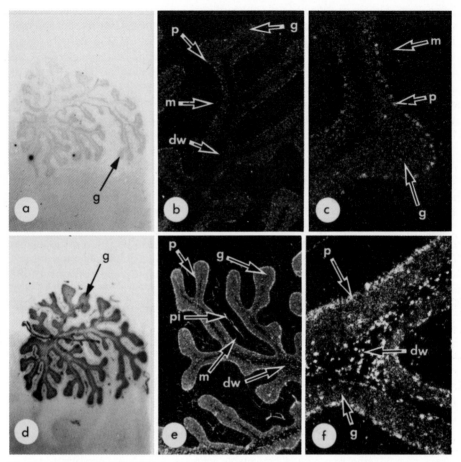

Fig. 2a-f. Induction of *hsp70* mRNA in the rabbit cerebellum. In situ hybridization was carried out as in Fig. 1 with sagittal tissue sections of the cerebellum. *Upper panels* control animals; *lower panels* hyperthermic animals (body temperature elevated 2–3 °C above normal for 1 h). **a,d** X-ray film images, mag. 3.5x. **b,e** Dark-field images of emulsion autoradiograms of cerebellum, mag 16x and 10x, respectively. **c,f** Dark-field images showing deep white matter (fiber tracts), mag. 72x. *dw* deep white matter; *g* granule cell layer; *m* molecular cell layer; *p* Purkinje cell layer; *pi* pia mater (Sprang and Brown 1987)

6 Induction of an *Hsp70* Gene at the Site of Tissue Injury in the Brain

Recently it has been demonstrated that localized tissue injury induces expression of a gene encoding Hsp70 in the mammalian nervous system (Brown et al. 1989). A small surgical cut was made in the rat cerebral cortex. By 2 h postsurgery a dramatic and highly localized induction of *hsp70* mRNA was detected in cells proximal to the lesion site using in situ hybridization and X-ray film autoradiography (Fig. 3a,b). By 12 h the intensity of the signal had diminished, and by 24 h only a few cells along the walls of the cut demonstrated a high level of *hsp70* mRNA (Fig. 3c,d).

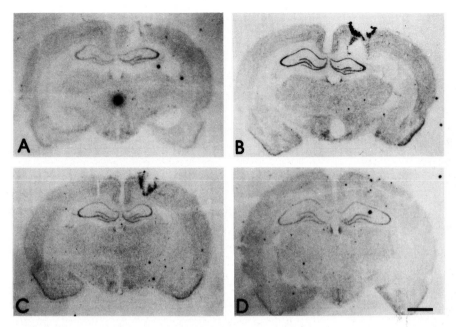

Fig. 3A-D. Time course of induction of an *hsp70* gene at the site of tissue injury in the rat brain detected by in situ hybridization. A surgical cut was made in the cerebral cortex of the adult rat. At various postlesion intervals, animals were sacrificed and brain sections were processed for in situ hybridization with ^{35}S-labeled *hsp70* antisense riboprobe. Survival times were **A** 1 h; **B** 2 h; **C** 12 h; and **D** 24 h. A highly localized and transient induction of *hsp70* mRNA which was apparent at the lesion site by 2 h was progressively diminished by 12 and 24 h. Bar 2mm (Brown et al. 1989)

To investigate the effect at the cellular level, brain tissue sections were coated with emulsion following hybridization with labeled *hsp70* antisense riboprobe. Cells at the glial surface of the surgical cut exhibited dramatic induction of heat shock mRNA at 2 h postlesion (Fig. 4a). By 3.5 h after injury, cells located deeper within the cut had also induced heat shock mRNA (Fig. 4b). This pattern was also evident at 12 h, although the intensity of signal had decreased (Fig. 4c). At 24 h there was a marked decrease in the apparent amount of heat shock mRNA at the lesion site; however, a few cells along the walls of the cut and primarily at the surface of the cerebral cortex exhibited a high level of heat shock mRNA (Fig. 4d). Both neurons and glial cells at the site of the surgical cut appeared to respond to tissue injury by induction of *hsp70* mRNA. The pattern of constitutive expression was not affected by the surgical procedure. This study indicated that induction of a member of the *hsp70* gene family is a physiologically relevant response which is activated at an early stage following tissue injury in the nervous system. The pattern of induction of *hsp70* mRNA can thus serve as a marker to identify a population of reactive cells that respond rapidly to trauma.

A second example which illustrates that tissue injury can induce the expression of a heat shock gene in the mammalian nervous system is a recent experiment which demonstrates that peripheral axotomy can act as a cell stressor to induce *hsp70* mRNA (New et al. 1989). Cutting the adult hamster facial nerve resulted in the induction of *hsp70* mRNA (detected by Northern blotting) in facial nerve nuclear groups which

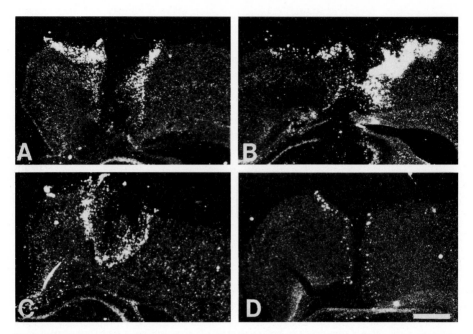

Fig. 4A-D. Analysis of *hsp70* gene induction at the cellular level at the injury site. The pattern of induction of an *hsp70* gene at the site of tissue injury was analyzed by in situ hybridization using liquid emulsion and dark-field microscopy for higher resolution. Survival times were **A** 2 h; **B** 3.5 h; **C** 12 h; and **D** 24 h. The relative number of densely labeled cells appeared to increase between 2 and 3.5 h postlesion, then decrease progressively by 12 and 24 h. Cells near the pial surface responded more intensively than cells located deeper in the cortex at 2 h, while at 3.5 h and 12 h, cells in all neocortical layers had responded. Bar 500 um (Brown et al. 1989)

were dissected from the brain stem. The time course of induction parallels that which was observed in the in situ hybridization study on wounding of the rat cerebral cortex (Brown et al. 1989).

7 Immunological Detection of Hsp70 in Brain Tissue

The pattern of induction of Hsp70 has been mapped out in the mammalian brain following such traumatic events as ischemia (Vass et al. 1988; Dwyer et al. 1989; Ferriero et al. 1990), induction of seizure activity (Gonzalez et al. 1989; Lowenstein et al. 1989; Vass et al. 1989), spinal cord injury (Gower et al. 1989), and hyperthermia (Barbe et al. 1988; Tytell et al. 1989a,b; Marini et al. 1990). These studies have been facilitated by the availability of antibodies directed against Hsp70 (Welch et al. 1989; Welch. 1990). In agreement with in situ hybridization studies (Sprang and Brown 1987; Masing and Brown 1989), immunocytochemical observations have revealed a strong induction of Hsp70 in glial cells either in vivo or in vitro following temperature elevation (Nishimura et al. 1988,1991; Marini et al. 1990).

A neuronal pattern of induction of Hsp70 is apparent following ischemia (Vass et al. 1988) or kainic acid-induced seizures (Gonzalez et al. 1989; Vass et al. 1989). Results suggest that Hsp70 immunocytochemistry may serve as a convenient marker for neuronal circuitry involved in excitotoxic mechanisms after ischemia and other stresses (Vass et al. 1988). It has been suggested that heat shock proteins may be useful early markers of cellular injury and may identify previously unrecognized areas of central nervous system vulnerability (Gonzalez et al. 1989; Ferriero et al. 1990). Whether Hsp70 is engaged in a vital reactive and-or repair process in response to neural trauma, or is only a useful marker for cell injury, is an important question which awaits future investigation.

Different types of neural trauma induce characteristic cellular responses in the mammalian brain with regard to the type of brain cell that responds by inducing Hsp70 and the timing of the induction response. Following transient ischemia in the gerbil brain, immunocytochemical studies suggest that Hsp70 is induced in specific hippocampal neurons but not in glial cells (Vass et al. 1988). Accumulation of Hsp70 was minimal in hippocampal CA1 neurons which die after brief ischemia but was pronounced in dentate granule cells and CA3 neurons which survive. The peak of CA3 immunoreactivity did not occur until 48 h postischemia suggesting that Hsp70 induction is a response to delayed hippocampal pathophysiology rather than a direct response to the initial ischemic insult. A recent study has confirmed that Hsp70 is not detectable by immunocytochemistry in CA1 neurons after ischemia (Nowak 1989). However, a response at the transcriptional level was noted in these cells by in situ hybridization in this study.

8 Tissue-Protective Effects of Heat Shock in the Nervous System

Functional Hsp70 or expression of *hsp70* genes appears to be required if mammalian cells are to survive traumatic insults such as temperature shock (Johnston and Kucey 1988; Riabowol et al. 1988). Brief heat shock has been found to confer tissue protection in the nervous system against subsequent traumatic events. For example, if rat embryos are exposed to heat shock at 43 °C at a specific phase of neural development, major regions of the brain fail to develop. A brief, nonteratogenic heat shock at 42 °C administered prior to the defect-inducing temperature shock confers complete tissue protection against the teratogenic effects (Walsh et al. 1987,1989; see Chap. 4). Protection against light-induced degeneration of rat retinal photoreceptors is also conferred by prior whole-body hyperthermia (Barbe et al. 1988; Tytell et al. 1989a,b).

Identification of the critical factor(s) which are responsible for these heat-induced tissue protective effects in the nervous system remains to be a carried out. An important question to be addressed is the determination of whether induction of Hsp70 in a particular brain cell or import of the protein into that cell really improve chances for survival, and if so, what is the operative mechanism? In *Drosophila* cells grown in tissue culture, heat shock proteins which are induced by prior mild heat treatment have been reported to alleviate the disruption of RNA splicing that results from severe heat shock (Yost and Lindquist 1986, 1988). The *Drosophila* heat shock proteins are thought to exert a protective effect on an early step in the pathway of the removal of intervening sequences from mRNA precursors.

9 Conclusion

Induction of *hsp70* genes in specific populations of neural cells following hyperthermia and other traumatic events, including tissue injury, suggests that Hsp70 may play a role in reactive and perhaps protective mechanisms in the nervous system. In the future, experiments involving either the induction of heat shock proteins or microinjection of purified Hsp70 into the brain may open up new avenues for enhancing cellular repair in the nervous system, and reducing neuron loss after tissue injury which is induced by various traumatic conditions.

The ability of brain cells to survive hyperthermia and other stresses may be related to their capacity for induced Hsp synthesis. It would be interesting to explore whether this induction response is hampered or fails in selected neural cells during aging or neurodegenerative diseases.

Different types of neural trauma appear to induce characteristic cellular responses in the mammalian brain with regard to the type of brain cell that reacts by inducing Hsp70 synthesis and the timing of the induction response. The differential sensitivity of various brain regions to particular types of metabolic stress may, in part, be a function of the relative vigor of the Hsp70 response.

Following hyperthemia, rather unexpected areas of sensitivity or reactivity have been revealed such as the dramatic induction of Hsp70 in glial cells. Production of Hsp70 by glial cells following hyperthermia and subsequent transfer of the protein to neurons may turn out to be part of a basic mechanisms by which glia and neurons interact at a molecular level. In the nervous system, the pattern of induction of Hsp70 following various traumatic conditions has proved to be useful early marker of cellular injury.

Acknowledgements. Studies on brain Hsp70s in this laboratory are supported by grants from the Medical Research Council of Canada.

References

Barbe MF, Tytell M, Gower DJ, Welch WJ (1988) Hyperthermia protects against light damage in the rat retina. Science 241:1817–1820

Blake MJ, Nowak TS, Holbrook NJ (1990) In vivo hyperthermia induces expression of HSP70 mRNA in brain regions controlling the neuroendocrine response to stress. Mol Brain Res 8:89–92

Brown IR (1983) Hyperthermia induces the synthesis of a heat shock protein by polysomes isolated from the fetal and neonatal mammalian brain. J Neurochem 40:1490–1493

Brown IR (1985a) Effect of hyperthermia and LSD on gene expression in the mammalian brain and other organs. In: Atkinson BG, Walden CB (eds) Changes in eukaryotic gene expression in response to environment stress. Academic Press, Orlando, pp 211–225

Brown IR (1985b) Modification of gene expression in the mammalian brain after hyperthermia. In: Zomzely-Neurath C, Walker WA (eds) Gene expression in brain. John Wiley, New York, pp 157–171

Brown IR, Rush SJ (1990) Expression of heat shock genes (hsp70) in the mammalian brain: distinguishing constitutively expressed and hyperthermia-inducible species. J Neurosci Res 25:14–19

Brown IR, Heikkila JJ, Cosgrove JW (1982) Analysis of protein synthesis in the mammalian brain using LSD and hyperthermia as experimental probes. In: Brown IR (ed) Molecular approaches to neurobiology. Academic Press, New York, pp 221–253

Brown IR, Lowe DG, Moran LA (1985) Expression of a heat shock gene in fetal and maternal rabbit brain. Neurochem Res 10:1277–1284

Brown IR, Rush SJ, Ivy GO (1989) Induction of a heat shock gene at the site of tissue injury in the rat brain. Neuron 2:1559–1564

Clark BD, Brown IR (1982) Protein synthesis in the mammalian retina following the intravenous administration of LSD. Brain Res 247:97–104

Clark BD, Brown IR (1985) Axonal transport of a heat shock protein in the rabbit visual system. Proc Natl Acad Sci USA 82:1281–1285

Clark BD, Brown IR (1986a) Induction of a heat shock protein in the isolated mammalian retina. Neurochem Res 11:269–279

Clark BD, Brown IR (1986b) A retinal heat shock protein is associated with elements of the cytoskeleton and binds to calmodulin. Biochem Biophys Res Commun 139:974–981

Clark BD, Brown IR (1987) Altered expression of a heat shock protein in the mammalian nervous system in the presence of agents which effect microtubule stability. Neurochem Res 12:819–823

Cosgrove JW, Brown IR (1983) Heat shock protein in the mammalian brain and other organs following a physiologically relevant increase in body temperature induced by LSD. Proc Natl Acad Sci USA 80:569–573

Craig EA (1985) The heat shock response. CRC Crit Rev Biochem 18:239–280

Currie RW, White FP (1981) Trauma-induced protein in rat tissues: a physiological role for a "heat shock" protein? Science 214:72–73

de Waegh S, Brady ST (1989) Axonal transport of a clathrin uncoating ATPase (HSC70): a role for HSC70 in the modulation of coated vesicle assembly in vivo. J Neurosci Res 23:433–440

Dienel GA, Kiessling M, Jacewicz M, Pulsinelli WA (1986) Synthesis of heat shock proteins in rat brain cortex after transient ischemia. J Cereb Blood Flow Metab 6:505–510

Dwyer BE, Nishimura RN, Brown IR (1989) Synthesis of the major inducible heat shock protein in rat hippocampus after neonatal hypoxia-ischemia. Exp Neurol 104:28–31

Ferriero DM, Soberano HQ, Simon RP, Sharp FR (1990) Hypoxia-ischemia induces heat shock protein-like (HSP72) immunoreactivity in neonatal rat brain. Dev Brain Res 53:145–150

Freedman MS, Clark BD, Cruz TF, Gurd JW, Brown IR (1981) Selective effects of LSD and hyperthermia on the synthesis of synaptic proteins and glycoproteins. Brain Res 207:129–145

Gonzalez MF, Shiraishi K, Hisanaga K, Sagar SM, Mandabach M, Sharp FR (1989) Heat shock proteins as markers of neural injury. Mol Brain Res 6:93–100

Gower DJ, Hollman C, Lee KS, Tytell MT (1989) Spinal cord injury and the stress protein response. J Neurosurg 70:605–611

Greenberg SC, Lasek RJ (1985) Comparison of labelled heat shock proteins in neuronal and non-neuronal cells of *Aplysia californica*. Neuroscience 5:1239–1245

Hightower LE, Guidon PT (1989) Selective release from cultured mammalian cells of heat-shock (stress) proteins that resemble glia-axon transfer proteins. J Cell Physiol 138:257–266

Jacewicz MJ, Kiessling M, Pulsinelli WA (1986) Selective gene expression in focal cerebral ischemia. J Cereb Blood Flow Metab 6:263–272

Johnston RN, Kucey BL (1988) Competitive inhibition of hsp70 gene expression causes thermosensitivity. Science 242:1551–1554

Kiessling M, Dienel GA, Jacewicz M, Pulsinelli WA (1986) Protein synthesis in postischemic rat brain: a two dimensional electrophoretic analysis. J Cereb Blood Flow Metab 6:642–649

Lim L, Hall C, Leung T, Whatley S (1984) The relationship of the rat brain 68kDa microtubule-associated protein with synaptosomal plasma membranes and with the *Drosophila* 70kDa heat-shock protein. Biochem J 224:677–680

Lindquist S (1986) The heat-shock response. Annu Rev Biochem 55:1151–1191

Lindquist S, Craig EA (1988) The heat shock proteins. Annu Rev Genet 22:631–677

Lowenstein D, Gonzalez M, Simon R, Sharp F (1989) The pattern of 72kd heat shock protein-like immunoreactivity in the rat brain following generalized status epilepticus. Soc Neurosci Abstr 15:1030

Manrezza P, Brown IR (1990) Time course of induction of a heat shock gene (hsp70) in the rabbit cerebellum after LSD in vivo: involvement of drug-induced hyperthermia. Neurochem Res 15:53–59

Marini AM, Kozuka M, Lipsky RH, Nowak TS (1990) 70-kilodalton heat shock protein induction in cerebellar astrocytes and cerebellar granule cells in vitro: comparison with immunocytochemical localization after hyperthermia in vivo. J Neurochem 54:1509–1516

Masing TE, Brown IR (1989) Cellular localization of heat shock gene expression in rabbit cerebellum by in situ hybridization with plastic-embedded tissue. Neurochem Res 14:725–731

Miller EK, Raese JD, Morrison-Bogorad MR (1989) The family of heat shock protein 70 mRNAs are differentially induced in rat cerebellum, cortex, and non-neuronal tissues. Soc Neurosci Abstr 15:1127

Moran LA, Chauvin M, Kennedy MD, Korri M, Lowe DG, Nicholson RC, Perry MD (1983) The major heat-shock protein (hsp70) gene family: related sequences in mouse, *Drosophila* and yeast. Can J Biochem Cell Biol 61:488–499

Morrison-Bogorad MR, Groshan K, Miller EK, Raese JD (1989) In situ quantitation of heat shock 70 mRNAs in rat cerebellum after amphetamine-induced hyperthermia. Soc Neurosci Abstr 15:1127

New GA, Hendrickson BR, Jones KJ (1989) Induction of heat shock protein 70 mRNA in adult hamster facial nuclear groups following axotomy of the facial nerve. Metab Brain Dis 4:273–279

Nishimura RN, Dwyer BE, Welch W, Cole R, de Vellis J, Liotta K (1988) The induction of the major heat-stress protein in purified rat glial cells. J Neurosci Res 20:12–18

Nishimura RN, Dwyer BE, Clegg K, Cole R, de Vellis J (1991) Comparison of the heat shock response in cultured cortical neurons and astrocytes. Mol Brain Res 9:39–45

Nowak TS (1985) Synthesis of a stress protein following transient ischemia in the gerbil. J Neurochem 45:1635–1641

Nowak TS (1988) Effects of amphetamine on protein synthesis and energy metabolism in mouse brain: role of drug-induced hyperthermia. J Neurochem 50:285–294

Nowak TS (1989) Heat shock response in gerbil brain after ischemia- in situ hybridization analysis. Soc Neurosci Abstr 15:1128

Nowak TS, Bond U, Schlesinger MJ (1990) Heat shock RNA levels in brain and other tissues after hyperthermia and transient ischemia. J Neurochem 54:451–458

Pardue ML, Feramisco JR, Lindquist S (1989) Stress-induced proteins. UCLA Symp Mol Cell Biol New Series v 96, Alan R Liss, New York

Pelham HRB (1989) Heat shock and the sorting of luminal proteins. EMBO J 8:3171–3176

Riabowol KT, Mizzen LA, Welch WJ (1988) Heat shock is lethal to fibroblasts microinjected with antibodies against hsp70. Science 242:433-436

Schlesinger MJ (1986) Heat shock proteins: the search for functions. J Cell Biol 103:321–325

Schlesinger MJ, Ashburner M, Tissieres A (eds) (1982) Heat shock: from bacteria to man. Cold Spring Harbor Laboratory Press, New York

Sprang GK, Brown IR (1987) Selective induction of a heat shock gene in fibre tracts and cerebellar neurons of the rabbit brain detected by in situ hybridization. Mol Brain Res 3:89–93

Stevenson MA, Calderwood SK (1990) Members of the 70-kilodalton heat shock protein family contain a highly conserved calmodulin-binding domain. Mol Cell Biol 10:1234–1238

Subjeck JR, Shyy TT (1986) Stress protein systems of mammalian cells. Am J Physiol 250 (Cell Physiol 19): C1–C17

Tytell M, Barbe MF (1987) Synthesis and axonal transport of heat shock proteins. In: Smith RS, Bisby MA (eds) Axonal transport. Neurology and neurobiology, Vol 25. Alan R Liss, New York, pp 473–492

Tytell M, Greenberg SG, Lasek RJ (1986) Heat shock-like protein is transferred from glial to axon. Brain Res 363:161–164

Tytell M, Barbe MF, Gower DJ, Brown IR (1989a) Localization of retinal heat shock (stress) protein and mRNA induced by hyperthemia. Soc Neurosci Abstr 15:116

Tytell M, Barbe MF, Gower DJ (1989b) Photoreceptor protection from light damage by hyperthermia. In: LaVail ML (ed) Inherited and environmentally induced retinal degenerations. Alan R Liss, New York, pp 523–538

Uhl GR (1986) In situ hybridization in brain. Plenum, New York

Uney JB, Leigh PN, Marsden CD, Lees A, Anderton BH (1988) Stereotaxic injection of kainic acid into the striatum of rats induces synthesis of mRNA for heat shock protein 70. FEBS Lett 235:215–218

Vass K, Welch WJ, Nowak TS (1988) Localization of 70 kDa stress protein induction in gerbil brain after ischemia. Acta Neuropathol (Berl) 77:128–135

Vass K, Berger ML, Nowak TS, Welch WJ, Lassmann H (1989) Induction of stress protein hsp70 in nerve cells after status epilepticus in the rat. Neurosci Lett 100:254–259

Walsh DA, Klein NW, Hightower LE, Edwards MJ (1987) Heat shock and thermotolerance during early rat embryo development. Teratology 36:181–191

Walsh DA, Li K, Speirs J, Crowther CE, Edwards MJ (1989) Regulation of the inducible heat-shock 71 genes in early neural development of cultured rat embryos. Teratology 40:321–334

Watowich SS, Morimoto R (1988) Complex regulation of heat shock- and glucose-responsive genes in human cells. Mol Cell Biol 8:393–405

Welch WJ (1990) The mammalian stress response: cell physiology and biochemistry of stress proteins. In: Morimoto R, Georgopoulos C, Tissieres, A. (eds) Role of the heat shock or stress protein response in human disease and medicine. Cold Spring Harbor Laboratory Press, New York, pp 223–278

Welch WJ, Feramisco JR (1982) Purification of the major mammalian heat shock proteins. J Biol Chem 257:14949–14959

Welch WJ, Mizzen LA, Arrigo AP (1989) Structure and function of mammalian stress proteins. In: Pardue ML, Feramisco JR, Lindquist S (eds) Stress-induced proteins. UCLA Symp Molec Cell Biol New Series v 96. Alan R Liss, New York, pp 187–202

Whatley SA, Leung T, Hall C, Lim L (1986) The brain 68-kilodalton microtubule-associated protein is a cognate form of the 70-kilodalton mammalian heat-shock protein and is present as a specific isoform in synaptosomal membranes. J Neurochem 47:1576–1583

Wu B, Hunt C, Morimoto R (1985) Structure and expression of the human gene encoding major heat shock protein HSP70. Mol Cell Biol 5:330-341

Yost HJ, Lindquist S (1986) RNA splicing is interrupted by heat shock and is rescued by heat shock protein synthesis. Cell 45:185–193

Yost HJ, Lindquist S (1988) Translation of unspliced transcripts after heat shock. Science 242:1544–1548